CELLULAR THERAPY FOR NEUROLOGICAL INJURY

GENE AND CELL THERAPY SERIES

Series Editors

Anthony Atala & Nancy Templeton

PUBLISHED TITLES

Cellular Therapy for Neurological Injury
Charles S. Jr., Cox

Placenta: The Tree of Life
Ornella Parolini

GENE AND CELL THERAPY

CELLULAR THERAPY FOR NEUROLOGICAL INJURY

Edited by

Charles S. Cox, Jr.

CRC Press
Taylor & Francis Group
Boca Raton London New York

CRC Press is an imprint of the
Taylor & Francis Group, an **informa** business

CRC Press
Taylor & Francis Group
6000 Broken Sound Parkway NW, Suite 300
Boca Raton, FL 33487-2742

Printed on acid-free paper
Version Date: 20160114

International Standard Book Number-13: 978-1-4822-2591-4 (Hardback)

Visit the Taylor & Francis Web site at
http://www.taylorandfrancis.com

and the CRC Press Web site at
http://www.crcpress.com

Contents

Series Preface

Cell and gene therapies can provide useful therapeutics for numerous diseases and disorders, particularly for those that have no other effective treatments. This *Cell and Gene Therapy* book series covers all current topics in gene and cell therapies and their supporting disciplines, including basic research discoveries through to clinical applications. Because these fields are continually evolving to produce advanced treatments based on scientific breakthroughs, each book in the series provides a timely in-depth coverage of its focused topic.

This first volume of the series covers topics currently relevant to the field of cellular therapy for neurological injuries. A remarkable group of authors comprehensively cover vital neurologically related tissue areas such as the brain and spinal cord. Major topics are covered, including stroke, traumatic brain injury, spinal cord injury, and myelomeningocele. The volume includes the whole continuum of development, from basic biology, to preclinical models and clinical trials. This volume is a comprehensive body of work that also addresses current barriers in the field, as well as opportunities moving forward.

We would like to thank the volume editor, Dr. Charles Cox, and the authors, all experts in their field, for their valuable contributions. We also would like to thank our senior acquisitions editor, Dr. C. R. Crumly, and members of his CRC Press staff, for their efforts and dedication to the *Cell and Gene Therapy* book series.

Anthony Atala, MD
Wake Forest Institute for Regenerative Medicine

Nancy Smyth Templeton, PhD
Optimal Expressions LLC

Preface

The field of regenerative medicine, in general, and cellular therapies, in particular, has grown rapidly over the past five years. A particular area of interest has been in the area of neurological injury—mainly because there are so few reparative therapeutic options available in current clinical practice. This includes diseases such as stroke, traumatic brain injury, spinal cord injury as well as congenital defects such as myelomeningocele. The use of cell-based therapies ranges from immunomodulatory strategies to tissue replacement/augmentation—depending upon the cell type and injury type. This book seeks to explore these approaches, offering the background material that puts current advances and literature into context as the field is moving rapidly. Leaders in specific areas provide this perspective relative to those neurological diseases/injuries/conditions to give a broad overview of the common themes in cell therapy.

Editor

Dr. Charles S. Cox, Jr., is professor of pediatric surgery, and the George and Cynthia Mitchell Distinguished chair in neuroscience, directing the Pediatric Surgical Translational Laboratories and Pediatric Program in Regenerative Medicine at the University of Texas Medical School at Houston. He is codirector of the Texas Trauma Institute and directs the Pediatric Trauma Program at the University of Texas-Houston/Children's Memorial Hermann Hospital in the Texas Medical Center.

A native of Texas, Dr. Cox earned his undergraduate degree from the University of Texas at Austin in the Plan II Liberal Arts Honors Program. Upon graduating from the University of Texas Medical Branch, he completed his surgery residency at the University of Texas Medical School at Houston. Further postgraduate fellowships were completed in pediatric surgery at the University of Michigan, an NIH T32–sponsored clinical and research fellowship in cardiopulmonary support/circulatory support devices/bio-hybrid organs at the Shriners Burns Institute, and surgical critical care/trauma at the University of Texas Medical School at Houston. He is certified by the American Board of Surgery in Surgery, with added qualifications in pediatric surgery and surgical critical care. He served in Afghanistan with the 82nd Airborne in the 909th Forward Surgical Team in 2002.

Dr. Cox has served on scientific study sections/review groups for the National Institutes of Health, California Institute for Regenerative Medicine, American Heart Association, Veterans Affairs MERIT Awards, Department of Defense, Congressionally Directed Medical Research Programs, as well as National Research Programs in Canada, Singapore, Hong Kong, Spain, and the Czech Republic. He is the author of over 150 scientific publications, 20 book chapters, and is the editor of a text titled, *Progenitor Cell Therapy for Neurological Injury.*

Contributors

Benjamin Aertker
Department of Pediatric Surgery
University of Texas–Houston Medical School
Houston, Texas

Qi Lin Cao
Vivian L. Smith Department of Neurosurgery
University of Texas Health Science Center at Houston
and
Center for Stem Cell and Regenerative Medicine
The Brown Foundation Institution of Molecular Medicine
University of Texas Health Science Center at Houston
Houston, Texas

Charles S. Cox, Jr.
Department of Pediatric Surgery
University of Texas–Houston Medical School
Houston, Texas

Pramod K. Dash
Departments of Neurobiology and Anatomy, and Neurosurgery
The University of Texas Medical School at Houston
Houston, Texas

Megan N. Evilsizor
Translational Neurotrauma Research Program
Barrow Neurological Institute
Phoenix, Arizona

Stuart L. Gibb
Blood Systems Research Institute
San Francisco, California

Daniel R. Griffiths
Translational Neurotrauma Research Program
Barrow Neurological Institute
Phoenix, Arizona

James Huh
Departments of Anesthesiology, Critical Care, and Pediatrics
Children's Hospital of Philadelphia
Philadelphia, Pennsylvania

Michael Hylin
Departments of Neurobiology and Anatomy, and Neurosurgery
The University of Texas Medical School at Houston
Houston, Texas

George P. Liao
Department of Pediatric Surgery
University of Texas–Houston Medical School
Houston, Texas

Jonathan Lifshitz
Translational Neurotrauma Research Program
Barrow Neurological Institute
Phoenix, Arizona

Geoffrey S.F. Ling
Department of Neurology
Uniformed Services University of the Health Sciences
Bethesda, Maryland

Joseph B. Long
Center for Military Psychiatry and Neuroscience
Walter Reed Army Institute of Research
Silver Spring, Maryland

Tracy K. McIntosh
Media NeuroConsultants, Inc.
Media, Pennsylvania

Vivek Misra
Department of Neurology
University of Texas Health Science Center at San Antonio
San Antonio, Texas

Anthony N. Moore
Departments of Neurobiology and Anatomy, and Neurosurgery
The University of Texas Medical School at Houston
Houston, Texas

Barclay Morrison III
Department of Biomedical Engineering
Columbia University
New York, New York

Linda J. Noble-Haeusslein
Departments of Neurological Surgery and Physical Therapy and Rehabilitation Sciences
University of California–San Francisco
San Francisco, California

Scott D. Olson
University of Texas Health Science Center at Houston
Houston, Texas

Shibani Pati
Blood Systems Research Institute
San Francisco, California

Darwin J. Prockop
Institute for Regenerative Medicine
Texas A&M University College of Medicine
Temple, Texas

Luke R. Putnam
Department of Pediatric Surgery
The University of Texas Medical School at Houston
Houston, Texas

Ramesh Raghupathi
Department of Anatomy and Neuroscience
Drexel University College of Medicine
Philadelphia, Pennsylvania

David Ritzel
Dyn-FX Consulting, Ltd.
Amherstburg, Ontario, Canada

Sean I. Savitz
University of Texas Health Science Center at Houston
Houston, Texas

Bridgette D. Semple
Departments of Neurological Surgery and Physical Therapy and Rehabilitation Sciences
University of California–San Francisco
San Francisco, California

Sushil Sharma
University of Texas Health Science Center at Houston
Houston, Texas

Richard L. Sutton
Department of Neurosurgery
University of California
Los Angeles, California

Theresa C. Thomas
Translational Neurotrauma Research Program
Barrow Neurological Institute
Phoenix, Arizona

KuoJen Tsao
Department of Pediatric Surgery
The University of Texas Medical School at Houston
Houston, Texas

Edward W. Vogel III
Department of Biomedical Engineering
Columbia University
New York, New York

Bing Yang
University of Texas Health Science Center at Houston
Houston, Texas

1 Acute Neurological Injury *Pathophysiology as Related to Therapeutic Targets*

Charles S. Cox, Jr., Benjamin Aertker, and George P. Liao

CONTENTS

1.1 INTRODUCTION

Traumatic brain injury (TBI) is estimated to result in 275,000 annual hospitalizations as well as 17,000 deaths (Faul et al. 2010) and leave 1.1% of the total U.S. population with long-term disabilities (Zaloshnja et al. 2008). Data from clinical trials show that less than 50% of patients who present with severe TBI will have a favorable outcome a year out from their injury (Myburgh et al. 2008). This large health burden is associated with significant financial ramifications for patients and society. Indirect costs of TBI have been calculated to total $51.2 billion ($66.3 billion in 2014) in an analysis of data from 2003 (Rutland-Brown et al. 2006). In addition, long-term mortality following TBI has remained relatively unchanged for the past 20 years (Brooks et al. 2013). In children, the deficits as a result of injury may be especially important, as they can result in prolonged alterations in cognitive development (Ewing-Cobbs et al. 1997).

Current management of TBI in the acute setting is primarily dependent on control of intracranial hypertension, cerebral perfusion pressure, and supportive care (Brain Trauma Foundation et al. 2007; Grände 2006). Attempts to limit secondary injury following TBI based on interventions successful in animal models and small clinical studies have failed in phase III clinical trials (McConeghy et al. 2012). Confirming the efficacy of new treatments is also complicated by the heterogeneous nature of TBI, the multiple pathophysiologic pathways that result in injury, and the need for improved clinical endpoints (Saatman et al. 2008). Through this review, we aim to provide the reader with an understanding of past therapeutic avenues, the mechanism of action for currently proposed interventions, their development status, and the challenges limiting their success. Other chapters focus on the use of cell therapies for spinal cord injury (SCI), and some comparisons are made between TBI and SCI in this chapter to highlight similarities/differences in strategies for regenerative medicine approaches.

1.2 MECHANISMS OF INJURY

1.2.1 Primary Traumatic Brain Injury

The primary injury refers to the kinetic energy transfer that disrupts brain tissue and results in organ dysfunction. The direct impact of parenchymal tissue against bone leads to neuronal and vascular injury. Intracranial bleeding is a component of the primary injury and can be intra- or extra-axial. The initial cell and structural damage following TBI, or primary injury, is the result of the mechanical stress or shear force on tissues, ischemia from cerebrovascular damage, and hematoma formation (Das et al. 2012). Additional injury may then occur over a period of months in biphasic fashion. This secondary injury is a result of several mechanisms, including neuronal excitotoxicity, mitochondrial dysfunction, and a hyperinflammatory response to injury (Morganti-Kossmann et al. 2007).

Four different pathoanatomic classifications of injury may be seen following TBI: contusions, subarachnoid hemorrhage, hematomas, and diffuse axonal injury (Saatman et al. 2008). The initial mechanical injury is also thought to result in microvascular dysfunction in the brain parenchyma and may be the mechanism for delayed intracerebral hemorrhage that is seen in approximately one half of the patients presenting with intra-axial or traumatic subarachnoid hemorrhage. This "blossoming" is typically seen within 12 hours of injury but may occur as far out as 3–4 days (Kurland et al. 2012). Evolution of the initial injury was first reported following the advent of x-ray computed tomography in the late 1970s (Gudeman et al. 1979). Even though certain factors such as decreased platelet counts and platelet dysfunction may be associated with this process (Kurland et al. 2012; Schnüriger et al. 2010), current treatment options to prevent the evolution of traumatic intracranial hemorrhage are limited. Tranexamic acid was evaluated in a subset of patients in the Clinical Randomisation of an Antifibrinolytic in Significant Haemorrhage (CRASH-2) trial and showed a trend but no significant difference toward reduced mortality and intracranial hemorrhage growth at 24–48 hours (Perel et al. 2012). The results from the CRASH-2 trial were consistent with a subsequent randomized controlled trial evaluating tranexamic acid (Yutthakasemsunt et al. 2013). A systematic review assessing the use of hemostatic agents also showed the potential for positive effect in the setting of isolated TBI (Ker et al. 2015). Clinical evaluation oftranexamic acid remains ongoing in two randomized clinical trials: the Clinical Randomisation of an Antifibrinolytic in Significant Head Injury (CRASH-3) trial (clinicaltrials. gov Identifier: NCT01402882) (Dewan et al. 2012) and the Prehospital Tranexamic Acid Use for Traumatic Brain Injury trial (ClinicalTrials.gov Identifier: NCT01990768).

In addition to the focal injuries discussed, patients may suffer from more dispersed lesions referred to as diffuse axonal injury or traumatic axonal injury (TAI). This injury pattern was first described in a postmortem case series analysis of five patients who were left comatose after suffering from closed head injuries (Strich 1956). Later a classification system was developed that graded the injury from one to three on the basis of the depth of the lesion location (Adams et al. 1989). Axonal injury following TBI appears to occur by two different mechanisms: membrane microperforations that lead to changes in axolemma permeability or termination of axonal transport, which results in axonal swelling (Andriessen et al. 2010). The former is typically the result when axons are subjected to greater than 20% strain and cause a primary axotomy. The latter pathway is the result of less severe strain and causes a secondary axotomy over the course of hours to days (Maxwell et al. 1997). With the advent of more modern imaging techniques, this pattern of white matter injury has been found to occur in greater than 70% of patients who survive moderate to severe head injuries (Skandsen et al. 2010), and pathologic evidence is seen ubiquitously in fatal cases of TBI (Gentleman et al. 1995).

In survivors of TBI, the ramifications may be seen as alterations in fiber tract integrity as shown by changes in fractional anisotropy and mean diffusivity with diffusion tensor imaging (see Figure 1.1). Decreases in fiber tract integrity as shown by fractional anisotropy have been shown in multiple white matter tracts as a long-term result of TBI in both children and adults following mild to severe TBI (Palacios et al. 2013; Wilde et al. 2012; Wu 2010). In contrast to long-term effects of injury for all severities, mild TBI may result in a significant increase in fractional anisotropy in the first month following injury, which is posited to be the result of axonal swelling following injury (Roberts 2014). This fiber tract disruption has been associated with a number of neuropsychological or cognitive outcomes (Ewing-Cobbs 2008; Kraus 2007; Palacios 2013). In the long term, these injuries appear to diminish maturation of the corpus callosum in pediatric patients, particularly the isthmus and splenium (Beauchamp et al. 2011; Ewing-Cobbs 2008).

1.2.2 Secondary Traumatic Brain Injury

1.2.2.1 Neuronal Excitotoxicity

Neuronal excitotoxicity following TBI occurs through both glutamate and calcium-dependent mechanisms (Arundine and Tymianski 2004). The postinjury state is marked by an increase in the release of glutamate and other excitatory amino acids that occurs following depletion of cellular ATP. Subsequent activation of N-methyl-D-aspartate (NMDA), α-amino-3-hydroxy-5-methyl-4-isoxazolepropionic acid (AMPA), and kainate receptors then results in an increase in intracellular calcium and cell death (Lee et al. 1999). Additionally, studies have shown that the rise in intracellular calcium may play a role in mitochondrial dysfunction and free radical production (Starkov et al. 2004). In an effort to mitigate this path of secondary

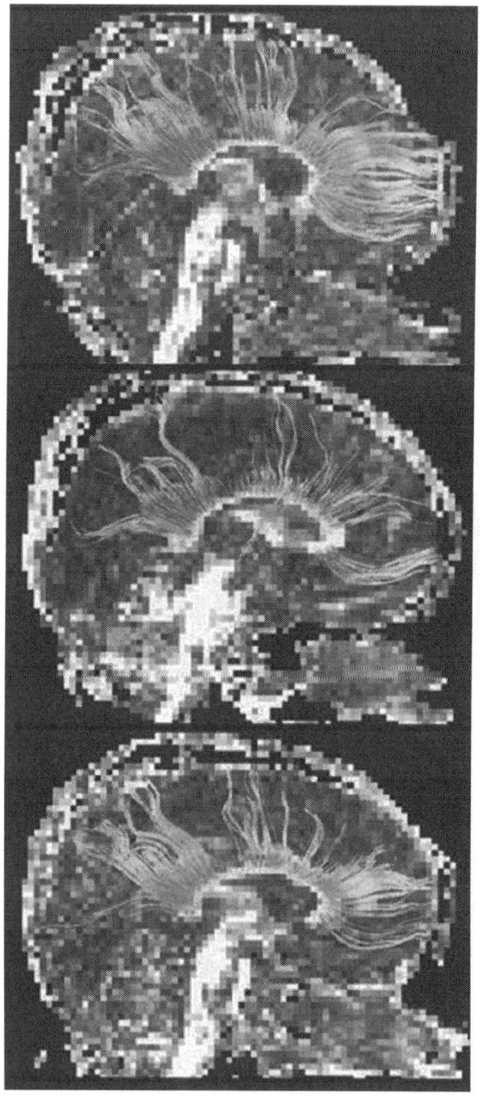

FIGURE 1.1 These images were obtained from a patient acutely following a severe TBI (top), 1 month post-injury (middle), and 6 months out from injury (bottom). These images show a reduction in fiber tract density from baseline to 1 month post-injury as well as continued decreases in fractional anisotropy at 6 months post-injury.

injury, multiple NMDA antagonists have been investigated, but none have proven to be successful, as covered in a 2006 review (Muir 2006). Since that review, additional randomized control trials with dexanabinol and magnesium sulfate have been completed, but neither was demonstrated to be efficacious (Maas et al. 2006; Temkin et al. 2007).

1.2.2.2 Inflammatory Response to Injury (TBI)

The inflammatory response to injury is composed of both local and systemic components, which can further be divided into humoral and cellular responses. Investigations into the role of infiltrating inflammatory cells, as well as local production of cytokines by glial cells, continue to provide new targets for improved neuroprotection (Morganti-Kossmann et al. 2007). Immediately following injury, alterations in the production of multiple chemokines and cytokines can be found in the brain parenchyma. Animal models are often used to determine the spatial and temporal response of these markers to TBI. In rodents, controlled cortical impact (CCI) injuries have been shown to result in elevations in interleukin 1 alpha (IL-1α), interleukin 1 beta (IL-1β), interleukin 6 (IL-6), and tumor necrosis factor alpha (TNFα) that occur at the primary lesion and penumbra. This initial cytokine response resolves over approximately 48 hours (Harting et al. 2009). A murine model of TBI via a cryogenic injury shows elevations of the pro-inflammatory cytokines IL-1β, IL-6, interleukin 17 (IL-17), and TNFα in addition to the anti-inflammatory cytokines interleukin 4 (IL-4) and interleukin 10 (IL-10) (Galindo et al. 2011). Studies in humans have also shown elevations in many of the same cytokines, including IL-1β, interleukin 8 (IL-8), IL-10, and TNFα (Woodcock and Morganti-Kossmann 2013). This cytokine response to injury appears to be more marked in traumatic injuries than ischemic events (Williams et al. 2007). Investigations of the individual role of cytokines on injury severity have shown some of their effects to be time dependent. This temporal dependency has been demonstrated with investigation into the effects of TNFα. Inhibition of TNFα synthesis or inhibition through the use of binding proteins has correlated with improved functional outcomes, blood-brain barrier integrity, and lesion size in animal models when tested acutely following brain injury due to multiple mechanisms, including closed head injury, ischemia, and CCI. Alternatively, models of TBI in TNFα knockout mice demonstrate that TNFα provides benefit in the chronic response to injury and subsequent recovery. These models show larger lesion volumes and reduced functional recovery when compared to wild-type animals (Scherbel et al. 1999; Shohami et al. 1999). Although continued research is needed in the cytokine response to injury, measurements of these values remain a promising avenue for monitoring therapeutic efficacy of new therapies (Woodcock and Morganti-Kossmann 2013). The approach of targeting individual cytokines as a treatment is unlikely to yield an effective agent due to redundancies in the pro-inflammatory cascade and the negative ramifications that manifest as a result of inhibiting the protective functions of these mediators.

The postinjury state also leads to alteration in the production of chemokines with acute increases in CCL2, CCL3, CCL11, CXCL1, and G-CSF (Galindo et al. 2011). Chemokines may act as chemoattractants and also help control the expression of integrins and selectins that contribute to the migration of peripheral inflammatory cells (Worthylake and Burridge 2001). The invasion of the central nervous system by inflammatory cells is partially dependent on blood-brain barrier breakdown. Polymorphonuclear cells can be found to invade the brain parenchyma as early as 6 hours after injury, peaking at approximately 24 hours and remaining present for 3–5 days (Royo et al. 1999; Williams et al. 2007). Although T cells have a less well-delineated role in the pathophysiology of TBI, T lymphocyte invasion is also found to peak around 24 hours following a CCI insult (Clausen et al. 2007). Penetrating and CCI models show this response is followed by a macrophage and microglial response that reaches its peak intensity over the 3–7 days following injury and may still be observed at 1 month following injury (Turtzo et al. 2014; Williams et al. 2007). This response may be even more prolonged. Postmortem evaluation of humans who have suffered moderate to severe TBI shows the presence of activated macrophages/microglia in approximately 25% of patients at 1 year from injury with continuation as far out as 18 years (Johnson et al. 2013). The macrophage/microglial populations are shown to result in a mix of pro-inflammatory M1 and anti-inflammatory M2 microglia/macrophage populations following CCI (Bedi et al. 2013a; Turtzo et al. 2014). It is thought that M1 microglia/macrophage populations are responsible for the production of oxidative species and much of the phagocytic activity. As a result, they may contribute to additional damage of healthy parenchyma. The M2 populations are believed to play a role in angiogenesis, remodeling of the extracellular matrix, and support regeneration following injury (Kumar and Loane 2012). Another potential benefit of the M2 response was demonstrated in a mechanical impact model of spinal cord injury where it has been noted to stimulate axonal growth at sites of central nervous system (CNS) injury (Kigerl et al. 2009).

1.2.2.3 Inflammatory Response to Injury (Spinal Cord Injury)

There is growing appreciation that there are regional differences in the inflammatory response to traumatic injury within the CNS according to both physical location and temporal influence. Previously, there was the thought that all CNS injuries produced similar inflammatory responses to injury. However, it has become clear that SCI results in a greater inflammatory response compared to TBI in terms of any cell type examined or time point postinjury (Batchelor 2008). In terms of cell populations, SCI activated and recruited more microglia/macrophages compared to focal and less dense accumulation of microglia around the site of a TBI injury. However, these studies were relatively short term (7 days), and we and others have demonstrated that there is prolonged and chronic microglial activation in both animal models of TBI as well as patients (Caplan, in press; Ramlackhansingh 2011). A prominent difference in the cellular inflammatory response to injury is the pronounced polymorphonuclear neutrophil (PMN) infiltration after SCI with a lesser contribution to TBI (Schnell 1999). However, there has been less of a focus on the neutrophil in both TBI and SCI since depletion experiments in both models do not prevent blood-brain barrier (BBB) permeability. In terms of the adaptive cellular immune response, T and B cell infiltration is much greater in SCI compared to TBI. Taken collectively, injuries of similar mechanical magnitude create a much greater inflammatory response compared to TBI.

Like many in the field, our group has correlated BBB permeability closely with the degree of inflammation after TBI—using it as a surrogate in experimental models. The brain–spinal cord barrier (BSCB) serves a similar function, but many studies suggest that the BSCB is more permeable and susceptible to inflammatory responses than the BBB, and it may be related to relatively lower expression of tight junctional proteins (Ge 2006; Prockop 1995). This permeability may influence CNS antigen presentation and the development of an autoimmune-type chronic inflammatory reaction. In that regard, the spinal cord appears to be more susceptible to this type of injury.

The discussion in the previous section highlighted the role of the infiltrating monocyte that transforms into a microglial/macrophage phenotype versus the endogenous CNS microglial cells. The M1-M2 phenotypic description probably represents an oversimplification of the continuum of inflammatory responses that variegate between phagocytotic and trophic. What is clear is that SCI results in greater infiltration of the peripheral immune cells compared to TBI, and that the infiltrating cells have a greater propensity to the M1 phenotype. There is some speculation that the greater macrophage inflammatory phenotypic expression in SCI is related to the anatomic concentration of white matter (and thus myelin) in the spinal cord, which is highly antigenic in terms of generating a pathological macrophage response. Regardless, much work remains in deciphering the role of microglia and infiltrating macrophages in response to TBI/SCI. Although the tools are limited in this area, some depletion experiments use liposomal clodronate as well as CD11b conditional CNS knockout mice to explore the relative contributions of each approach (Heppner 2005; van Rooijen and Sanders 1994). In conclusion, comparative studies indicate that the relative inflammatory response to SCI is greater than TBI in terms of innate and adaptive immune effects and that this is facilitated by increased BSCB permeability. Although these differences are significant, they do not preclude the development of similar approaches in the neuroinflammatory response to injury. However, detailed targets and dosing regimens are important in the design of any potential therapeutic as noted that the magnitude and timing of these inflammatory responses vary according to the anatomic site of injury.

1.2.2.4 Clinical Neuro-Intensive Care

1.2.2.4.1 Traumatic Brain Injury

The current management of the injured brain following TBI remains largely supportive, with much of the focus on optimizing cerebral perfusion by managing cerebral edema and intracranial pressures (Clausen and Bullock 2001; Frattalone and Ling 2013). Clinical care guidelines have been developed and are associated with improved outcomes when used by critical care teams (Adelson et al. 2003; Suarez et al. 2004; Varelas et al. 2004). However, these practices lack robust randomized control trials. Currently, the management strategy escalates in intensity in a tiered fashion with first-tier treatments typically including sedation, establishing an intracranial pressure (ICP) threshold, monitoring cerebral perfusion, using neuromuscular blockade, cerebrospinal fluid (CSF) drainage, and hyperosmolar therapy (Sakellaridis et al. 2011). Second-tier treatments include hyperventilation, barbiturates for pharmacological coma with electroencephalogram monitoring for burst suppression, hypothermia, and surgical decompression.

First-tiered therapy begins with sedation, which is a combination of anesthetics and analgesics. Propofol is a commonly used anesthetic for quick clearance required for frequent neurological tests, but it can cause myocardial depression as part of an infusion syndrome and may not reduce the cerebral ischemic burden (Johnston et al.

2003). Norepinephrine and phenylephrine are commonly used pressors to maintain adequate cerebral perfusion pressure, because they have the least effect on cerebral vasomotor tone, but overaggressive hypertension may increase the risk of acute respiratory distress syndrome (Frattalone and Ling 2013). Fever increases the cerebral metabolic burden and increases the ICP and is treated aggressively with a low index of suspicion for infection and atelectasis. Hyperosmolar therapy includes mannitol, which is administered at 0.5 g–1g/kg and produces an effect within 15–30 min. This can be administered every 6 hours to a target serum osmolarity of 310–320 Osm/L. In addition to lowering the intracranial pressure, mannitol also has been shown to improve cerebral blood flow (CBF) (Scalfani et al. 2012). For hyperacute ICP elevations and for herniation syndromes, 23% hypertonic saline can be used and can reduce the ICP by up to 50% within minutes and produce a durable response over hours. Sodium chloride and sodium acetate can be used as a mixture to minimize hyperchloremic metabolic acidosis.

Second-tier options include barbiturates and surgical decompression. Barbiturate coma reduces the cerebral metabolic rate as well as the ICP but also has numerous systemic risks, including hypotension, hypocalcemia, hepatic renal dysfunction, sepsis, and ileus. Furthermore, the long-term outcome for barbiturate coma is unknown (Bochicchio et al. 2006; Eisenberg et al. 1988). The use of mild induced hypothermia (body temperature between 32°C and 35°C) has produced mixed results with some studies suggesting no benefit while others suggesting modest benefit (Sydenham et al. 2009). Hypothermia does exacerbate electrolyte disorders, arrhythmia, and infections. The DECRA (Early Decompressive Craniectomy in Patients with Severe Traumatic Brain Injury) trial examined outcomes for patients undergoing decompressive craniectomies for elevated intracranial pressures. Although decompression lowers intracranial pressure and decreases length of stay, the investigators cited worse long-term outcomes. However, the trial did not include a significant population undergoing craniectomies, namely those with space occupying hematomas or those undergoing unilateral craniectomies (Cooper et al. 2011). The international multicenter RESCUEicp (Randomised Evaluation of Surgery with Craniectomy for Uncontrollable Elevation of intracranial pressure) has been designed to compare surgical decompression versus medical management alone.

Experimental therapies aim to improve the monitoring and therapeutic response to the post-injured brain. A controversial issue in neurocritical care is in regard to the practice of directed therapy to maintain ICP levels below 20 mm Hg. The multicentered randomized Benchmark Evidence from South American Trials: Treatment of Intracranial Pressures (BEST TRIP) study reported no difference in functional/cognitive outcome, mortality, median intensive care unit (ICU) stay, and serious adverse events between maintaining ICP at or below 20 mm Hg to imaging and clinical examination alone (Chesnut et al. 2012; Melhem et al. 2014). Proponents of continued ICP and cerebral perfusion pressure (CPP) monitoring suggest that the study used practices that varied from established guidelines and did not specifically look into ICP monitor use for the management of intracranial hypertension, thereby limiting external validity and generalizability. Intracranial pressure is an indicator of injury severity, but the operational process of measuring, interpreting, and making treatment decisions is complex, and outcome measures such as mortality fail to address the specific contribution of ICP-directed care (Le Roux 2014). Recent evidence looking specifically at large databases and studies following the Brain Trauma Foundation (BTF) guidelines suggest that ICP monitoring contributed to improved outcomes (Alali et al. 2013; Gerber et al. 2013; Talving et al. 2013).

The debate regarding ICP monitoring and outcome has led investigators to seek additional, multimodal approaches to assess the physiological status of the injured brain. Multimodal monitoring includes brain oxygen monitoring (currently considered a level III clinical practice guideline recommendation) and microdialysis (not yet endorsed as a guideline). Poor short-term outcome is associated with hypoxia measured by $pBrO_2$ (partial pressure of oxygen in brain tissue) independent of elevated ICP, low CPP, and injury severity (Oddo et al. 2011). The multicentered phase II BOOST 2 (Brain Tissue Oxygen Monitoring in Traumatic Brain Injury) trial (clinicaltrials.gov Identifier: NCT00974259), estimated to complete in 2014, will evaluate whether $pBrO_2$ levels below the critical threshold of 20 mm Hg can be reduced with monitoring, in addition to the evaluation of safety, feasibility, and GOSE (Glasgow Outcome Scale-Extended) scores 6 months after injury.

Microdialysis has the ability to provide information regarding the metabolic status of penumbral brain tissue, and includes real-time glucose, lactate, glycerol, and glutamate measurements, although robust randomized clinical trials have not yet been pursued. Studies have suggested that metabolic derangements can be detected by microdialysis prior to increases in ICP (Belli et al. 2008). Investigators have also demonstrated that metabolic crisis, defined by brain glucose <0.8 mmol/L and lactate/pyruvate ratio >25, can occur at an incidence of 74% despite adequate resuscitation and controlled ICP (Stein et al. 2012). In rodent models, lactate levels are elevated at the site of injury after TBI (Chen et al. 2000). Furthermore, microdialysis has been used to detect inadequate glucose levels in the brain with the use of strict systemic glycemic control (Oddo et al. 2008).

Although the use of $pBrO_2$ monitoring and microdialysis has not been widely adopted in clinical use, these two devices provide investigators valuable tools beyond simple ICP measurements when evaluating emerging therapeutics. Combined microdialysis and positron emission tomography in patients following severe TBI demonstrated that metabolic crisis can even be present without cerebral ischemia as measured by oxygen extraction fraction and cerebral venous oxygen content (Vespa et al. 2005).

The long-term sequelae of TBI are difficult to measure in terms of outcomes. Outcome measures of mortality and function are often used in clinical trials but are often nonspecific in regard to the therapeutic strategy under investigation. The use of the GOS has contributed to the failure of many clinical trials when used as a primary dichotomized outcome measure. This is due to the inability to move patients with treatment from the extremes of the scale, such that studies are habitually underpowered. In recent years, imaging has become an important outcome measure for TBI. Several regions of the brain are sensitive to TBI and include the hippocampus (Colicos et al. 1996; Colicos and Dash 1996; Williams et al. 2001). Long-term changes in hippocampal areas such as the dentate gyrus and CA1 can stem from newborn neuron death after TBI, which affects memory and causes learning deficits

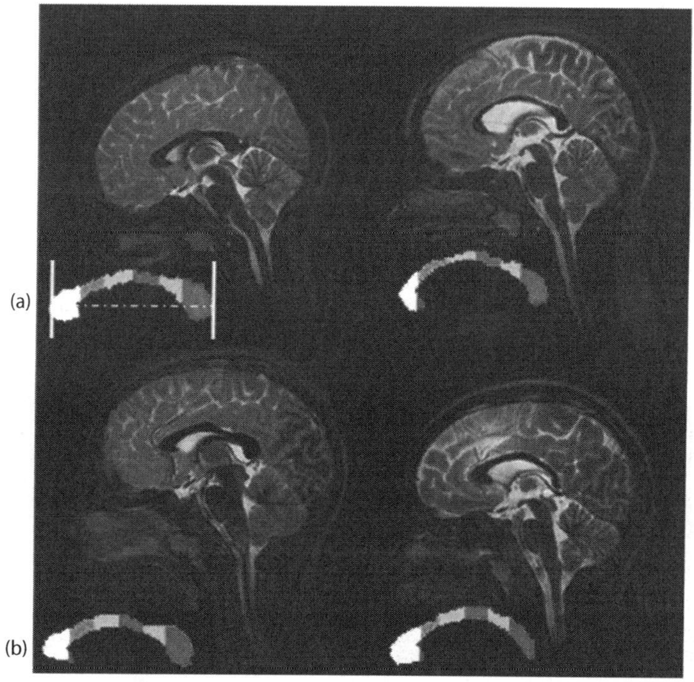

FIGURE 1.2 Alterations in corpus callosum growth may be seen in pediatric patients following TBI. In the age matched comparisons above (control on left, injured on right), a reduction in cross sectional area may be noted in the splenium) of both examples. (a) and (b) represent the volume of the segments of the corpus callosum demonstrated on the sagittal MRI scans. The splenium is the posterior portion of the corpus callosum that is attenuated after injury. (Reproduced from Ewing-Cobbs, L. et al., 2008. *Neuroimage* 42:1305–1315.)

(Atkins 2011; Gao et al. 2008a). Corpus callosum volume loss has also been demonstrated in humans and has been topographically correlated with neuropsychological outcomes (Figure 1.2) (Ewing-Cobbs et al. 2008). Many studies seek to determine if early clinical intervention can translate to long-term improvements. The average ICP during the acute neurointensive period has been used as an early target for therapy in hopes that this indicator can correlate with long-term outcome. Studies have reported that average ICPs during the first 48 hours do not correlate with 6-month functional or neuropsychological outcomes (Badri et al. 2012). However, these studies generally do not reflect continuous monitoring trends, number of spikes, and waveforms, and are thus likely limited by design.

1.2.2.4.2 Spinal Cord Injury

Similar to TBI, the American Association of Neurological Surgeons published guidelines that identify the best care strategies based on the available clinical evidence. The strategies are quite similar to TBI. Although there is little class 1 evidence to support many clinical practices, there is a consensus that the avoidance of hypotension/hypoxia by using cardiopulmonary monitoring and interventions is warranted.

There was previous enthusiasm for the use of early, high-dose methylprednisolone, as the biological rationale for an anti-inflammatory is quite strong. Despite early promising results, this approach has fallen out of favor because the rate of complications due to the drug outweighed the relatively small/incremental improvements in functional outcomes.

1.2.2.5 Therapeutic Targets

1.2.2.5.1 Excitotoxicity

Glutamate has been identified as a major neuroexcitatory amino acid that exacerbates cell injury, such as astrocyte swelling following TBI (Bullock et al. 1998; Koura et al. 1998; Zauner et al. 1996). Elevated levels can be identified in the CSF after TBI (Yamamoto et al. 1999). Research regarding glutamate antagonism began in the 1990s (Bullock et al. 1992; Maxwell et al. 1994; Myseros and Bullock 1995). Recently, the adenosine A2A receptor, found in cells such as bone marrow–derived cells, has been associated with increased glutamate levels following TBI. After TBI, glutamate levels were reduced by adenosine A2A receptor inactivation or in knockout mouse models, along with reduced proinflammatory cytokines such as IL-1 and TNFα (Dai et al. 2010). Valproate is an antiepileptic drug that has multiple targets including the GABA, sodium channel, glycogen, and histone pathways, and has been demonstrated in rats to protect the BBB, reduce neural damage, and improve cognition (Dash et al. 2010). Topiramate is another antiepileptic drug that has been clinically used to reduce glutamate release after TBI in humans, and a phase II clinical trial designed to determine whether this drug can prevent epilepsy after injury is currently ongoing (Alves et al. 2003). Investigators have even explored the benefit of caffeine and alcohol with their potential neuroexcitatory modulating mechanisms (Dash et al. 2004; Li et al. 2008).

1.2.2.5.2 Oxidative Stress

Neuroprotection with improvements in behavior has been demonstrated in rodent models for antioxidants such as deferoxamine, selenium, α-phenyl-tert-*N*-butyl nitrone (PBN), and NXY-059 (Clausen et al. 2008; Long et al. 1996; Marklund et al. 2001; Yeo and Kang 2007). More recently, (–)-epigallocatechin-3-gallate (EGCG) in green tea has been shown in post-TBI rats to preserve neuronal stem cells (Itoh et al. 2012). In mice, CAPE (caffeic phenol acid ester) improved the BBB via an antioxidant oxidant pathway (Zhao et al. 2012). Edaravone, an antioxidant and neuroprotectant, scavenges NO, protects the BBB, and reduced CA3 neuronal loss, apoptosis, and astrocyte/glial activation in rats (Dohi et al. 2006, 2007; Itoh et al. 2009; Miyamoto et al. 2013a,b; Satoh et al. 2002; Wang et al. 2011). When given to a small number of TBI patients ($n = 17$), jugular bulb measurements demonstrated decreased reactive oxidative species (Dohi et al. 2006). Edaravone has been investigated in controlled trials for stroke, but the optimal dose and therapeutic window have not yet been established (Lapchak 2010). Determining the optimal dosage and therapeutic window will undoubtedly be crucial in the design of a human TBI trial.

1.2.2.5.3 Blood-Brain Barrier

The BBB has been a target of interest in TBI due to its participation in the neuroinflammatory process and the development of cerebral edema. Postinjury supplementation of the endogenously expressed cyclophilin A (a protein involved in endothelial cell activation and inflammation) reduced BBB permeability 24 hours after injury in rats (Redell et al. 2007). Tissue inhibitor of metalloproteinase (TIMP) metallopeptidase inhibitor 3 (TIMP-3) is a MMP inhibitor that stabilizes and improves BBB integrity in animal models (Menge et al. 2012; Tejima et al. 2009). Progesterone has been demonstrated in preclinical studies to promote endothelial progenitor cell (EPC)–mediated vascular remodeling, downregulate the inflammatory cascade, and decrease cerebral edema (Li et al. 2012). Sulforaphane (isothiocyanate) in cruciferous vegetables attenuates aquaporin-4 (AQP4) loss and improves the BBB (Zhao et al. 2005). Cannabinoid type 2 receptor agonists have been shown in mice models of TBI to improve BBB permeability and reduce macrophage/microglial activation and neuronal degeneration (Amenta et al. 2012). Citicoline, a naturally endogenous compound found to be effective in preclinical trials of BBB protection, was not found to significantly improve function or cognitive outcomes in human phase III trials (Baskaya et al. 2000; Zafonte et al. 2012).

1.2.2.5.4 Signaling Pathways

Many signaling pathways have been implicated in TBI. The Erk pathway has been described as an important extracellular signal pathway in preclinical models of TBI (Dash et al. 2002). Animal models have associated increased transcription factors such as CREB (cAMP) following TBI with changes in behavior (Dash et al. 1995). Strategies involving histones can preserve Akt signaling, decreasing apoptosis, and have been shown to increase nestin expression (Lu et al. 2013; Wang et al. 2013a). Phosphodiesterase targeting strategies have also been explored in TBI research. Phosphodiesterase (PDE)-4 treatment targeting the cAMP pathway improves histopathological outcomes and decreases inflammation (Atkins et al. 2007; Titus et al. 2013). The transforming growth factor-beta (TGF-β) pathway has been implicated in mice TBI models to be linked with the Runt-related transcription factor-1 (Runx1) to promote activation and proliferation in the dentate gyrus (Logan et al. 2013). Progesterone has been shown to regulate apoptotic protein expression in the dentate gyrus in rats but also has been shown to increase vasculogenesis (Barha et al. 2011; Li et al. 2012; Yao et al. 2005). Despite preclinical success, the phase III ProTECT trial using intravenous progesterone in the acute post-TBI period was halted for futility in late 2013 (Clinicaltrials.gov Identifier: NCT00822900). The international SyNAPSe study is another Phase III progesterone clinical trial for acute TBI trial that finished enrollment in late 2013 with results now pending (Clinicaltrials.gov Identifier: NCT01143064).

1.2.2.5.5 Growth Factors

Growth factors are an attractive target to be either endogenously augmented or delivered exogenously to aid in the regenerative process. Intraventricular infusion of fibroblast growth factor (FGF) into rats after TBI has been shown to increase neurogenesis in the subventricular zone (SVZ) (Sun et al. 2009). Nerve growth factor (NGF) has been used to promote astrocyte migration and is associated with reduced apoptosis following TBI in rats. Intraventricular delivery of epidermal growth factor (EGF) has been shown to be neuroprotective in rats (Sun et al. 2010). Vascular endothelial growth factor (VEGF) was shown to increase de novo hippocampal neurogenesis more than proliferation of resident neuroblasts in rats (Lee and Agoston 2010). In mice, intraventricular delivery augmented neurogenesis and angiogenesis and reduced post-TBI lesion volume (Thau-Zuchman et al. 2010). Gold salts have been used via local perilesion injections to reduce inflammation and apoptosis via VEGF and FGF (Larsen et al. 2008; Pedersen et al. 2009). The neutrophic/mitogenic protein S100B made by astrocytes has been demonstrated to increase hippocampal neurogenesis as well as improve cognition in post-TBI rats but has classically been correlated with poor outcomes when found in human CSF (Kleindienst et al. 2005; Kleindienst and Ross Bullock 2006). Recently, P75, a small molecule ligand, has been shown to bind neurotrophin receptors on neuronal precursors and enhance their regenerative properties in rats (Shi et al. 2013). A randomized double-blinded pilot study by Liu and colleagues is currently underway investigating repeated intranasal delivery of NGF for acute TBI (Clinicaltrials.gov Identifier: NCT01212679).

1.2.2.5.6 Neuronal Architecture

Cyclosporin A was shown in the 1990s to protect axons following TBI in rats by inhibiting calcium-induced mitochondrial damage (Buki et al. 1999; Okonkwo et al. 1999). Preinjury inhibition of calpain, a calcium influx–mediated cysteine protease, preserves axonal integrity following TBI in rats (Buki et al. 2003). Inhibitory myelin molecules such as Nogo-A have been implicated in axonal sprouting following injury, but monoclonal antibodies against this protein did not appear to act through sprouting or cell loss protection, but still improved cognition in animal models (Lenzlinger et al. 2005). Myelin-associated glycoprotein is another inhibitor of axonal growth, and investigators found improved sensorimotor function following intraventricular administration in rats (Thompson et al. 2006). Cyclosporin is currently being investigated in a phase II trial.

1.2.2.5.7 Post-TBI Neuroinflammation

The neuroinflammatory state following TBI is complex and likely an evolving process of dynamic cytokine expression. For example, TNF has been shown to be toxic to the neuronal stem cell proliferation phase, but not during differentiation. In fact, IFNγ enhances neuronal stem cell differentiation and neurite outgrowth (Wong et al. 2004). This may explain why anti-TNF or anti-IL-6 strategies have not been found to improve acute edema or motor or cognitive function in rats (Marklund et al. 2005). In preclinical studies, strategies to neutralize IL-1 have been shown to improve edema, tissue loss, and cognition in mice (Clausen et al. 2009, 2011). Studies have shown the local milieu in the rat brain in the first 48 hours following TBI is highly proinflammatory, with elevated levels of IL-1b, IL-6, and TNFα, along with the presence of microglia and macrophages (Harting et al. 2008).

The proinflammatory environment can serve as a target of cell therapy but may limit drug efficacy or even exacerbate injury. Purely pharmacologic anti-inflammatory strategies may also interfere with complex reparative inflammatory pathways. Bortezomib, a selective proteasome inhibitor used in multiple sclerosis, was found to have neuroprotective properties in animal models and was associated with decreased NFkB expression (Qu et al. 2010). A multifunctional immunomodulator, TSG-6, reduces neutrophil extravasation and BBB leakage in animal models (Watanabe et al. 2013). Ibuprofen has been demonstrated to improve the outcome of transplanted stem cells in animal models (Wallenquist et al. 2012). A protein involved in the generation of inflammation-mediating prostaglandins, COX-2, was found to be exclusively expressed in rat neurons but not astroglia and may have protective roles in only certain neurons (Dash et al. 2000; Kunz et al. 2002). The administration of COX-2 inhibitors in a rat TBI model was found to improve cognition but worsen motor function. Thus, the use of selective anti-inflammatory drugs may not be specific enough to truly target neuroprotection. Anti-inflammatory drugs such as rolipram may unfortunately increase bleeding in animal models (Atkins et al. 2012). Inhibition of single proinflammatory cascades of overlapping/redundant signal transduction pathways has not proven successful.

1.2.2.5.8 Neurovasculature

Investigators have also explored strategies of optimizing the neurovascular niche for resident stem cell activation and function in response to hypoxia following TBI. The hypoxia-induced factor (HIF-1α) pathway has downstream components that include VEGF, SDF-1, brain-derived neurotrophic factor (BDNF), tyrosine kinase receptor TrkB and associated co-receptor Nrp-1, as well as chemokine receptor CXCR4 and nitric oxide (NO) (Madri 2009). The NO donor DETA/NONOate delivered via intraperitoneal injection in rats, improved proliferation, survival, and differentiation of resident neuronal stem cells (Lu et al. 2003). Statins have also been shown to induce angiogenesis, reduce neurologic deficits, increase neuronal survival, and increase hippocampal synaptogenesis–induced angiogenesis in rats (Lu et al. 2004, 2007). In human studies, statin therapy for 10 days following moderate to severe TBI was found to reduce TNFα levels at 72 hours after injury as well as disability scores at time points up to 6 months (Sánchez-Aguilar et al. 2013). Erythropoietin is being investigated in a phase III trial as a subcutaneous injection in patients with severe TBI under the hypothesis that secondary injury can be improved through optimizing oxygen delivery.

1.2.2.6 Cell Therapy

Multiple cell therapy approaches are currently being investigated for their therapeutic potential in the treatment of TBI. These include mesenchymal stromal cells, multipotent adult progenitor cells (MAPCs), and bone marrow mononuclear cells (BM-MNCs), all of which may be procured from harvesting bone marrow. Pluripotent stem cells in the form of embryonic stem cells or induced pluripotent stem cells, neural stem cells, and umbilical cord blood cells have been evaluated as well. Additionally, research into adipose stromal cells for TBI is being explored. These cell therapies have been proposed to work through primarily three different mechanisms: engraftment with transdifferentiation, loco-regional effects, and systemic modulation of the immune response (Walker et al. 2011).

This review of cell therapies for TBI will focus primarily on cell lines harvested from bone marrow as current clinical trials are primarily being evaluated with these lineages. Both mesenchymal stromal cells and MAPC are plastic adherent cell lines (Roobrouck et al. 2011). Mesenchymal stem or stromal cells (MSCs) were originally characterized by their ability to differentiate into adipocytes, chondrocytes, and osteoblasts in addition to being positive for CD105, CD73 and negative for CD14, CD34, CD45 (Pittenger et al. 1999). The MAPCs share similarities with mesenchymal stem cells but also offer greater differentiation ability and may be expanded in culture for longer periods of time (Jiang et al. 2002). Finally, BM-MNCs consist of the nonadherent cell populations from bone marrow. This fraction includes the CD34+ endothelial progenitor cell population, which makes up approximately 1% of the harvested cells (Dedeepiya et al. 2012).

1.2.2.6.1 Modulation of Local and Systemic Inflammatory Response

A change in the inflammatory response following direct transplantation of stem cells and systemic administration has been evaluated in preclinical experiments. These data support the ability of stem cell therapies to alter cytokine production and modulate inflammatory cell populations following TBI. Intraparenchymal transplantation of MSC in animal models of TBI has shown variable effects. In the acute phase, MSC transplantation has demonstrated both elevations (Walker et al. 2010a) and decreases (Galindo et al. 2011) in anti-inflammatory cytokines. An acute increase in IL-6 production from baseline following MSC treatment, however, is also supported by in vitro evidence of NFkB activation in neural stem cells leading to increased synthesis of IL-6 (Walker et al. 2010a). These conflicting results may be due to variation in the time following injury and could even represent evidence of increased inflammation as a result of the transplantation. Additionally, by 30 days out from injury, there is evidence of continued IL-6 elevation (Galindo et al. 2011). An elevation in IL-10 and reduction in TNFα expression in response to MSC intraventricular transplantation in a stroke model is also reported (Liu et al. 2009). Intravenous administration of bone marrow–derived stem cells has been shown to modulate the cytokine response to TBI with more consistent results in inducing an anti-inflammatory response (Walker et al. 2010b; Zhang et al. 2013). The MSC administration shows extensive effects on cytokines with decreases in IL-1β, IL-6, IL-17, TNFα, and interferon gamma (IFNγ) in the acute phase of injury in addition to increases in IL-10 and TGF-β1 (Zhang et al. 2013).

Some recent evidence indicates that the occurrence of a cell therapy–induced anti-inflammatory response may be mediated by production of TNFα-stimulated gene/protein 6 (TSG-6) as shown in experiments utilizing MSC. The TSG-6 was initially discovered through the analysis of a cDNA library created by TNFα stimulated fibroblasts (Lee et al. 1992). The MSC administration has been shown to increase TSG-6 levels following experimental models of myocardial infarction (Lee et al. 2009), TBI (Zhang et al. 2013), anoxic brain injury (Lin et al. 2013), and peritoneal injury (Wang et al. 2012). Furthermore, its role has been demonstrated in vitro with the use of siRNA for TSG-6, which is able to prevent any MSC-related decrease in IL-1β, IL-6, TNFα, and iNOS expression (Liu et al. 2014b). In an experiment performed by Lee et al., the in vitro analysis of MSC indicated that TSG-6 transcription is actually increased in response to TNFα in excess of what is typically seen from fibroblasts when stimulated under similar conditions. This expression of TSG-6 is also associated with decreased pro matrix metalloproteinase activity and decreased infarct size in rodent models of myocardial infarction (Lee et al. 2009). Moreover,

TSG-6 administration alone has been investigated in the acute phase of a murine TBI model and shown to be associated with improvements in memory as well as lesion size (Watanabe et al. 2013). Data are currently lacking on whether other cell lines that are being investigated for their therapeutic use in TBI may operate through this pathway.

The ability of stem cells to bring about modulation in the inflammatory response may also be due to their direct interaction with tissue away from the brain parenchyma. Evidence of decreased splenic mass following acute stroke and TBI has led to the evaluation of the spleen in its role in the inflammatory response to injury. This reaction is thought to be due to increases in sympathetic nervous system–mediated release of splenocytes (Ajmo Jr. et al. 2009; Vendrame et al. 2006). Further support for stem cell–spleen interaction in brain injuries is provided by an experiment performed by Lee et al. with NSC in a rat model of intracerebral hemorrhage. This experiment was able to show that rat status after splenectomy did not derive any further benefit from IV NSC administration (Lee et al. 2008). The MAPCs, when cultured with splenocytes in vitro, have been shown to increase the production of IL-4 and IL-10 as well as increase the expression of $CD4^+$ T cells. These results corresponded with reduced splenic apoptosis and preserved spleen mass in vivo as illustrated in Figure 1.3 (Walker et al. 2010b, 2012a). A contact-mediated decrease in TNFα production by macrophages in co-culture with NSC has also been reported in vitro and could occur in vivo due to interactions at the splenic marginal zone (Lee et al. 2008). This is consistent with earlier results where cord blood administration in animal models of middle cerebral artery occlusion resulted in an increase in Th2 type response when compared to controls, which may play a role in immunologic modulation and functional improvement (Vendrame et al. 2006).

Another factor that has been found to mediate the beneficial effects of MSC administration is tissue inhibitor of matrix metalloproteinase 3 (TIMP3). Evidence of TIMP3 attenuating increased blood-brain barrier permeability after injury has been shown by knockout of TIMP3 production with the use of siRNA and with systemic administration of recombinant TIMP3. Experiments performed in animal models and in vitro show that MSCs increase the production of TIMP3 via a contact-mediated interaction with pulmonary endothelial cells. Increased production is noted to occur in the spleen as well (Menge et al. 2012). This reduction in permeability following injury is found to coincide with increases in the adherens junction protein cadherin and tight junction proteins occludin and claudin-5 (Pati et al. 2011).

Reduction of total microglia/macrophage, neutrophil (MPO^+), and lymphocyte ($CD3^+$) number may occur with MSC treatment after TBI when compared to controls (Zhang et al. 2013). Additional animal models have shown that administration of bone marrow–derived stem cells is able to illicit an increase in the regulatory T cell response following injury (see Figure 1.4) and increase the M2:M1 ratio of macrophages and microglia in the acute phase (Walker et al. 2012a). This alteration has also been shown with BM-MNC and may be due to a proapoptotic CC3-dependent pathway that is selective for the activated microglial population (Bedi et al. 2013b). Although attenuation of the inflammatory response occurs acutely, the duration of this change is unknown, and other experiments have shown the prolonged M1 response found in TBI to be associated with white matter injury (Wang et al. 2013a).

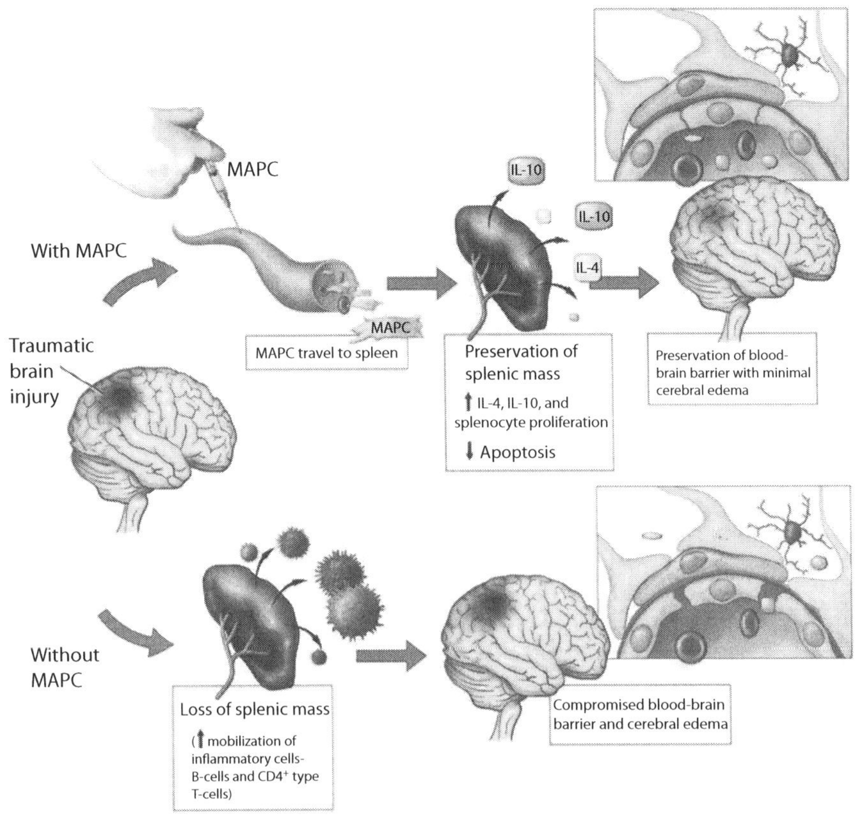

FIGURE 1.3 A potential mechanism by which systemically administered MAPC may produce beneficial effects on blood-brain barrier permeability following TBI. (Reproduced from Walker, P.A. et al. 2010. *Stem Cells Dev.* 19:867–876.)

Finally, preservation or restoration of the blood-brain barrier by cell therapies could have beneficial effects on the inflammatory response to injury. Multiple studies in animal models have shown intravenous administration of BM-MNC, MAPC, and MSC to preserve blood-brain barrier integrity (Bedi et al. 2013b; Menge et al. 2012; Walker et al. 2010b, 2012a).

1.2.2.6.2 Growth Factor Production

Intravenously administered hMSCs have been shown to increase the levels of NGF, NT-3, and BDNF in the brain parenchyma (Kim et al. 2010; Mahmood et al. 2004). In vitro analyses have established that MSC produce multiple growth factors in response to normal and injured brain environments including BDNF, NGF, VEGF, and hepatocyte growth factor (HGF) (Chang et al. 2013; Chen et al. 2005). In the instance of NGF, there is evidence that it may remain elevated for as long as 7 weeks when rodent models of TBI are treated with intraventricular transplantation of MSC (Chen et al. 2005).

However, the mechanism for this increase in neurotrophic factors has been disputed. It is argued that local synthesis by MSC in the perilesional area is unlikely

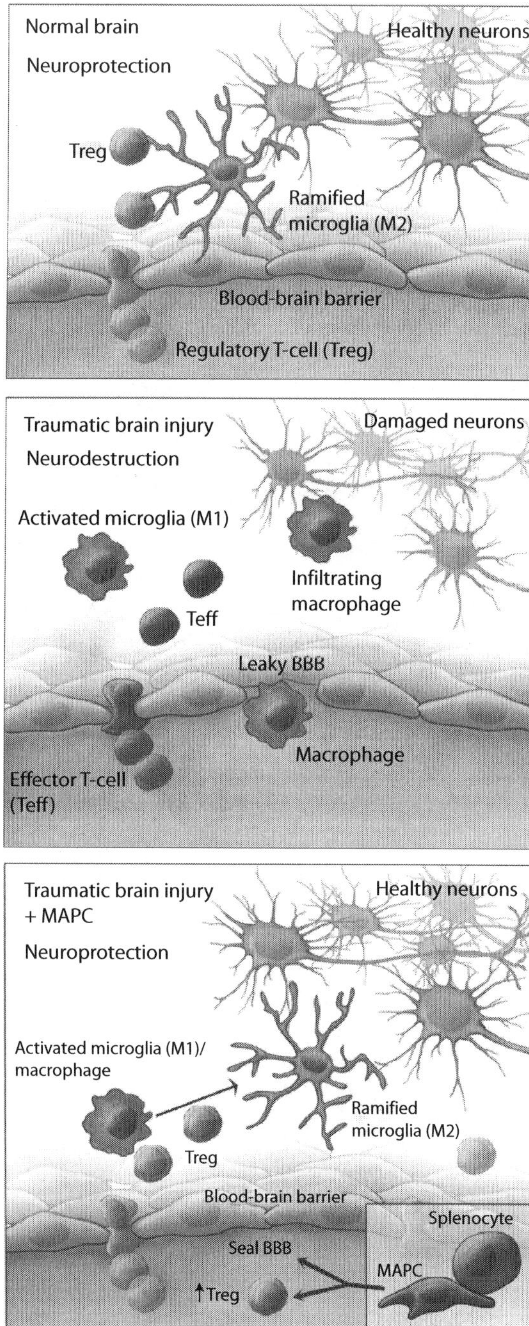

FIGURE 1.4 Administration of MAPC systemically have been demonstrated to increase the proportion of microglia/macrophage with an M2 phenotype by way of interactions with splenocytes. (Reproduced from Walker, P.A. et al., 2012. *J Neuroinflammation.* 9:228.)

to be sufficient in quantity to bring about any functional improvement (Walker et al. 2012b). When cell therapies are administered systemically, the number of cells reaching the injury site is severely limited by a first-pass effect in the pulmonary vasculature as a result of its narrow diameter in the capillary bed (Fischer et al. 2009; Harting et al. 2009; Schrepfer et al. 2007). Cell sequestration in tissue remote from the injured brain parenchyma is dependent on the exact cell lineage administered. The MAPCs, however, have been shown to double the number of cells that are able to avoid sequestration in the pulmonary capillary beds when compared to MSCs, and BM-MNCs avoid this sequestration even further (Fischer et al. 2009).

1.2.2.6.3 Engraftment

Even with direct implantation, durable survival of transplanted cells occurs at a low frequency. Complete rejection of all transplanted MSCs, which are thought to be poorly immunogenic, has been shown as early as 14 days in an animal model that did not undergo a prior TBI (Coyne et al. 2006). Survival of any NSC at 3 and 14 weeks following transplantation has been shown to occur in <60% in animal models of TBI (Riess et al. 2002). To improve cell survival following transplantation, neurotrophic factors such as brain-derived neurotrophic factor and neural growth factor have been used successfully (Mahmood et al. 2002). Direct implantation with the use of scaffolds is another potential avenue for improved cell survival (Mahmood et al. 2013). Additionally, co-grafting of stem cells with another cell type has also been performed. Transplantation of NSC with olfactory ensheathing cells has been shown to result in a 3.4-fold increase in cell survival compared to NSC alone (Liu et al. 2014a). Co-grafting BM-MSCs with Schwann cells has been reported to have effects separate from just changes in survival. This combination has shown improvement in cell migration of transplanted cells toward lesions in animal models of TBI as seen with suprapara-magnetic iron oxide (SPIO)-labeled cells (Xu et al. 2014). Evidence now exists that the degree of engraftment has a correlation with functional outcomes. This has been demonstrated by behavioral improvements with intracarotid administration of NSC (Guzman et al. 2008).

The success of cell engraftment into brain parenchyma may be variable upon the cell line used due to its constitutive surface markers and have a temporal dependency. The CD49d expression may be required for adequate cell engraftment as supported by absence of engraftment of hNPC with minimal CD49d expression (Lundberg et al. 2012). Alternatively, selection for CD49d-positive neural stem cells has shown to enhance engraftment in a murine stroke model (Guzman et al. 2008). Some data exist that IV hMSC administration allows for trace engraftment near the injury site with differentiation toward astrocytic ($GFAP^+$) and neuronal lineages (NeuN) (Kim et al. 2010; Mahmood et al. 2001, 2004). However, differentiation into a neural lineage has been disputed (Breunig et al. 2007; Castro et al. 2002).

1.2.2.6.4 Replacement or Regeneration

Both MSC and human umbilical cord blood cells have been reported to transdifferentiate into neural and glial cell lineages. The MSC transdifferentiation into neural cells was first shown in 2000 after proliferation in vitro (Woodbury et al. 2000). This study was followed shortly by reports of MSC transdifferentiation into astrocytes and neurons in vivo (Lu et al. 2001). Reports of spontaneous migration of transplanted bone marrow cells into the brain of animals and humans have also been published. The vast majority of these cells were noted to differentiate into nonneuronal cells, but some of the cells found in the hippocampus and cortex displayed markers of neuronal lineage (Mezey et al. 2000, 2003). However, it has been argued that these results are due to cell fusion of marrow cells with a recipient cell in the brain (Terada et al. 2002). Subsequent studies cast doubt on in vivo differentiation of MSC into neuronal elements. One such paper did so by demonstrating that cell labels such as bromodeoxyuridine and benzamide may transfer from graft cells to host cells after elimination by an immune-mediated response (Coyne et al. 2006). In light of results showing minimal engraftment in addition to rare or even questionable transdifferentiation, neural replacement has been posited to have limited ability as a current therapeutic mechanism by some authors (Walker et al. 2010b).

Although direct neuronal cell replacement with progenitor cell therapies remains audacious, other avenues remain for improving neuronal function following TBI. Axonal sprouting and cortical rewiring have been found in animal models following stroke and TBI. Evidence from animal models of TBI and stroke demonstrates that this process may be amplified by treatment with intraparenchymal transplantation of MSC (Liu et al. 2008; Mahmood et al. 2013). The MSC transplantation appears to have some limited ability to promote axonal regeneration across lesions and provide neuroprotection to neurons that have undergone axotomy in a model of spinal cord injury (Novikova et al. 2011). These histopathologic changes following treatment appear to arise with associated changes in function. In an animal model of TBI, the degree to which corticospinal tract axonal sprouting occurs within the spinal cord has been shown to have a correlation with improvement in motor function (Mahmood et al. 2013).

The central nervous system is known to have a limited ability for both axonal sprouting from uninjured neurons and regeneration from damaged neurons when compared to the peripheral nervous system. One of the reasons for this dichotomy is the differences in local environment and presence of myelin-associated inhibitors such as Nogo-A, MAG, and OMgp (Geoffroy and Zheng 2014). In vitro MSCs have been demonstrated to mitigate these effects through multiple mechanisms: stimulation of neurite growth by soluble factors and contact-mediated reductions in the effect of these inhibitors (Wright et al. 2007b). In the case of MSC transplantation with collagen scaffolds following TBI, there is evidence of decreased Nogo-A synthesis by oligodendrocytes. This same study showed a decrease in Nogo-A levels with MSC transplantation alone, but not to the same degree as those transplanted with the collagen scaffold (Mahmood et al. 2014). An increase in axonal sprouting with anti Nogo-A antibodies has also been shown in a stroke model. Furthermore, this increase in axonal sprouting from the contralateral corticospinal tract occurs in concert with reorganization of the undamaged motor cortex (Lindau et al. 2013). The use of scaffolds with cell therapies, which has shown improvement in cell engraftment, may also aid in functional neurorestoration by way of axonal sprouting in major white matter tracts such as the corpus callosum and corticospinal tracts (Mahmood et al. 2013; Xiong et al. 2009). This ability to produce signs of therapeutic effect diffusely throughout the CNS is promising and may be necessary for functional recovery in severe TBI due to the high incidence of diffuse axonal injury. Further experimentation is necessary to determine if these loco-regional effects are scalable to larger primates. In vitro analysis of the effect of MSC-derived factors on NSC have been able to show evidence of increased neurite outgrowth even independent of glial interaction, which could be beneficial for repair of neuronal circuitry (Croft and Przyborski 2009). Increased axonal sprouting has also been reported following intravenous administration of MSC in rodents following middle cerebral artery occlusion (Liu et al. 2010). However, no current study appears to have investigated whether or not systemically administered progenitor cell therapy is able to mediate the axonal sprouting or regeneration in TBI.

Alternatively, replacement/repair at the site of the injured brain parenchyma may be possible without the need of cell transplantation. Modulation of the constitutive cell populations at the site of injury is currently being investigated for treatment of CNS injuries. Neuronal replacement may be achievable by cellular reprogramming of glial or neural progenitor cells in the brain. One pathway that has been targeted with gene transfer via viral transfection is the Notch-Hes1 signaling pathway. Stereotactical infusion of Hes1 short interfering RNA prior to lateral fluid percussion injury has been shown to result in increased differentiation of hippocampal neural progenitor cells into NeuN+ neurons and is associated with improvements in spatial memory following injury (Zhang et al. 2014). Downregulation of Hes1 is also known to result in an inverse relationship to neurogenesis and cell proliferation (Baek et al. 2006). Furthermore, this downregulation may be a natural response to injury as it has been observed to occur spontaneously following TBI in rodent models to a limited degree (Yang et al. 2009).

Areas without a ready supply of neural progenitor cells may benefit from reprogramming of local glial cells. Conversion of astrocytes in the striatum and cortex has been reported in vivo. This has been achieved by viral transfection with the transcription factor SOX2 in uninjured animals resulting in doublecortin + neuroblasts, but few mature neurons as shown by staining for NeuN. Replication of this effect with SOX2 transfection is also observable in vivo following a stab wound model of TBI. In this model, no differentiation of astrocytes was noted on the nonlesioned hemisphere. However, the authors in this study reported the development of action potentials and synapse formation (Heinrich et al. 2014). A similar process has also been seen with animal models of spinal cord injury. Transfection of astrocytes with SOX2 containing lentiviruses following spinal cord hemisection resulted in induction to form neuroblasts as shown by staining for doublecortin. This transformation was observable at 4 weeks following injury, and mature neurons appeared later around 8 weeks following injury. Additionally, this method of neural replacement resulted in synaptic formation with preexisting motor neurons (Su et al. 2014). Further increases in cell survival and maturation toward neuronal lineages have been achieved following SOX2 transfection with the use of histone deacetylase inhibitors or transfection to express noggin/BDNF or Ascl1 (Heinrich et al. 2014; Niu et al. 2013; Su et al. 2014).

1.2.2.6.5 Does Route Matter?

Multiple avenues of delivery for progenitor cell therapies have been evaluated in both animal models and in humans. Three of the most common approaches from least to most invasive are intravenous (IV), intra-arterial (IA), intrathecal, and intraparenchymal delivery. Each of these means of delivery has its own advantages. Over the past several years as phase I/II trials have commenced for both stroke and TBI patients, with safety data now available for both intravenous and intra-arterial delivery, harvest and administration BM-MNC for the acute treatment of TBI has now been shown to be safe in pediatric patients (Cox Jr. et al. 2011). These results are consistent with the safe administration of BM-MNC in the acute (24–72 hours) treatment of strokes in the adult population (Savitz et al. 2011). Intra-arterial administration of BM-MNC has also been performed in stroke patients with infusion into the middle cerebral artery (Friedrich et al. 2012; Moniche et al. 2012). Additional trials evaluating the safety of intra-arterial infusion of $CD34^+$ cell populations and MSC have been completed (Banerjee et al. 2014; Bang et al. 2005). Together these studies help show that autologous progenitor cell therapies are logistically feasible. The absence of severe adverse events resulting from worsening cerebral ischemia with intra-arterial administration or end organ dysfunction as a result of sequestration with intravenous administration is an important finding during the translational period. Further study will be needed to determine the ideal means of administration in TBI patients as comorbid conditions, the presence of polytrauma, and need for continued resuscitation may limit some avenues of treatment.

Studies performed in animal models are available to compare differences in efficacy between arterial and venous delivery of cells. However, the results from these studies have varied. Analysis of cell distribution following intravenous administration by Harting et al. showed that less than 4% of cells reached the arterial circulation and only 0.295% of the infused cells were able to reach the carotid artery in a rodent TBI model (Harting et al. 2009). Lundberg and associates compared the cell engraftment of multiple cell lines (human MSC, human NPC, rat NPC) when given intra-arterially versus intravenously and found significantly higher rates of engraftment with IA administration of hMSC, but no evidence of human NPC migration to the site of injury via either administration method. The authors in this study determined that this was likely due to the lack of VCAM-1 to CD49d interaction in the latter cell population (Lundberg et al. 2012). Rodent stroke models have also been able to demonstrate higher rates of engraftment by IA administration of BM-MNC (Kamiya et al. 2008; Yang et al. 2013), MSC (Byun et al. 2013), and with NSC (Pendharkar et al. 2010) when compared to IV. However, the effect of increased engraftment by IA administration remains unclear as currently available data vary. Kamiya and colleagues were able to show decreased lesion size and motor function when compared to control animals with intra-arterial but not with intravenous administration (Kamiya et al. 2008). Alternatively, other studies comparing intra-arterial with intravenous administration of BM-MNC have shown no difference in behavioral/functional or structural outcomes in several animal models (Vasconcelos-dos-Santos et al. 2012; Yang et al. 2013). In addition, intravenous administration was found to cause potentially beneficial attenuation of cytokines with greater decreases in IL-1β and increases in IL-10 when compared to intra-arterial administration. No current studies report differences in behavioral, motor, or structural outcomes with intravenous versus intra-arterial administration of stem cells in animal TBI models.

Ultimately, the need for intra-arterial administration will likely depend on the degree to which functional outcomes are dependent on loco-regional versus systemic therapeutic mechanisms. Supporting this idea are data from intravenous administration of the secretome produced by MSC on culture media. This secretome has been found to result in decreased perilesional neuronal apoptosis as well as lesion volume and possibly increased neurogenesis as shown by increased numbers of BrdU/NeuN/DAPI staining neurons at the lesion site (Chang et al. 2013). This provides some evidence that engraftment and differentiation may not be necessary to achieve neuroprotection or repair with progenitor cell therapies.

Pathways that have been implicated in the protective effects of treatment with progenitor cells and MSC in particular include paracrine-mediated activation of the MAPK/Erk 1, 2, and PI3-K/Akt. In vitro analysis has demonstrated that secretory factors from MSC are able to illicit enhanced phosphorylation of MAPK/Erk 1, 2 in neurons and lead to decreased neuronal apoptosis. This decrease in apoptosis is prevented with the addition of inhibitors of MEK1 thus preventing MAPK/Erk 1, 2 activation and inhibition of PI3-K (Isele et al. 2007). These two pathways have also been shown to be protective toward astrocytes (Gao et al. 2005). Activation of the PI3-K/Akt pathway following treatment with MSC also appears to result in increased axonal outgrowth. The activation of this pathway using cells in a transwell coculture provides further evidence that this outcome is secondary to yet undiscovered secretory factors or paracrine effects (Liu et al. 2013).

1.2.2.6.6 Clinical Investigation

In review of the available literature, there is one published case series where direct autologous MSC transplantation was performed in humans following TBI. In this report, seven patients (ages 6–55) underwent bone marrow harvest with subsequent isolation and expansion of their MSC population. The cells were then administered locally and later intravenously. No serious adverse effects were reported during a 6-month follow-up period (Zhang et al. 2008). Intrathecal administration of MSC has also been performed in humans. In a series of 97 patients, Tian and associates were able to demonstrate that they could safely harvest and administer autologous MSC using as much as 5 mL (1×10^6 cells/mL) via lumbar puncture in patients with persistent vegetative state or other deficits 1 month after injury. In this series of patients, the greatest improvements were seen in younger patients who underwent treatment closer to the time of injury (Tian et al. 2013). The use of allogenic, umbilical cord MSC was evaluated in patients with chronic deficits following TBI. Twenty of the 40 enrolled patients underwent four injections of 1×10^7 cells (2 mL) over a period of 5–7 days. As a result of the intervention, 20% of the displayed symptoms of low intracranial pressure but no other events were reported. The investigators reported statistically significant improvements in functional assessments 6 months out from treatment which was not seen in controls (Wang et al. 2013b). However, due to the single-blinded nature of the study, the results should be interpreted with caution as a placebo effect cannot be eliminated.

Currently, there are phase II clinical trials that are recruiting adult and pediatric patients to evaluate the use of BM-MNCs in acute, severe TBI (clinicaltrials.gov Identifiers: NCT01575470 and NCT01851083). A phase I clinical trial has also started to enroll patients with chronic TBI symptoms for treatment with intrathecal administration of BM-MNC (Clinicaltrials.gov Identifier: NCT02028104).

1.2.2.6.7 Spinal Cord Injury and Cell Therapies

Other chapters in this monograph will explore the issues of cell therapy for SCI in greater detail. However, it is worth noting that the local injury characteristics of SCI differ from the majority of TBIs. Specifically, in TBI there are often multifocal, bihemispheric, multiple lesions that make direct implantation of cells or cells/scaffolds directly much more problematic. In contrast, SCI often has a lesion cavity that results from a contusion, and this has proven amenable to cell/scaffold approaches. Further, the inflammatory component of SCI makes it a target for a combination-type approach using a systemic cell infusion followed by, or in conjunction with, a direct implantation.

1.3 DISCUSSION

Cell-based therapies have become a new avenue of investigation in a multitude of disorders that affect the CNS. Current research shows that these therapies may work through multiple pathways with increasing evidence for both systemic alteration of the inflammatory response and loco-regional effects following injury. Systemically administered cell therapies may provide a way for CNS repair primarily through attenuation of an inflammatory response that may prove disproportionate and vestigial in nature when modern clinical care is provided. This effect, in concert with alterations or amplification of endogenous neurotrophin production, may be more readily available as an avenue for treatment when compared to earlier goals of cell replacement. Determination of the optimal timing for treatment will likely play a significant role with regard to the source and cell lineage used. This is due to the restraints that adequate cell expansion would have on acute treatment. In these instances, allogenic cell sources would be ideal.

Although improved functional recovery is of utmost importance with any new therapy, the development of more specific clinical analyses, imaging modalities, and biomarkers may provide greater direction on which therapeutic avenues show the most potential. The lack of such data prior to the initiation of larger phase III trials has been argued to be one of several culprits plaguing efforts in treatments for TBI (Schwamm 2014). Improvements in study design and analysis of collected data are also ongoing and may provide future benefit (Maas and Lingsma 2008; Panczykowski et al. 2012). Secondary measures of effect may be especially necessary for therapies with pleiotropic mechanisms of action as is the case with cell therapies. A wide variety of cell lines are currently being evaluated in animal models, and it will be important to delineate accurately the mechanisms that these cell lines act through.

Research continues on more traditional therapeutic avenues, such as surgical management of intracranial hypertension, where limited data from prospective, randomized trials exist. For instance, the efficacy of performing a decompressive bilateral fronto-temporal craniectomy was examined in the DECRA trial but showed no improvement in long-term outcome (Cooper et al. 2011). Another trial, RESCUEicp

(Randomised Evaluation of Surgery with Craniectomy for Uncontrollable Elevation of Intra-Cranial Pressure), is currently ongoing (Hutchinson et al. 2006).

In addition to the continued evaluation of therapeutic agents, treatment protocols, and procedures, the development of improved measures of disease severity and recovery should play a large role in future TBI research. It has been argued that the variability of treatment protocols between centers provides another obstacle to evaluating new therapies in large, multicenter trials (www.adapttrial.org). It is possible that all of these factors may need to be improved upon in order to show success in a large, multicenter trial.

REFERENCES

Adams, J.H., Doyle, D., Ford, I., Gennarelli, T., Graham, D., McLellan, D., 1989. Diffuse axonal injury in head injury: Definition, diagnosis and grading. *Histopathology*. 15:49–59.

Adelson, P.D., Bratton, S.L., Carney, N.A., Chesnut, R.M., du Coudray, H.E., Goldstein, B., Kochanek, P.M. et al., American Association for Surgery of Trauma, Child Neurology Society, International Society for Pediatric Neurosurgery, International Trauma Anesthesia and Critical Care Society, Society of Critical Care Medicine, World Federation of Pediatric Intensive and Critical Care Societies, 2003. Guidelines for the acute medical management of severe traumatic brain injury in infants, children, and adolescents. Chapter 1: Introduction. *Pediatr Crit Care Med*. 4 (3 Suppl):S2–4.

Ajmo, Jr., C.T., Collier, L.A., Leonardo, C.C., Hall, A.A., Green, S.M., Womble, T.A., Cuevas, J., Willing, A.E., Pennypacker, K.R., 2009. Blockade of adrenoreceptors inhibits the splenic response to stroke. *Exp Neurol*. 218:47–55.

Alali, A.S., Fowler, R.A., Mainprize, T.G., Scales, D.C., Kiss, A., de Mestral, C., Ray, J.G., Nathens, A.B., 2013. Intracranial pressure monitoring in severe traumatic brain injury: Results from the American College of Surgeons Trauma Quality Improvement Program. *J Neurotrauma*. 30 (20):1737–1746.

Amenta, P.S., Jallo, J.I., Tuma, R.F., Elliott, M.B., 2012. A cannabinoid type 2 receptor agonist attenuates blood-brain barrier damage and neurodegeneration in a murine model of traumatic brain injury. *J Neurosci Res*. 90 (12):2293–2305.

Andriessen, T.M., Jacobs, B., Vos, P.E., 2010. Clinical characteristics and pathophysiological mechanisms of focal and diffuse traumatic brain injury. *J Cell Mol Med*. 14:2381–2392.

Arundine, M., Tymianski, M., 2004. Molecular mechanisms of glutamate-dependent neurodegeneration in ischemia and traumatic brain injury. *Cell Mol Life Sci*. 61:657–668.

Atkins, C.M. 2011. Decoding hippocampal signaling deficits after traumatic brain injury. *Transl Stroke Res*. 2 (4):546–555.

Atkins, C.M., Kang, Y., Furones, C., Truettner, J.S., Alonso, O.F., Dietrich, W.D., 2012. Postinjury treatment with rolipram increases hemorrhage after traumatic brain injury. *J Neurosci Res*. 90 (9):1861–1871.

Atkins, C.M., Oliva, A.A., Jr., Alonso, O.F., Pearse, D.D., Bramlett, H.M., Dietrich, W.D., 2007. Modulation of the cAMP signaling pathway after traumatic brain injury. *Exp Neurol*. 208 (1):145–158.

Badri, S., Chen, J., Barber, J., Temkin, N.R., Dikmen, S.S., Chesnut, R.M., Deem, S., Yanez, N.D., Treggiari, M.M., 2012. Mortality and long-term functional outcome associated with intracranial pressure after traumatic brain injury. *Intensive Care Med*. 38 (11):1800–1809.

Baek, J.H., Hatakeyama, J., Sakamoto, S., Ohtsuka, T., Kageyama, R., 2006. Persistent and high levels of Hes1 expression regulate boundary formation in the developing central nervous system. *Development* (Cambridge, England). 133:2467–2476.

Banerjee, S., Bentley, P., Hamady, M., Marley, S., Davis, J., Shlebak, A., Nicholls, J., Williamson, D.A., Jensen, S.L., Gordon, M. et al. 2014. Intra-arterial immunoselected CD34+ stem cells for acute ischemic stroke. *Stem Cells Transl Med.* 3:1322–1330.

Bang, O.Y., Lee, J.S., Lee, P.H., Lee, G., 2005. Autologous mesenchymal stem cell transplantation in stroke patients. *Ann Neurol.* 57:874–882.

Barha, C.K., Ishrat, T., Epp, J.R., Galea, L.A., Stein, D.G., 2011. Progesterone treatment normalizes the levels of cell proliferation and cell death in the dentate gyrus of the hippocampus after traumatic brain injury. *Exp Neurol.* 231 (1):72–81.

Baskaya, M.K., Dogan, A., Rao, A.M., Dempsey, R.J., 2000. Neuroprotective effects of citicoline on brain edema and blood-brain barrier breakdown after traumatic brain injury. *J Neurosurg.* 92 (3):448–452.

Batchelor, P.E., Tan, S., Wills, T.E., Porritt, M.J., Howells, D.W., 2008. Comparison of inflammation in the brain and spinal cord following mechanical injury. *J Neurotrauma.* 25:1217–1225.

Beauchamp, M.H., Ditchfield, M., Catroppa, C., Kean, M., Godfrey, C., Rosenfeld, J.V., Anderson, V., 2011. Focal thinning of the posterior corpus callosum: Normal variant or post-traumatic? *Brain Inj.* 25:950–957.

Bedi, S.S., Smith, P., Hetz, R.A., Xue, H., Cox, C.S., 2013a. Immunomagnetic enrichment and flow cytometric characterization of mouse microglia. *J Neurosci Methods.* 219:176–182.

Bedi, S.S., Walker, P.A., Shah, S.K., Jimenez, F., Thomas, C.P., Smith, P., Hetz, R.A., Xue, H., Pati, S., Dash, P.K., 2013b. Autologous bone marrow mononuclear cells therapy attenuates activated microglial/macrophage response and improves spatial learning after traumatic brain injury. *J Trauma Acute Care Surg.* 75:410–416.

Belli, A., Sen, J., Petzold, A., Russo, S., Kitchen, N., Smith, M., 2008. Metabolic failure precedes intracranial pressure rises in traumatic brain injury: A microdialysis study. *Acta Neurochirurgica.* 150 (5):461–469; discussion 470.

Bochicchio, G.V., Bochicchio, K., Nehman, S., Casey, C., Andrews, P., Scalea, T.M., 2006. Tolerance and efficacy of enteral nutrition in traumatic brain-injured patients induced into barbiturate coma. *JPEN J Parenter Enteral Nutr.* 30 (6):503–506.

Brain Trauma Foundation, AANS, CNS, AANS/CNS Joint Section on Neurotrauma and Critical Care, 2007. Guidelines for the management of severe traumatic brain injury, 3rd edition. *J Neurotrauma.* 24:S1–S106.

Breunig, J.J., Arellano, J.I., Macklis, J.D., Rakic, P., 2007. Everything that glitters isn't gold: A critical review of postnatal neural precursor analyses. *Cell Stem Cell.* 1:612–627.

Brooks, J.C., Strauss, D.J., Shavelle, R.M., Paculdo, D.R., Hammond, F.M., Harrison-Felix, C.L., 2013. Long-term disability and survival in traumatic brain injury: Results from the National Institute on Disability and Rehabilitation Research Model Systems. *Arch Phys Med Rehabil.* 94:2203–2209.

Buki, A., Farkas, O., Doczi, T., Povlishock, J.T., 2003. Preinjury administration of the calpain inhibitor MDL-28170 attenuates traumatically induced axonal injury. *J Neurotrauma.* 20 (3):261–268.

Buki, A., Okonkwo, D.O., Povlishock, J.T., 1999. Postinjury cyclosporin A administration limits axonal damage and disconnection in traumatic brain injury. *J Neurotrauma.* 16 (6):511–521.

Bullock, R., Kuroda, Y., Teasdale, G.M., McCulloch, J., 1992. Prevention of post-traumatic excitotoxic brain damage with NMDA antagonist drugs: A new strategy for the nineties. *Acta Neurochir Suppl.* 55:49–55.

Bullock, R., Zauner, A., Woodward, J.J., Myseros, J., Choi, S.C., Ward, J.D., Marmarou, A., Young. H.F., 1998. Factors affecting excitatory amino acid release following severe human head injury. *J Neurosurg.* 89 (4):507–518.

Byun, J.S., Kwak, B.K., Kim, J.K., Jung, J., Ha, B.C., Park, S., 2013. Engraftment of human mesenchymal stem cells in a rat photothrombotic cerebral infarction model: Comparison of intra-arterial and intravenous infusion using MRI and histological analysis. *J Korean Neurosurg Soc.* 54:467–476.

Caplan, H.W., Mandy, F., Mandy, F., Zelnick, P., Pavuluri, Y., Mitchell, M.B., Smith, P., Cox, C.S., Bedi, S.S. Spatio-temporal distribution of microglia/macrophages after traumatic brain injury. *Brain Res.* (forthcoming).

Castro, R.F., Jackson, K.A., Goodell, M.A., Robertson, C.S., Liu, H., Shine, H.D., 2002. Failure of bone marrow cells to transdifferentiate into neural cells in vivo. *Science.* 297:1299.

Chang, C.-P., Chio, C.-C., Cheong, C.-U., Chao, C.-M., Cheng, B.-C., Lin, M.-T., 2013. Hypoxic preconditioning enhances the therapeutic potential of the secretome from cultured human mesenchymal stem cells in experimental traumatic brain injury. *Clin Sci.* 124:165–176.

Chen, Q., Long, Y., Yuan, X., Zou, L., Sun, J., Chen, S., Perez-Polo, J.R., Yang, K., 2005. Protective effects of bone marrow stromal cell transplantation in injured rodent brain: Synthesis of neurotrophic factors. *J Neurosci Res.* 80:611–619.

Chen, T., Qian, Y.Z., Rice, A., Zhu, J.P., Di, X., Bullock, R., 2000. Brain lactate uptake increases at the site of impact after traumatic brain injury. *Brain Res.* 861 (2):281–287.

Chesnut, R.M., Temkin, N., Carney, N., Dikmen, S., Rondina, C., Videtta, W., Petroni, G. et al., Global Neurotrauma Research Group, 2012. A trial of intracranial-pressure monitoring in traumatic brain injury. *N Engl J Med.* 367 (26):2471–2481.

Clausen, T., Bullock, R., 2001. Medical treatment and neuroprotection in traumatic brain injury. *Curr Pharm Design.* 7 (15):1517–1532.

Clausen, F., Hanell, A., Bjork, M., Hillered, L., Mir, A.K., Gram, H., Marklund, N., 2009. Neutralization of interleukin-1β modifies the inflammatory response and improves histological and cognitive outcome following traumatic brain injury in mice. *Eur J Neurosci.* 30 (3):385–396.

Clausen, F., Hanell, A., Israelsson, C., Hedin, J., Ebendal, T., Mir, A.K., Gram, H., Marklund, N., 2011. Neutralization of interleukin-1β reduces cerebral edema and tissue loss and improves late cognitive outcome following traumatic brain injury in mice. *Eur J Neurosci.* 34 (1):110–123.

Clausen, F., Lorant, T., Lewén, A., Hillered, L., 2007. T lymphocyte trafficking: A novel target for neuroprotection in traumatic brain injury. *J Neurotrauma.* 24:1295–1307.

Clausen, F., Marklund, N., Lewen, A., Hillered, L., 2008. The nitrone free radical scavenger NXY-059 is neuroprotective when administered after traumatic brain injury in the rat. *J Neurotrauma.* 25 (12):1449–1457.

Colicos, M.A., Dash, P.K., 1996. Apoptotic morphology of dentate gyrus granule cells following experimental cortical impact injury in rats: Possible role in spatial memory deficits. *Brain Res.* 739 (1–2):120–131.

Colicos, M.A., Dixon, C.E., Dash, P.K., 1996. Delayed, selective neuronal death following experimental cortical impact injury in rats: Possible role in memory deficits. *Brain Res.* 739 (1–2):111–119.

Cooper, D.J., Rosenfeld, J.V., Murray, L., Arabi, Y.M., Davies, A.R., D'Urso, P., Kossmann, T., Ponsford, J., Seppelt, I., Reilly, P., 2011. Decompressive craniectomy in diffuse traumatic brain injury. *N Engl J Med.* 364:1493–1502.

Cox Jr., C.S., Baumgartner, J.E., Harting, M.T., Worth, L.L., Walker, P.A., Shah, S.K., Ewing-Cobbs, L., Hasan, K.M., Day, M.-C., Lee, D., 2011. Autologous bone marrow mononuclear cell therapy for severe traumatic brain injury in children. *Neurosurgery.* 68:588–600.

Coyne, T.M., Marcus, A.J., Woodbury, D., Black, I.B., 2006. Marrow stromal cells transplanted to the adult brain are rejected by an inflammatory response and transfer donor labels to host neurons and glia. *Stem Cells.* 24:2483–2492.

Croft, A.P., Przyborski, S.A., 2009. Mesenchymal stem cells expressing neural antigens instruct a neurogenic cell fate on neural stem cells. *Exp Neurol.* 216:329–341.

Dai, S.S., Zhou, Y.G., Li, W., An, J.H., Li, P., Yang, N., Chen, X.Y. et al., 2010. Local glutamate level dictates adenosine A2A receptor regulation of neuroinflammation and traumatic brain injury. *J Neurosci.* 30 (16):5802–5810.

Das, M., Mohapatra, S., Mohapatra, S.S., 2012. New perspectives on central and peripheral immune responses to acute traumatic brain injury. *J Neuroinflammation.* 9:236.

Dash, P.K., Mach, S.A., Moore, A.N., 2000. Regional expression and role of cyclooxygenase-2 following experimental traumatic brain injury. *J Neurotrauma.* 17 (1):69–81.

Dash, P.K., Mach, S.A., Moore, A.N., 2002. The role of extracellular signal-regulated kinase in cognitive and motor deficits following experimental traumatic brain injury. *Neuroscience* 114 (3):755–767.

Dash, P.K., Moore, A.N., Dixon, C.E., 1995. Spatial memory deficits, increased phosphorylation of the transcription factor CREB, and induction of the AP-1 complex following experimental brain injury. *J Neurosci.* 15 (3, pt 1):2030–2039.

Dash, P.K., Moore, A.N., Moody, M.R., Treadwell, R., Felix, J.L., Clifton, G.L., 2004. Post-trauma administration of caffeine plus ethanol reduces contusion volume and improves working memory in rats. *J Neurotrauma.* 21:1573–1583.

Dash, P.K., Orsi, S.A, Zhang, M., Grill, R.J., Pati, S., Zhao, J., Moore, A.N., 2010. Valproate administered after traumatic brain injury provides neuroprotection and improves cognitive function in rats. *PLoS One* 5:e11383.

Dedeepiya, V.D., Rao, Y.Y., Jayakrishnan, G.A., Parthiban, J.K., Baskar, S., Manjunath, S.R., Senthilkumar, R., Abraham, S.J., 2012. Index of CD34+ cells and mononuclear cells in the bone marrow of spinal cord injury patients of different age groups: A comparative analysis. *Bone Marrow Res.* 2012:1–8.

Dewan, Y., Komolafe, E.O., Mejía-Mantilla, J.H., Perel, P., Roberts, I., Shakur, H., 2012. CRASH-3-tranexamic acid for the treatment of significant traumatic brain injury: Study protocol for an international randomized, double-blind, placebo-controlled trial. *Trials.* 13:87.

Dohi, K., Satoh, K., Mihara, Y., Nakamura, S., Miyake, Y., Ohtaki, H., Nakamachi, T., Yoshikawa, T., Shioda, S., Aruga, T., 2006. Alkoxyl radical-scavenging activity of edaravone in patients with traumatic brain injury. *J Neurotrauma.* 23 (11):1591–1599.

Dohi, K., Satoh, K., Nakamachi, T., Yofu, S., Hiratsuka, K., Nakamura, S., Ohtaki, H., Yoshikawa, T., Shioda, S., Aruga, T., 2007. Does edaravone (MCI-186) act as an antioxidant and a neuroprotector in experimental traumatic brain injury? *Antioxid Redox Signal.* 9 (2):281–287.

Eisenberg, H.M., Frankowski, R.F., Contant, C.F., Marshall, L.F., Walker, M.D., 1988. High-dose barbiturate control of elevated intracranial pressure in patients with severe head injury. *J Neurosurg.* 69 (1):15–23.

Ewing-Cobbs, L., Fletcher, J.M., Levin, H.S., Francis, D.J., Davidson, K., Miner, M.E., 1997. Longitudinal neuropsychological outcome in infants and preschoolers with traumatic brain injury. *J Int Neuropsychol Soc.* 3:581–591.

Ewing-Cobbs, L., Prasad, M.R., Swank, P., Kramer, L., Cox, C.S., Jr., Fletcher, J.M., Barnes, M., Zhang, X., Hasan, K.M., 2008. Arrested development and disrupted callosal microstructure following pediatric traumatic brain injury: Relation to neurobehavioral outcomes. *Neuroimage.* 42:1305–1315.

Faul, M., Xu, L., Wald, M., Coronado, V.G., 2010. *Traumatic Brain Injury in the United States: Emergency Department Visits, Hospitalizations and Deaths 2002–2006.* Atlanta, GA: Centers for Disease Control and Prevention, National Center for Injury Prevention and Control, 2–70.

Fischer, U.M., Harting, M.T., Jimenez, F., Monzon-Posadas, W.O., Xue, H., Savitz, S.I., Laine, G.A., Cox Jr., C.S., 2009. Pulmonary passage is a major obstacle for intravenous stem cell delivery: The pulmonary first-pass effect. *Stem Cells Dev.* 18:683–692.

Frattalone, A.R., Ling, G.S., 2013. Moderate and severe traumatic brain injury: Pathophysiology and management. *Neurosurg Clin N Am.* 24 (3):309–319.

Friedrich, M.A., Martins, M.P., Araújo, M.D., Klamt, C., Vedolin, L., Garicochea, B., Raupp, E.F., Sartori El Ammar, J., Machado, D.C., da Costa, J.C., 2012. Intra-arterial infusion of autologous bone marrow mononuclear cells in patients with moderate to severe middle cerebral artery acute ischemic stroke. *Cell Transplant.* 21:S13–S21.

Galindo, L.T., Filippo, T.R., Semedo, P., Ariza, C.B., Moreira, C.M., Camara, N.O., Porcionatto, M.A., 2011. Mesenchymal stem cell therapy modulates the inflammatory response in experimental traumatic brain injury. *Neurol Res Int.* 2011.

Gao, Q., Li, Y., Chopp, M., 2005. Bone marrow stromal cells increase astrocyte survival via upregulation of phosphoinositide 3-kinase/threonine protein kinase and mitogen-activated protein kinase kinase/extracellular signal-regulated kinase pathways and stimulate astrocyte trophic factor gene expression after anaerobic insult. *Neuroscience.* 136:123–134.

Gao, X., Deng-Bryant, Y., Cho, W., Carrico, K.M., Hall, E.D., Chen, J., 2008. Selective death of newborn neurons in hippocampal dentate gyrus following moderate experimental traumatic brain injury. *J Neurosci Res.* 86 (10):2258–2270.

Ge, S., Pachter, J.S., 2006. Isolation and culture of microvascular endothelial cells from murine spinal cord. *J Neuroimmunol.* 177: 209–214.

Gerber, L.M., Chiu, Y.L., Carney, N., Hartl, R., Ghajar, J., 2013. Marked reduction in mortality in patients with severe traumatic brain injury. *J Neurosurg.* 119:1583–1590.

Gentleman, S., Roberts, G., Gennarelli, T.A., Maxwell, W., Adams, J., Kerr, S., Graham, D., 1995. Axonal injury: A universal consequence of fatal closed head injury? *Acta Neuropathol.* 89:537–543.

Geoffroy, C.G., Zheng, B., 2014. Myelin-associated inhibitors in axonal growth after CNS injury. *Curr Opin Neurobiol.* 27:31–38.

Gould, E., Tanapat, P., McEwen, B.S., Flügge, G., Fuchs, E., 1998. Proliferation of granule cell precursors in the dentate gyrus of adult monkeys is diminished by stress. *Proc Natl Acad Sci.* 95:3168–3171.

Grände, P.O., 2006. The "Lund Concept" for the treatment of severe head trauma—Physiological principles and clinical application. *Intensive Care Med.* 32:1475–1484.

Gudeman, S.K., Kishore, P., Miller, D.J., Girevendulis, A.K., Lipper, M.H., Becker, D.P., 1979. The genesis and significance of delayed traumatic intracerebral hematoma. *Neurosurgery.* 5:309–313.

Guzman, R., De Los Angeles, A., Cheshier, S., Choi, R., Hoang, S., Liauw, J., Schaar, B., Steinberg, G., 2008. Intracarotid injection of fluorescence activated cell-sorted CD49d-positive neural stem cells improves targeted cell delivery and behavior after stroke in a mouse stroke model. *Stroke.* 39:1300–1306.

Harting, M.T., Jimenez, F., Adams, S.D., Mercer, D.W., Cox Jr., C.S., 2008. Acute, regional inflammatory response after traumatic brain injury: Implications for cellular therapy. *Surgery* 144 (5):803–813.

Harting, M.T., Jimenez, F., Xue, H., Fischer, U.M., Baumgartner, J., Dash, P.K., Cox Jr., C.S., 2009. Intravenous mesenchymal stem cell therapy for traumatic brain injury: Laboratory investigation. *J Neurosurg.* 110:1189.

Heinrich, C., Bergami, M., Gascon, S., Lepier, A., Vigano, F., Dimou, L., Sutor, B., Berninger, B., Gotz, M., 2014. Sox2-mediated conversion of NG2 glia into induced neurons in the injured adult cerebral cortex. *Stem Cell Reports.* 3:1000–1014.

Heppner, F.L., Greter, M., Marino, D. et al., 2005. Experimental autoimmune encephalomyelitis repressed by heral and central nervous systems. *Results Probl Cell Differ.* 48:339–351.

Hutchinson, P.J., Corteen, E., Czosnyka, M., Mendelow, A.D., Menon, D.K., Mitchell, P., Murray, G. et al. 2006. Decompressive craniectomy in traumatic brain injury: The randomized multicenter RESCUEicp study (www.RESCUEicp.com). *Acta Neurochir Suppl.* 96:17–20.

Isele, N.B., Lee, H.-S., Landshamer, S., Straube, A., Padovan, C.S., Plesnila, N., Culmsee, C., 2007. Bone marrow stromal cells mediate protection through stimulation of PI3-K/Akt and MAPK signaling in neurons. *Neurochem Int.* 50:243–250.

Itoh, T., Satou, T., Nishida, S., Tsubaki, M., Hashimoto, S., Ito, H., 2009. The novel free radical scavenger, edaravone, increases neural stem cell number around the area of damage following rat traumatic brain injury. *Neurotox Res.* 16 (4):378–389.

Jiang, Y., Jahagirdar, B.N., Reinhardt, R.L., Schwartz, R.E., Keene, C.D., Ortiz-Gonzalez, X.R., Reyes, M., Lenvik, T., Lund, T., Blackstad, M., 2002. Pluripotency of mesenchymal stem cells derived from adult marrow. *Nature.* 418:41–49.

Johnson, V.E., Stewart, J.E., Begbie, F.D., Trojanowski, J.Q., Smith, D.H., Stewart, W., 2013. Inflammation and white matter degeneration persist for years after a single traumatic brain injury. *Brain.* 136:28–42.

Johnston, A.J., Steiner, L.A., Chatfield, D.A., Coleman, M.R., Coles, J.P., Al-Rawi, P.G., Menon, D.K., Gupta, A.K., 2003. Effects of propofol on cerebral oxygenation and metabolism after head injury. *Br J Anaesth.* 91 (6):781–786.

Kamiya, N., Ueda, M., Igarashi, H., Nishiyama, Y., Suda, S., Inaba, T., Katayama, Y., 2008. Intra-arterial transplantation of bone marrow mononuclear cells immediately after reperfusion decreases brain injury after focal ischemia in rats. *Life Sci.* 83:433–437.

Ker, K., Roberts, I., Shakur, H., Coats, T.J., 2015. Antifibrinolytic drugs for acute traumatic injury. *Cochrane Database Sys Rev.* (5): CD004896.

Kigerl, K.A., Gensel, J.C., Ankeny, D.P., Alexander, J.K., Donnelly, D.J., Popovich, P.G., 2009. Identification of two distinct macrophage subsets with divergent effects causing either neurotoxicity or regeneration in the injured mouse spinal cord. *J Neurosci.* 29:13435–13444.

Kim, H.-J., Lee, J.-H., Kim, S.-H., 2010. Therapeutic effects of human mesenchymal stem cells on traumatic brain injury in rats: Secretion of neurotrophic factors and inhibition of apoptosis. *J Neurotrauma.* 27:131–138.

Kleindienst, A., McGinn, M.J., Harvey, H.B., Colello, R.J., Hamm, R.J., Bullock, M.R., 2005. Enhanced hippocampal neurogenesis by intraventricular S100B infusion is associated with improved cognitive recovery after traumatic brain injury. *J Neurotrauma.* 22 (6):645–655.

Kleindienst, A., Bullock, M.R., 2006. A critical analysis of the role of the neurotrophic protein S100B in acute brain injury. *J Neurotrauma.* 23 (8):1185–1200.

Koura, S.S., Doppenberg, E.M., Marmarou, A., Choi, S., Young, H.F., Bullock, R., 1998. Relationship between excitatory amino acid release and outcome after severe human head injury. *Acta Neurochir* Suppl. 71:244–246.

Kraus, M.F., Susmaras, T., Caughlin, B.P., Walker, C.J., Sweeney, J.A., Little, D.M., 2007. White matter integrity and cognition in chronic traumatic brain injury: A diffusion tensor imaging study. *Brain.* 130:2508–2519.

Kumar, A., Loane, D.J., 2012. Neuroinflammation after traumatic brain injury: Opportunities for therapeutic intervention. *Brain Behav Immun.* 26:1191–1201.

Kunz, T., Marklund, N., Hillered, L., Oliw, E.H., 2002. Cyclooxygenase-2, prostaglandin synthases, and prostaglandin H2 metabolism in traumatic brain injury in the rat. *J Neurotrauma.* 19 (9):1051–1064.

Kurland, D., Hong, C., Aarabi, B., Gerzanich, V., Simard, J.M., 2012. Hemorrhagic progression of a contusion after traumatic brain injury: A review. *J Neurotrauma.* 29:19–31.

Lapchak, P.A., 2010. A critical assessment of edaravone acute ischemic stroke efficacy trials: Is edaravone an effective neuroprotective therapy? *Expert Opin Pharmacother.* 11 (10):1753–1763.

Larsen, A., Kolind, K., Pedersen, D.S., Doering, P., Pedersen, M.O., Danscher, G., Penkowa, M., Stoltenberg, M., 2008. Gold ions bio-released from metallic gold particles reduce inflammation and apoptosis and increase the regenerative responses in focal brain injury. *Histochem Cell Biol.* 130 (4):681–692.

Lee, C., Agoston, D.V., 2010. Vascular endothelial growth factor is involved in mediating increased de novo hippocampal neurogenesis in response to traumatic brain injury. *J Neurotrauma*. 27 (3):541–553.

Lee, J.-M., Zipfel, G.J., Choi, D.W., 1999. The changing landscape of ischaemic brain injury mechanisms. *Nature*. 399:A7–A14.

Lee, R.H., Pulin, A.A., Seo, M.J., Kota, D.J., Ylostalo, J., Larson, B.L., Semprun-Prieto, L., Delafontaine, P., Prockop, D.J., 2009. Intravenous hMSCs improve myocardial infarction in mice because cells embolized in lung are activated to secrete the anti-inflammatory protein TSG-6. *Cell Stem Cell*. 5:54–63.

Lee, S.-T., Chu, K., Jung, K.-H., Kim, S.-J., Kim, D.-H., Kang, K.-M., Hong, N.H., Kim, J.-H., Ban, J.-J., Park, H.-K., 2008. Anti-inflammatory mechanism of intravascular neural stem cell transplantation in haemorrhagic stroke. *Brain*. 131:616–629.

Lee, T.H., Wisniewski, H.G., Vilcek, J., 1992. A novel secretory tumor necrosis factor-inducible protein (TSG-6) is a member of the family of hyaluronate binding proteins, closely related to the adhesion receptor CD44. *J Cell Biol*. 116:545–557.

Lenzlinger, P.M., Shimizu, S., Marklund, N., Thompson, H.J., Schwab, M.E., Saatman, K.E., Hoover, R.C. et al., 2005. Delayed inhibition of Nogo-A does not alter injury-induced axonal sprouting but enhances recovery of cognitive function following experimental traumatic brain injury in rats. *Neuroscience*. 134 (3):1047–1056.

Le Roux, P., 2014. Intracranial pressure after the BEST TRIP trial: A call for more monitoring. *Curr Opin Crit Care*. 20 (2):141–147.

Li, W., Dai, S., An, J., Li, P., Chen, X., Xiong, R., Liu, P. et al., 2008. Chronic but not acute treatment with caffeine attenuates traumatic brain injury in the mouse cortical impact model. *Neuroscience*. 151 (4):1198–1207.

Li, Z., Wang, B., Kan, Z., Zhang, B., Yang, Z., Chen, J., Wang, D., Wei, H., Zhang, J.N., Jiang, R., 2012. Progesterone increases circulating endothelial progenitor cells and induces neural regeneration after traumatic brain injury in aged rats. *J Neurotrauma*. 29 (2):343–353.

Lin, Q.M., Zhao, S., Zhou, L.L., Fang, X.S., Fu, Y., Huang, Z.T., 2013. Mesenchymal stem cells transplantation suppresses inflammatory responses in global cerebral ischemia: Contribution of TNF-alpha-induced protein 6. *Acta Pharmacol Sin*. 34:784–792.

Lindau, N.T., Bänninger, B.J., Gullo, M., Good, N.A., Bachmann, L.C., Starkey, M.L., Schwab, M.E., 2013. Rewiring of the corticospinal tract in the adult rat after unilateral stroke and anti-Nogo-A therapy. *Brain*. 139:739–756.

Liu, N., Chen, R., Du, H., Wang, J., Zhang, Y., Wen, J., 2009. Expression of IL-10 and TNF-alpha in rats with cerebral infarction after transplantation with mesenchymal stem cells. *Cell Mol Immunol*. 6:207–213.

Liu, S.J., Zou, Y., Belegu, V., Lv, L.-Y., Lin, N., Wang, T.-Y., McDonald, J.W., Zhou, X., Xia, Q.-J., Wang, T.-H., 2014a. Co-grafting of neural stem cells with olfactory ensheathing cells promotes neuronal restoration in traumatic brain injury with an anti-inflammatory mechanism. *J Neuroinflammation*. 11:66.

Liu, Y., Zhang, Y., Lin, L., Lin, F., Li, T., Du, H., Chen, R., Zheng, W., Liu, N., 2013. Effects of bone marrow-derived mesenchymal stem cells on the axonal outgrowth through activation of PI3K/AKT signaling in primary cortical neurons followed oxygen-glucose deprivation injury. *PloS One*. 8:e78514.

Liu, Y., Zhang, R., Yan, K., Chen, F., Huang, W., Lv, B., Sun, C., Xu, L., Li, F., Jiang, X., 2014b. Mesenchymal stem cells inhibit lipopolysaccharide-induced inflammatory responses of BV2 microglial cells through TSG-6. *J Neuroinflammation*. 11:135.

Liu, Z., Li, Y., Zhang, Z.G., Cui, X., Cui, Y., Lu, M., Savant-Bhonsale, S., Chopp, M., 2010. Bone marrow stromal cells enhance inter- and intracortical axonal connections after ischemic stroke in adult rats. *J Cereb Blood Flow Metab*. 30:1288–1295.

Liu, Z., Li, Y., Zhang, X., Savant-Bhonsale, S., Chopp, M., 2008. Contralesional axonal remodeling of the corticospinal system in adult rats after stroke and bone marrow stromal cell treatment. *Stroke.* 39:2571–2577.

Logan, T.T., Villapol, S., Symes, A.J., 2013. TGF-beta superfamily gene expression and induction of the Runx1 transcription factors in adult neurogenic regions after brain injury. *PLoS One* 8, E59250.

Long, D.A., Ghosh, K., Moore, A.N., Dixon, C.E., Dash, P.K., 1996. Deferoxamine improves spatial memory performance following experimental brain injury in rats. *Brain Res.* 717 (1–2):109–117.

Lu, D., Goussev, A., Chen, J., Pannu, P., Li, Y., Mahmood, A., Chopp, M., 2004. Atorvastatin reduces neurological deficit and increases synaptogenesis, angiogenesis, and neuronal survival in rats subjected to traumatic brain injury. *J Neurotrauma.* 21 (1):21–32.

Lu, D., Mahmood, A., Wang, L., Li, Y., Lu, M., Chopp, M., 2001. Adult bone marrow stromal cells administered intravenously to rats after traumatic brain injury migrate into brain and improve neurological outcome. *Neuroreport.* 12:559–563.

Lu, D., Mahmood, A., Zhang, R., Copp, M., 2003. Upregulation of neurogenesis and reduction in functional deficits following administration of DEtA/NONOate, a nitric oxide donor, after traumatic brain injury in rats. *J Neurosurg.* 99 (2):351–361.

Lu, D., Qu, C., Goussev, A., Jiang, H., Lu, C., Schallert, T., Mahmood, A., Chen, J., Li, Y., Chopp, M., 2007. Statins increase neurogenesis in the dentate gyrus, reduce delayed neuronal death in the hippocampal CA3 region, and improve spatial learning in rat after traumatic brain injury. *J Neurotrauma.* 24 (7):1132–1146.

Lu, J., Frerich, J.M., Turtzo, L.C., Li, S., Chiang, J., Yang, C., Wang, X. et al., 2013. Histone deacetylase inhibitors are neuroprotective and preserve NGF-mediated cell survival following traumatic brain injury. *Proc Natl Acad Sci USA.* 110 (26):10747–10752.

Lundberg, J., Södersten, E., Sundström, E., Le Blanc, K., Andersson, T., Hermanson, O., Holmin, S., 2012. Targeted intra-arterial transplantation of stem cells to the injured CNS is more effective than intravenous administration: Engraftment is dependent on cell type and adhesion molecule expression. *Cell Transplant.* 21:333–343.

Maas, A.I., Lingsma. H.F., 2008. New approaches to increase statistical power in TBI trials: Insights from the IMPACT study. *Acta Neurochir Suppl.* 101:119–124.

Maas, A.I., Murray, G., Henney III, H., Kassem, N., Legrand, V., Mangelus, M., Muizelaar, J.-P., Stocchetti, N., Knoller, N., 2006. Efficacy and safety of dexanabinol in severe traumatic brain injury: Results of a phase III randomised, placebo-controlled, clinical trial. *Lancet Neurol.* 5:38–45.

Madri, J.A., 2009. Modeling the neurovascular niche: Implications for recovery from CNS injury. *J Physiol Pharmacol.* 60 (Suppl 4):95–104.

Mahmood, A., Lu, D., Chopp, M., 2004. Intravenous administration of marrow stromal cells (MSCs) increases the expression of growth factors in rat brain after traumatic brain injury. *J Neurotrauma.* 21:33–39.

Mahmood, A., Lu, D., Wang, L., Chopp, M., 2002. Intracerebral transplantation of marrow stromal cells cultured with neurotrophic factors promotes functional recovery in adult rats subjected to traumatic brain injury. *J Neurotrauma.* 19:1609–1617.

Mahmood, A., Lu, D., Wang, L., Li, Y., Lu, M., Chopp, M., 2001. Treatment of traumatic brain injury in female rats with intravenous administration of bone marrow stromal cells. *Neurosurgery.* 49:1196–1204.

Mahmood, A., Wu, H., Qu, C., Mahmood, S., Xiong, Y., Kaplan, D., Chopp, M., 2014. Down-regulation of Nogo-A by collagen scaffolds impregnated with bone marrow stromal cell treatment after traumatic brain injury promotes axonal regeneration in rats. *Brain Res.* 1542:41–48.

Mahmood, A., Wu, H., Qu, C., Xiong, Y., Chopp, M., 2013. Effects of treating traumatic brain injury with collagen scaffolds and human bone marrow stromal cells on sprouting of corticospinal tract axons into the denervated side of the spinal cord. *J Neurosurg.* 118:381–389.

Marklund, N., Clausen, F., Lewen, A., Hovda, D.A., Olsson, Y., Hillered, L., 2001. Alpha-phenyl-tert-N-butyl nitrone (PBN) improves functional and morphological outcome after cortical contusion injury in the rat. *Acta Neurochir.* 143:73–81.

Marklund, N., Keck, C., Hoover, R., Soltesz, K., Millard, M., LeBold, D., Spangler, Z., Banning, A., Benson, J., McIntosh, T.K., 2005. Administration of monoclonal antibodies neutralizing the inflammatory mediators tumor necrosis factor alpha and interleukin-6 does not attenuate acute behavioral deficits following experimental traumatic brain injury in the rat. *Restor Neurol Neurosci.* 23 (1):31–42.

Maxwell, W.L., Bullock, R., Landholt, H., Fujisawa, H., 1994. Massive astrocytic swelling in response to extracellular glutamate—A possible mechanism for post-traumatic brain swelling? *Acta Neurochir Suppl.* 60:465–467.

Maxwell, W.L., Povlishock, J.T., Graham, D.L., 1997. A mechanistic analysis of nondisruptive axonal injury: A review. *J Neurotrauma.* 14:419–440.

McConeghy, K.W., Hatton, J., Hughes, L., Cook, A.M., 2012. A review of neuroprotection pharmacology and therapies in patients with acute traumatic brain injury. *CNS Drugs.* 26:613–636.

Melhem, S., Shutter, L., Kaynar, A., 2014. A trial of intracranial pressure monitoring in traumatic brain injury. *Crit Care.* 18 (1):302.

Menge, T., Zhao, Y., Zhao, J., Wataha, K., Geber, M., Zhang, J., Letourneau, P., Redell, J., Shen, L., Wang, J., 2012. Mesenchymal stem cells regulate blood-brain barrier integrity in traumatic brain injury through production of the soluble factor TIMP3. *Sci Transl Med.* 4 (161): 161ra150.

Mezey, E., Chandross, K.J., Harta, G., Maki, R.A., McKercher, S.R., 2000. Turning blood into brain: Cells bearing neuronal antigens generated in vivo from bone marrow. *Science.* 290:1779–1782.

Mezey, É., Key, S., Vogelsang, G., Szalayova, I., Lange, G.D., Crain, B., 2003. Transplanted bone marrow generates new neurons in human brains. *Proc Natl Acad Sci.* 100:1364–1369.

Miyamoto, K., Ohtaki, H., Dohi, K., Tsumuraya, T., Nakano, H., Kiriyama, K., Song, D., Aruga, T., Shioda, S., 2013a. Edaravone increases regional cerebral blood flow after traumatic brain injury in mice. *Acta Neurochir Suppl.* 118:103–109.

Miyamoto, K., Ohtaki, H., Dohi, K., Tsumuraya, T., Song, D., Kiriyama, K., Satoh, K., Shimizu, A., Aruga, T., Shioda, S. 2013b. Therapeutic time window for edaravone treatment of traumatic brain injury in mice. *BioMed Res Int.* 2013:379206.

Moniche, F., Gonzalez, A., Gonzalez-Marcos, J.-R., Carmona, M., Piñero, P., Espigado, I., Garcia-Solis, D., Cayuela, A., Montaner, J., Boada, C., 2012. Intra-arterial bone marrow mononuclear cells in ischemic stroke: A pilot clinical trial. *Stroke.* 43:2242–2244.

Morganti-Kossmann, M.C., Satgunaseelan, L., Bye, N., Kossmann, T., 2007. Modulation of immune response by head injury. *Injury.* 38:1392–1400.

Muir, K.W., 2006. Glutamate-based therapeutic approaches: Clinical trials with NMDA antagonists. *Curr Opin Pharmacol.* 6:53–60.

Myburgh, J.A., Cooper, D.J., Finfer, S.R., Venkatesh, B., Jones, D., Higgins, A., Bishop, N., Higlett, T., Australian, Australiasian Traumatic Brain Injury Study Investigators for the Australian, 2008. Epidemiology and 12-month outcomes from traumatic brain injury in Australia and New Zealand. *J Trauma.* 64:854–862.

Myseros, J.S., Bullock, R., 1995. The rationale for glutamate antagonists in the treatment of traumatic brain injury. *Ann NY Acad Sci.* 765:262–271; discussion 298.

Niu, W., Zang, T., Zou, Y., Fang, S., Smith, D.K., Bachoo, R., Zhang, C.L., 2013. In vivo reprogramming of astrocytes to neuroblasts in the adult brain. *Nature Cell Biol.* 15:1164–1175.

Novikova, L.N., Brohlin, M., Kingham, P.J., Novikov, L.N., Wiberg, M., 2011. Neuroprotective and growth-promoting effects of bone marrow stromal cells after cervical spinal cord injury in adult rats. *Cytotherapy.* 13:873–887.

Oddo, M., Levine, J.M., Mackenzie, L., Frangos, S., Feihl, F., Kasner, S.E., Katsnelson, M. et al., 2011. Brain hypoxia is associated with short-term outcome after severe traumatic brain injury independently of intracranial hypertension and low cerebral perfusion pressure. *Neurosurgery.* 69 (5):1037–1045; discussion 1045.

Oddo, M., Schmidt, J.M., Carrera, E., Badjatia, N., Connolly, E.S., Presciutti, M., Ostapkovich, N.D., Levine, J.M., Le Roux, P., Mayer, S.A., 2008. Impact of tight glycemic control on cerebral glucose metabolism after severe brain injury: A microdialysis study. *Crit Care Med.* 36 (12):3233–3238.

Okonkwo, D.O., Buki, A., Siman, R., Povlishock, J.T., 1999. Cyclosporin A limits calcium-induced axonal damage following traumatic brain injury. *Neuroreport.* 10 (2):353–358.

Palacios, E.M., Sala-Llonch, R., Junque, C., Fernandez-Espejo, D., Roig, T., Tormos, J.M., Bargallo, N., Vendrell, P., 2013. Long-term declarative memory deficits in diffuse TBI: Correlations with cortical thickness, white matter integrity and hippocampal volume. *Cortex.* 49:646–657.

Panczykowski, D.M., Puciao, P.M., Scruggs, B.J., Bauer, J.S., Hricik, A.J., Beers, S.R., Okonkwo, D.O., 2012. Prospective independent validation of IMPACT modeling as a prognostic tool in severe traumatic brain injury. *J Neurotrauma.* 29:47–52.

Pati, S., Khakoo, A.Y., Zhao, J., Jimenez, F., Gerber, M.H., Harting, M., Redell, J.B. et al., 2011. Human mesenchymal stem cells inhibit vascular permeability by modulating vascular endothelial cadherin/beta-catenin signaling. *Stem Cells Dev.* 20:89–101.

Pedersen, M.O., Larsen, A., Pedersen, D.S., Stoltenberg, M., Penkova, M., 2009. Metallic gold treatment reduces proliferation of inflammatory cells, increases expression of VEGF and FGF, and stimulates cell proliferation in the subventricular zone following experimental traumatic brain injury. *Histol Histopathol.* 24 (5):573–586.

Pendharkar, A.V., Chua, J.Y., Andres, R.H., Wang, N., Gaeta, X., Wang, H., De, A., Choi, R., Chen, S., Rutt, B.K., 2010. Biodistribution of neural stem cells after intravascular therapy for hypoxic–ischemia. *Stroke.* 41:2064–2070.

Perel, P., Al-Shahi Salman, R., Kawahara, T., Morris, Z., Prieto-Merino, D., Roberts, I., Sandercock, P., Shakur, H., Wardlaw, J., 2012. CRASH-2 (Clinical Randomisation of an Antifibrinolytic in Significant Haemorrhage) intracranial bleeding study: The effect of tranexamic acid in traumatic brain injury—A nested randomised, placebo-controlled trial. *Health Technol Assess.* 16 (iii–xii):1–54.

Pittenger, M.F., Mackay, A.M., Beck, S.C., Jaiswal, R.K., Douglas, R., Mosca, J.D., Moorman, M.A., Simonetti, D.W., Craig, S., Marshak, D.R., 1999. Multilineage potential of adult human mesenchymal stem cells. *Science.* 284:143–147.

Prockop, L.D., Naidu, K.A., Binard, J.E., Ransohoff, J., 1995. Selective permeability of 3H-D-mannitol and 14C-carbolyl-inulin across the blood-brain barrier and blood-spinal cord barrier in the rabbit. *J Spinal Cord Med.* 18:221–226.

Qu, C., Mahmood, A., Ning, R., Xiong, Y., Zhang, L., Chen, J., Jiang, H., Chopp, M., 2010. The treatment of traumatic brain injury with velcade. *J Neurotrauma.* 27 (9):1625–1634.

Ramlackhansingh, A.F., Brooks, D.J., Greenwood, R.J., Bose, S.K., Turkheimer, F.E., Kinnunen, K.M., Gentleman, S. et al., 2011. Inflammation after trauma: Microglial activation and traumatic brain injury. *Ann Neurol.* 70:374–383.

Redell, J.B., Zhao, J., Dash, P.K., 2007. Acutely increased cyclophilin a expression after brain injury: A role in blood-brain barrier function and tissue preservation. *J Neurosi Res.* 85:1980–1988.

Riess, P., Zhang, C., Saatman, K.E., Laurer, H.L., Longhi, L.G., Raghupathi, R., Lenzlinger, P.M. et al., 2002. Transplanted neural stem cells survive, differentiate, and improve neurological motor function after experimental traumatic brain injury. *Neurosurgery.* 51:1043–1052; discussion 1052–1054.

Roberts, R.M., Mathias, J.L., Rose, S.E., 2014. Diffusion tensor imaging (DTI) findings following pediatric non-penetrating TBI: A meta-analysis. *Dev Neuropsychol.* 39, 600–637.

Roobrouck, V.D., Clavel, C., Jacobs, S.A., Ulloa-Montoya, F., Crippa, S., Sohni, A., Roberts, S.J. et al. 2011. Differentiation potential of human postnatal mesenchymal stem cells, mesoangioblasts, and multipotent adult progenitor cells reflected in their transcriptome and partially influenced by the culture conditions. *Stem Cells.* 29:871–882.

Royo, N.C., Wahl, F., Stutzmann, J.-M., 1999. Kinetics of polymorphonuclear neutrophil infiltration after a traumatic brain injury in rat. *Neuroreport* 10:1363–1367.

Rutland-Brown, W., Langlois, J.A., Thomas, K.E., Xi, Y.L., 2006. Incidence of traumatic brain injury in the United States, 2003. *J Head Trauma Rehabil.* 21:544–548.

Saatman, K.E., Duhaime, A.-C., Bullock, R., Maas, A.I., Valadka, A., Manley, G.T., 2008. Classification of traumatic brain injury for targeted therapies. *J Neurotrauma.* 25:719–738.

Sakellaridis, N., Pavlou, E., Karatzas, S., Chroni, D., Vlachos, K., Chatzopoulos, K., Dimopoulou, E., Kelesis, C., Karaouli, V., 2011. Comparison of mannitol and hypertonic saline in the treatment of severe brain injuries. *J Neurosurg.* 114 (2):545–548.

Sánchez-Aguilar, M., Tapia-Pérez, J.H., Sánchez-Rodríguez, J.J., Viñas-Ríos, J.M., Martínez-Pérez, P., de la Cruz-Mendoza, E., Sánchez-Reyna, M. et al. 2013. Effect of rosuvastatin on cytokines after traumatic head injury: Clinical article. *J Neurosurg.* 118:669–675.

Satoh, K., Ikeda, Y., Shioda, S., Tobe, T., Yoshikawa, T., 2002. Edarabone scavenges nitric oxide. *Redox Rep.* 7 (4):219–222.

Savitz, S.I., Misra, V., Kasam, M., Juneja, H., Cox, C.S., Alderman, S., Aisiku, I., Kar, S., Gee, A., Grotta, J.C., 2011. Intravenous autologous bone marrow mononuclear cells for ischemic stroke. *Ann Neurol.* 70:59–69.

Scalfani, M.T., Dhar, R., Zazulia, A.R., Videen, T.O., Diringer, M.N., 2012. Effect of osmotic agents on regional cerebral blood flow in traumatic brain injury. *J Crit Care.* 27 (5):526 e7–12.

Scherbel, U., Raghupathi, R., Nakamura, M., Saatman, K.E., Trojanowski, J.Q., Neugebauer, E., Marino, M.W., McIntosh, T.K., 1999. Differential acute and chronic responses of tumor necrosis factor-deficient mice to experimental brain injury. *Proc Natl Acad Sci.* 96:8721–8726.

Schnell, L., Fearn, S., Klassen, H., Schwab, M.E., Perry, V.H. 1999. Acute inflammatory responses to mechanical lesions in the CNS: Differences between brain and spinal cord. *Eur J Neurosci.* 11:3648–3658.

Schnüriger, B., Inaba, K., Abdelsayed, G.A., Lustenberger, T., Eberle, B.M., Barmparas, G., Talving, P., Demetriades, D., 2010. The impact of platelets on the progression of traumatic intracranial hemorrhage. *J Trauma.* 68:881–885.

Schrepfer, S., Deuse, T., Reichensupurner, H., Fischbein, M.P., Robbins, R.C., Pelletier, M.P., 2007. Stem cell transplantation: The lung barrier. *Tranplant Proc.* 39: 573–576.

Schwamm, L.H., 2014. Progesterone for traumatic brain injury—Resisting the sirens' song. *N Engl J Med.* 371:2522–2523.

Shi, J., Longo, F.M., Massa, S.M., 2013. A small molecule P75 ligand protects neurogenesis after traumatic brain injury. *Stem Cells.* 31:2561–2574.

Shohami, E., Ginis, I., Hallenbeck, J.M., 1999. Dual role of tumor necrosis factor alpha in brain injury. *Cytokine Growth Factor Rev.* 10:119–130.

Skandsen, T., Kvistad, K.A., Solheim, O., Strand, I.H., Folvik, M., Vik, A., 2010. Prevalence and impact of diffuse axonal injury in patients with moderate and severe head injury: A cohort study of early magnetic resonance imaging findings and 1-year outcome. *J Neurosurg.* 113:556–563.

Starkov, A.A., Chinopoulos, C., Fiskum, G., 2004. Mitochondrial calcium and oxidative stress as mediators of ischemic brain injury. *Cell Calcium*. 36:257–264.

Stein, N.R., McArthur, D.L., Etchepare, M., Vespa, P.M., 2012. Early cerebral metabolic crisis after TBI influences outcome despite adequate hemodynamic resuscitation. *Neurocrit Care*. 17:49–57.

Strich, S.J., 1956. Diffuse degeneration of the cerebral white matter in severe dementia following head injury. *J Neurol Neurosurg Psychiatry*. 19:163–185.

Su, Z., Niu, W., Liu, M.L., Zou, Y., Zhang, C.L., 2014. In vivo conversion of astrocytes to neurons in the injured adult spinal cord. *Nat Commun*. 5:3338.

Suarez, J.I., Zaidat, O.O., Suri, M.F., Feen, E.S., Lynch, G., Hickman, J., Georgiadis, A., Selman, W.R., 2004. Length of stay and mortality in neurocritically ill patients: Impact of a specialized neurocritical care team. *Crit Care Med*. 32 (11):2311–2317.

Sun, D., Bullock, M.R., McGinn, M.J., Zhou, Z., Altememi, N., Hagood, S., Hamm, R., Colello, R.J., 2009. Basic fibroblast growth factor-enhanced neurogenesis contributes to cognitive recovery in rats following traumatic brain injury. *Exp Neurol*. 216:56–65.

Sun, D., Bullock, M.R., Altememi, N., Zhou, Z., Hagood, S., Rolfe, A., McGinn, M.J., Hamm, R., Colello, R.J., 2010. The effect of epidermal growth factor in the injured brain after trauma in rats. *J Neurotrauma*. 27 (5):923–938.

Sydenham, E., Roberts, I., Alderson, P., 2009. Hypothermia for traumatic head injury. *Cochrane Database Syst Rev*. (2):CD001048.

Talving, P., Karamanos, E., Teixeira, P.G., Skiada, D., Lam, L., Belzberg, H., Inaba, K., Demetriades, D., 2013. Intracranial pressure monitoring in severe head injury: Compliance with Brain Trauma Foundation guidelines and effect on outcomes: A prospective study. *J Neurosurg*. 119 (5):1248–1254.

Tejima, E., Guo, S., Murata, Y., Arai, K., Lok, J., van Leyen, K., Rosell, A., Wang, X., Lo, E.H., 2009. Neuroprotective effects of overexpressing tissue inhibitor of metalloproteinase TIMP-1. *J Neurotrauma*. 26 (11):1935–1941.

Temkin, N.R., Anderson, G.D., Winn, H.R., Ellenbogen, R.G., Britz, G.W., Schuster, J., Lucas, T., Newell, D.W., Mansfield, P.N., Machamer, J.E., 2007. Magnesium sulfate for neuroprotection after traumatic brain injury: A randomised controlled trial. *Lancet Neurol*. 6:29–38.

Terada, N., Hamazaki, T., Oka, M., Hoki, M., Mastalerz, D.M., Nakano, Y., Meyer, E.M., Morel, L., Petersen, B.E., Scott, E.W., 2002. Bone marrow cells adopt the phenotype of other cells by spontaneous cell fusion. *Nature*. 416:542–545.

Thau-Zuchman, O., Shohami, E., Alexandrovich, A.G., Leker, R.R., 2010. Vascular endothelial growth factor increases neurogenesis after traumatic brain injury. *J Cereb Blood Flow Metab*. 30 (5):1008–1016.

Thompson, H.J., Marklund, N., LeBold, D.G., Morales, D.M., Keck, C.A., Vinson, M., Royo, N.C., Grundy, R., McIntosh, T.K., 2006. Tissue sparing and functional recovery following experimental traumatic brain injury is provided by treatment with an anti-myelin-associated glycoprotein antibody. *Eur J Neurosci*. 24 (11):3063–3072.

Tian, C., Wang, X., Wang, L., Wu, S., Wan, Z., 2013. Autologous bone marrow mesenchymal stem cell therapy in the subacute stage of traumatic brain injury by lumbar puncture. *Exp Clin Transplant*. 11:176–181.

Titus, D.J., Sakurai, A., Kang, Y., Furones, C., Jergova, S., Santos, R., Sick, T.J., Atkins, C.M., 2013. Phosphodiesterase inhibition rescues chronic cognitive deficits induced by traumatic brain injury. *J Neurosci*. 33 (12):5216–5226.

Turtzo, L.C., Lescher, J., Janes, L., Dean, D.D., Budde, M.D., Frank, J.A., 2014. Macrophagic and microglial responses after focal traumatic brain injury in the female rat. *J Neuroinflammation*. 11:82.

van Rooijen, N., Sanders, A., 1994. Liposome mediated depletion of macrophages: Mechanism of action, preparation of liposomes and applications. *J Immmunol Methods*. 174: 83–93.
Varelas, P.N., Conti, M.M., Spanaki, M.V., Potts, E., Bradford, D., Sunstrom, C., Fedder, W., Hacein Bey, L., Jaradeh, S., Gennarelli, T.A., 2004. The impact of a neurointensivist-led team on a semiclosed neurosciences intensive care unit. *Crit Care Med*. 32 (11):2191–2198.
Vasconcelos-dos-Santos, A., Rosado-de-Castro, P.H., Lopes de Souza, S.A., da Costa Silva, J., Ramos, A.B., Rodriguez de Freitas, G., Barbosa da Fonseca, L.M., Gutfilen, B., Mendez-Otero, R., 2012. Intravenous and intra-arterial administration of bone marrow mononuclear cells after focal cerebral ischemia: Is there a difference in biodistribution and efficacy? *Stem Cell Res*. 9:1–8.
Vendrame, M., Gemma, C., Pennypacker, K.R., Bickford, P.C., Davis Sanberg, C., Sanberg, P.R., Willing, A.E., 2006. Cord blood rescues stroke-induced changes in splenocyte phenotype and function. *Exp Neurol*. 199:191–200.
Vespa, P., Bergsneider, M., Hattori, N., Wu, H.M., Huang, S.C., Martin, N.A., Glenn, T.C., McArthur, D.L., Hovda, D.A., 2005. Metabolic crisis without brain ischemia is common after traumatic brain injury: A combined microdialysis and positron emission tomography study. *J Cereb Blood Flow Metab*. 25 (6):763–774.
Wallenquist, U., Brannvall, K., Clausen, F., Lewen, A., Hillered, L., Forsberg-Nilsson, K., 2009. Grafted neural progenitors migrate and form neurons after experimental traumatic brain injury. *Restor Neurol Neurosci*. 27 (4):323–334.
Wallenquist, U., Holmqvist, K., Hanell, A., Marklund, N., Hillered, L., Forsberg-Nilsson, K., 2012. Ibuprofen attenuates the inflammatory response and allows formation of migratory neuroblasts from grafted stem cells after traumatic brain injury. *Restor Neurol Neurosci*. 30: 9–19.
Walker, P.A., Bedi, S.S., Shah, S.K., Jimenez, F., Xue, H., Hamilton, J.A., Smith, P., Thomas, C.P., Mays, R.W., Pati, S., 2012. Intravenous multipotent adult progenitor cell therapy after traumatic brain injury: Modulation of the resident microglia population. *J Neuroinflammation*. 9:228.
Walker, P.A., Harting, M.T., Jimenez, F., Shah, S.K., Pati, S., Dash, P.K., Cox Jr., C.S., 2010a. Direct intrathecal implantation of mesenchymal stromal cells leads to enhanced neuroprotection via an NFkappaB-mediated increase in interleukin-6 production. *Stem Cells Dev*. 19:867–876.
Walker, P.A., Letourneau, P.A., Bedi, S., Shah, S.K., Jimenez, F., Cox Jr., C.S., 2011. Progenitor cells as remote "bioreactors": Neuroprotection via modulation of the systemic inflammatory response. *World J Stem Cells*. 3:9.
Walker, P.A., Shah, S.K., Jimenez, F., Aroom, K.R., Harting, M.T., Cox Jr., C.S., 2012b. Bone marrow–derived stromal cell therapy for traumatic brain injury is neuroprotective via stimulation of non-neurologic organ systems. *Surgery*. 152:790–793.
Walker, P.A., Shah, S.K., Jimenez, F., Gerber, M.H., Xue, H., Cutrone, R., Hamilton, J.A., Mays, R.W., Deans, R., Pati, S., 2010b. Intravenous multipotent adult progenitor cell therapy for traumatic brain injury: Preserving the blood brain barrier via an interaction with splenocytes. *Exp Neurol*. 225:341–352.
Wang, G.H., Jiang, Z.L., Li, Y.C., Li, X., Shi, H., Gao, Y.Q., Vosler, P.S., Chen, J., 2011. Free-radical scavenger edaravone treatment confers neuroprotection against traumatic brain injury in rats. *J Neurotrauma*. 28 (10):2123–2134.
Wang, G., Zhang, J., Hu, X., Zhang, L., Mao, L., Jiang, X., Liou, A.K.-F., Leak, R.K., Gao, Y., Chen, J., 2013a. Microglia/macrophage polarization dynamics in white matter after traumatic brain injury. *J Cereb Blood Flow Metab*. 33:1864–1874.
Wang, N., Li, Q., Zhang, L., Lin, H., Hu, J., Li, D., Shi, S., Cui, S., Zhou, J., Ji, J., 2012. Mesenchymal stem cells attenuate peritoneal injury through secretion of TSG-6. *PloS One*. 7: e43768.

Wang, S., Cheng, H., Dai, G., Wang, X., Hua, R., Liu, X., Wang, P., Chen, G., Yue, W., An, Y., 2013b. Umbilical cord mesenchymal stem cell transplantation significantly improves neurological function in patients with sequelae of traumatic brain injury. *Brain Res.* 1532: 76–84.

Watanabe, J., Shetty, A.K., Hattiangady, B., Kim, D.-K., Foraker, J.E., Nishida, H., Prockop, D.J., 2013. Administration of TSG-6 improves memory after traumatic brain injury in mice. *Neurobiol Dis.* 59:86–99.

Wilde, E.A., Ayoub, K.W., Bigler, E.D., Chu, Z.D., Hunter, J.V., Wu, T.C., McCauley, S.R., Levin, H.S., 2012. Diffusion tensor imaging in moderate-to-severe pediatric traumatic brain injury: Changes within an 18 month post-injury interval. *Brain Imaging Behav.* 6: 404–416.

Williams, A.J., Wei, H.H., Dave, J.R., Tortella, F.C., 2007. Acute and delayed neuroinflammatory response following experimental penetrating ballistic brain injury in the rat. *J Neuroinflammation.* 4:17.

Williams, S., Raghupathi, R., MacKinnon, M.A., McIntosh, T.K., Saatman, K.E., Graham, D.I., 2001. In situ DNA fragmentation occurs in white matter up to 12 months after head injury in man. *Acta Neuropathologica.* 102 (6):581–590.

Wong, G., Goldshmit, Y., Turnley. A.M., 2004. Interferon-gamma but not TNF alpha promotes neuronal differentiation and neurite outgrowth of murine adult neural stem cells. *Exp Neurol.* 187 (1):171–177.

Woodbury, D., Schwarz, E.J., Prockop, D.J., Black, I.B., 2000. Adult rat and human bone marrow stromal cells differentiate into neurons. *J Neurosci Res.* 61:364–370.

Woodcock, T., Morganti-Kossmann, M.C., 2013. The role of markers of inflammation in traumatic brain injury. *Front Neurol.* 4:18.

Worthylake, R.A., Burridge, K., 2001. Leukocyte transendothelial migration: Orchestrating the underlying molecular machinery. *Curr Opin Cell Biol.* 13:569–577.

Wright, K.T., El Masri, W., Osman, A., Roberts, S., Chamberlain, G., Ashton, B.A., Johnson, W.E., 2007b. Bone marrow stromal cells stimulate neurite outgrowth over neural proteoglycans (CSPG), myelin associated glycoprotein and Nogo-A. *Biochem Biophys Res Commun.* 354:559–566.

Wu, T.C., Wilde, E.A., Bigler, E.D., Li, X., Merkley, T.L., Yallampalli, R., McCauley, S.R. et al. 2010. Longitudinal changes in the corpus callosum following pediatric traumatic brain injury. *Dev Neurosci.* 32: 361–373.

Xiong, Y., Qu, C., Mahmood, A., Liu, Z., Ning, R., Li, Y., Kaplan, D.L., Schallert, T., Chopp, M., 2009. Delayed transplantation of human marrow stromal cell-seeded scaffolds increases transcallosal neural fiber length, angiogenesis, and hippocampal neuronal survival and improves functional outcome after traumatic brain injury in rats. *Brain Res.* 1263:183–191.

Xu, H., Ma, C., Cao, L., Wang, J., Fan, X., 2014. Study of co-transplantation of SPIO labeled bone marrow stromal stem cells and Schwann cells for treating traumatic brain injury in rats and in vivo tracing of magnetically labeled cells by MRI. *Eur Rev Med Pharmacol Sci.* 18:520–525.

Yamamoto, T., Rossi, S., Stiefel, M., Doppenberg, E., Zauner, A., Bullock, M.R., Marmarou, A., 1999 CSF and ECF glutamate concentrations in head injured patients. *Acta Neurochir Suppl.* 75:17–19.

Yang, B., Migliati, E., Parsha, K., Schaar, K., Xi, X., Aronowski, J., Savitz, S.I., 2013. Intra-arterial delivery is not superior to intravenous delivery of autologous bone marrow mononuclear cells in acute ischemic stroke. *Stroke.* 44:3463–3472.

Yang, X., Yang, S., Wang, J., Zhang, X., Wang, C., Hong, G., 2009. Expressive proteomics profile changes of injured human brain cortex due to acute brain trauma. *Brain Inj.* 23:830–840.

Yao, X.L., Liu, J., Lee, E., Ling, G.S., McCabe, J.T., 2005. Progesterone differentially regulates pro- and anti-apoptotic gene expression in cerebral cortex following traumatic brain injury in rats. *J Neurotrauma*. 22 (6):656–668.

Yeo, J.E., Kang, S.K., 2007. Selenium effectively inhibits ROS-mediated apoptotic neural precursor cell death in vitro and in vivo in traumatic brain injury. *Biochim Biophys Acta*. 1772 (11–12):1199–1210.

Yutthakasemsunt, S., Kittiwatanagul, W., Piyavechvirat, P., Thinkamrop, B., Phuenpathom, N., Lumbiganon, P., 2013. Tranexamic acid for patients with traumatic brain injury: A randomized, double-blinded, placebo-controlled trial. *BMC Emerg Med*. 13: 20.

Zafonte, R.D., Bagiella, E., Ansel, B.M., Novack, T.A., Friedewald, W.T., Hesdorffer, D.C., Timmons, S.D. et al., 2012. Effect of citicoline on functional and cognitive status among patients with traumatic brain injury: Citicoline Brain Injury Treatment Trial (COBRIT). *JAMA*. 308 (19):1993–2000.

Zaloshnja, E., Miller, T., Langlois, J.A., Selassie, A.W., 2008. Prevalence of long-term disability from traumatic brain injury in the civilian population of the United States, 2005. *J Head Trauma Rehabil*. 23:394–400.

Zauner, A., Bullock, R., Kuta, A.J., Woodward, J., Young, H.F., 1996. Glutamate release and cerebral blood flow after severe human head injury. *Acta Neurochir Suppl*. 67:40–44.

Zhang, R., Liu, Y., Yan, K., Chen, L., Chen, X.-R., Li, P., Chen, F.-F., Jiang, X.-D., 2013. Anti-inflammatory and immunomodulatory mechanisms of mesenchymal stem cell transplantation in experimental traumatic brain injury. *J Neuroinflammation*. 10:106.

Zhang, Z., Yan, R., Zhang, Q., Li, J., Kang, X., Wang, H., Huan, L. et al. 2014. Hes1, a Notch signaling downstream target, regulates adult hippocampal neurogenesis following traumatic brain injury. *Brain Res*. 1583:65–78.

Zhang, Z., Zhang, Z., Guan, L., Zhang, K., Zhang, Q., Dai, L., 2008. A combined procedure to deliver autologous mesenchymal stromal cells to patients with traumatic brain injury. *Cytotherapy*. 10:134–139.

Zhao, J., Moore, A.N., Clifton, G.L., Dash, P.K., 2005. Sulforaphane enhances aquaporin-4 expression and decreases cerebral edema following traumatic brain injury. *J Neurosc Res*. 82 (4):499–506.

Zhao, J., Pati, S., Redell, J.B., Zhang, M., Moore, A.N., Dash, P.K., 2012. Caffeic acid phenethyl ester protects blood-brain barrier integrity and reduces contusion volume in rodent models of traumatic brain injury. *J Neurotrauma*. 29:1209–1218.

2 Oligodendrocyte Precursor Cells in Spinal Cord Injury Repair

Qi Lin Cao

CONTENTS

2.1 INTRODUCTION

Despite extensive research, clinical advancements, and improved rehabilitation strategies, spinal cord injury (SCI) continues to be a significant cause of disability and mortality (Sekhon and Fehlings, 2001). About 11,000 new cases of SCI are reported each year, and 250,000 patients now live in the United States. Economic costs for SCI approach $10 billion a year (http://www.Sfn.org). Pathophysiology of SCI involves two broad chronological events: the primary injury and the secondary injury. The primary injury that encompasses the focal destruction of neural tissue caused by direct mechanical trauma then instigates a progressive wave of secondary injury, which via the activation of a barrage of noxious pathophysiological mechanisms exacerbates the injury and leads to greater functional loss after SCI (reviews, see Tator and Fehlings, 1991; Hulsebosch, 2002; Profyris et al., 2004). Oligodendrocytes (OLs) are particularly susceptible to oxidative stress, glutamate excitotoxicity, and the immune responses associated with the secondary injury cascade after SCI (Springer et al., 1999; Casha et al., 2001; Beattie et al., 2002; Nottingham et al., 2002; Park et al., 2004). Oligodendrocyte death and/or apoptosis occur at the injury center as early as a few hours following injury and significantly increase for several days thereafter (Crowe et al., 1997; Emery et al., 1998; Yong et al., 1998; Li et al., 1999). By 1 week, the number of apoptotic cells decreases at the injury center and there is an increase in apoptotic death away from the primary injury. This new apoptotic wave is predominantly localized in the white matter and can arise at large distances from the lesion center (Crowe et al., 1997; Liu et al., 1997; Li et al., 1999). The later phase of apoptosis lasts for at least a few weeks (Crowe et al., 1997; Li et al., 1999). Such delayed onset of apoptosis in OLs distant to the injury site appears to be unique to SCI and has important therapeutic implications. Because each OL myelinates multiple axons, their death leads to demyelination of many axons, which are left intact by the initial injury (Bunge et al., 1993; Totoiu and Keirstead, 2005; Cao et al., 2005b) (Figure 2.1). Consequently, the electrophysiological conductions of these axons are lost or delayed (Gledhill et al., 1973; Itoyama et al., 1983; Dusart et al., 1992). Dysfunction of these demyelinated axons in the injury epicenter and also in the areas distant to this epicenter may contribute further to long-term neurological deficits after SCI. Furthermore, OLs may provide trophic support to its myelinated axons by both contact-mediated and soluble mechanisms (Griffiths et al., 1998; Wilkins et al., 2001, 2003). Demyelinated axons are more vulnerable to the insults in the injured spinal cord and undergo secondary degeneration (Tsunoda and Fujinami, 2002; Compston, 2006). Importantly, preservation of a small portion of axons (5%–10%) in each individual tract can achieve significant meaningful recovery with locomotion (Blight, 1983b; Blight and DeCrescito, 1986). Numerous demyelinated, but otherwise intact, axons are observed after SCI in both experimental animals (Totoiu and Keirstead, 2005; Cao et al., 2005b) and humans (Bunge et al., 1993). Therefore, promoting remyelination is an important therapeutic strategy to enhance functional recovery by restoring the electrophysiological conduction of the demyelinated axons as well as preventing its degeneration. Oligodendrocyte precursor cells (OPCs) are the major cells responsible for

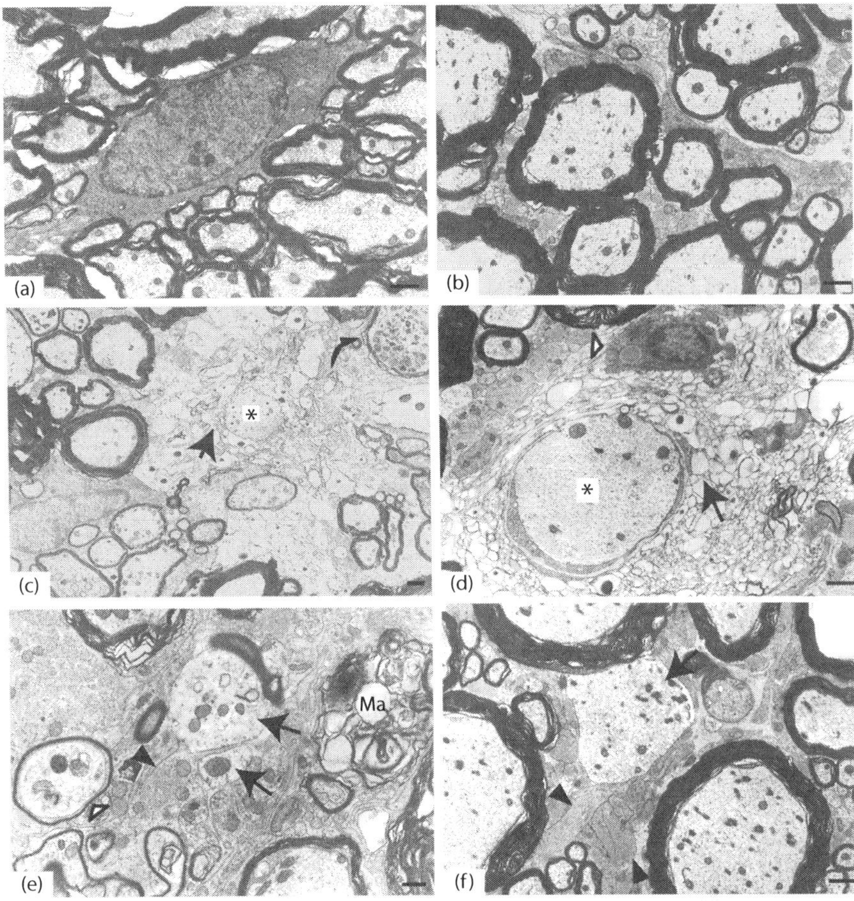

FIGURE 2.1 Demyelination after contusive SCI. (a,b) In the ventrolateral funiculus VLF of normal adult spinal cord, darkly stained compact myelin, the thickness of which is proportional to axon diameter, surrounded each axon. (c,d) In the VLF at 1 week after SCI, many myelin sheaths had degenerated, and numerous intramyelinic vacuoles were seen (arrows). (d) A macrophage was observed in close relation to degenerating myelin (open arrowhead). (c,d) The axons surrounded by degenerating myelin appeared morphologically normal (asterisks). (c) These axons could be distinguished from those that were undergoing degeneration (curved arrows). (e,f) Some demyelinated axons survived for at least 1 month after injury (arrows). (e,f) Healthy appearing demyelinated axons were observed in the VLF at the injury epicenter (e) and also a few millimeters away from the epicenter (f, arrow). These axons are adjacent to astrocytic processes (arrowheads). Scale bars, 2.5 μm. (Reproduced from Cao Q et al., 2005a. *J Neurosci* 25:6947–6957.)

the remyelination after SCI. In this chapter, we first discuss the origin of OPCs during development and the signals that regulate its development. We then briefly discuss the potential roles of OPCs in the normal central nervous system (CNS) and after SCI. Finally, we discuss the therapeutic potential of OPC transplantation after SCI and the concerns for its clinical translation.

2.2 DEFINITION AND MARKERS FOR OLIGODENDROCYTE PRECURSOR CELLS

Oligodendrocyte precursor cells (OPCs) are first identified in the low density culture of perinatal rat optical nerve by the expression of ganglioside moieties recognized by A2B5 antibody (Raff et al., 1983). They are first termed as *O-2A progenitors* because they could differentiate into OLs or type-2 astrocytes (astrocytes expressing both glial fibrillary acidic protein [GFAP] and A2B5), depending on culture conditions. Because attempts to identify type 2 astrocytes in the developing CNS in vivo have stalled, a consensus was reached that type 2 astrocyte may be an artifact of culture. The term *O-2A progenitors* has been gradually replaced by the term *OPCs* to reflect the prevailing view that these cells are dedicated mainly or exclusively to OL production during normal development. OPCs express a range of defining molecular markers including the proteoglycan nerve/glial antigen 2 (NG2) (Nishiyama et al., 1999) and the platelet-derived growth factor receptor alpha subunit (PDGFRα). They also express other markers such as O4 antigen and the transcription factors Olig2, Olig1, and Nkx2.2 (Fancy et al., 2004). Because some of these markers can also be expressed by other cell populations (such as NG2 and PDGFRα in pericytes) or alternatively maintained in later stages of OL development (such as Olig2 and Olig1 in mature OLs), a combination of markers is often suitable to unambiguously identify OPCs. Using these markers, previous studies show that OPCs generate in restricted areas during development and then spread to the developing CNS by proliferation and migration, becoming more or less uniformly distributed throughout the brain and spinal cord soon after birth in rodents (Miller, 1996; Richardson et al., 2006). After birth, OPCs differentiate into mature OLs to form myelin around axons.

Adult OPCs are first isolated from the adult optical nerve using the similar immunopanning approach for OPCs from the perinatal optical nerve (Ffrench-Constant and Raff, 1986; Wolswijk and Noble, 1989; Shi et al., 1998). OPCs may also be purified from adult spinal cord or other areas (Cheng et al., 2007). The adult OPCs closely resemble their perinatal counterpart in the expression of phenotypic marker antigens as well as their bi-potential to differentiate into OLs and type 2 astrocytes in vitro (Figure 2.2), although they are found to divide, migrate, and differentiate more slowly (Wolswijk and Noble, 1989; Wren et al., 1992; Psachoulia et al., 2009). The adult OPCs are thought to develop from their perinatal counterpart (Wren et al., 1992). Later studies using staining for NG2 or PGDFα show that adult OPCs are distributed throughout the mature brain and spinal cord including both gray and white matter (Pringle et al., 1992; Dawson et al., 2000). They are surprisingly numerous and compose around 5% of all cells in the CNS (Pringle et al., 1992; Dawson et al., 2003). Due to its abundance and ubiquitous distribution, adult OPCs are regarded as a novel "fifth neural cell type" after neurons, OLs, astrocytes, and microglia and are termed as *polydendrocytes* due to their morphology (Nishiyama et al., 2009). In addition, a spectrum of contrasting and often confusing names has emerged in an attempt both to describe and to classify this cell population, such as OPCs, oligodendrocyte progenitor cells, NG2-glia, glial precursor cells, and the original *O-2A progenitor*. The terms *precursor cell* and *progenitor cell* are often used interchangeably, although some consider a progenitor cell to have a greater developmental potential

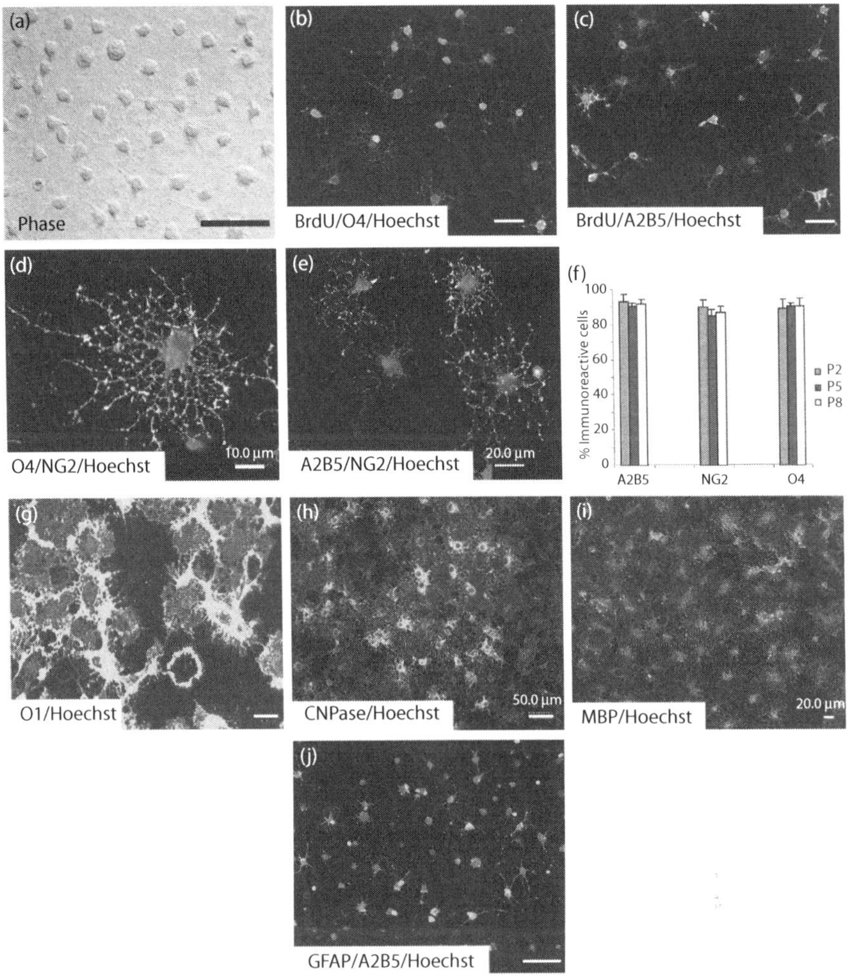

FIGURE 2.2 Isolation and differentiation of adult oligodendrocyte precursor cells (OPCs) in vitro. Adult OPCs were purified from the spinal cord of adult rats by immunopanning with the O4 antibody. The proliferating precursors displayed a characteristic adult OPC morphology with several small short processes emanating from a small round cell body (a). All cells expressed (b) O4, (b) A2B5, and (d,e) NG2. In the presence of fibroblast growth factor 2 (FGF2) and platelet-derived growth factor aa (PDGFaa), adult OPCs divided and are readily labeled by BrdU (b,c). These OPCs proliferated for multiple passages without changing phenotypes (f). Data in (f) were the mean ± SD of four independent experiments. Three days after withdrawal of FGF2 and PDGFaa, OPCs constitutively differentiated into OLs expressing O1 (g). Five days after differentiation, the majority of adult OPCs differentiated into mature OLs expressing CNPase (h) and MBP (i). When differentiated in the presence of 10% fetal bovine serum, more than 60% of the cells differentiated into GFAP+ astrocytes (j) that also expressed A2B5 and had a typical stellate morphology of type 2 astrocytes (j). Scale bars = 25 μm (a–c,g,j), 10 μm (d), 20 μm (e,i), and 50 μm (h). (BrdU, 5-bromo-2′-deoxyuridine; CNPase, 2′,3′-cyclic nucleotide phosphodiesterase; GFAP, glial fibrillary acidic protein; MBP, myelin basic protein.) (Reproduced from Cheng X et al., 2007. *Stem Cells* 25:3204–3214.)

than a precursor cell (Raff, 2003). Currently, no clear consensus has yet emerged on the most appropriate name for this cell population. In this review, we will use the term *OPCs* to refer to this cell population with all these different names.

2.3 ORIGIN OF OPCs

In contrast to their widespread distribution in the adult CNS, OPCs are generated from distinct areas in the embryonic nervous system. An early population of OPCs arises from neuroepithelial progenitor cells in the ventral neural tube around embryonic day 12.5 (E12.5). They are specified in response to the secreted morphogen sonic hedgehog (Shh) derived from the floor-plate of the neural tube (Rowitch, 2004; Richardson et al., 2006; Yu et al., 2013a). Secreted signals of fibroblast growth factors family (FGFs) also regulate early OPC development and loss of either FGF receptor 1 or 2 in ventral neuroepithelial progenitors blocks induction of OPCs (Furusho et al., 2011; Furusho et al., 2012). A later population of OPCs generates from the dorsal neural tube around E16.5. The contribution of this dorsal source is estimated to be 20%–30% of the total OPC number (Cai et al., 2005; Vallstedt et al., 2005). The specification of dorsal OPCs is not Shh-dependent. It remains unknown how dorsally derived OPCs are specified, because dorsally derived bone morphogenic proteins (BMPs) are typically thought to compete with Shh to inhibit OPC development. A third population of OPCs could arise from neural stem cells located in the subventricular zone (SVZ) after birth and in the adult brain (Menn et al., 2006). The neural stem cells in SVZ generate fast dividing transient amplifying cells (C cells) which mainly migrate tangentially along the rostral migratory stream toward the olfactory bulb and differentiate into interneurons (Lois and Alvarez-Buylla, 1993). A small fraction of C cells generate OPCs which migrate radially into the overlying white matter and cortex (Suzuki and Goldman, 2003; Menn et al., 2006). Their specification may rely on the same signals as during development, including Shh and FGF2 (Azim et al., 2012; Ferent et al., 2013). Other work suggests that additional signals may be also required, including epidermal growth factor (Gonzalez-Perez and Alvarez-Buylla, 2011) and pigment epithelium-derived factor (Sohn et al., 2012). Interestingly, the generation of OPC from SVZ is location dependent dorsoventrally and rostrocaudally. More OPCs are produced from the dorsal SVZ compared to the ventrolateral SVZ (Ortega et al., 2013). Furthermore, along the rostrocaudal axis of the brain, a higher number of OPCs are produced from the posterior SVZ than from rostral SVZ (Menn et al., 2006). The dorsal WNT and ventral BMP signals may contribute to the location difference of SVZ in OPC specification.

The developmental heterogeneity of OPCs has raised the interesting questions whether OPCs from different origins are functionally distinct subpopulations. The ventral OPCs arise first and then spread uniformly throughout the whole spinal cord. The dorsal OPCs appear later and remain mainly in dorsal and dorsolateral funiculi. During adulthood, corticospinal and rubrospinal tracts are myelinated mainly by OLs from dorsal OPCs although OLs from ventral OPCs dominate these tracts during early postnatal development (Tripathi et al., 2011). These results show that ventral and dorsal OPCs myelinate different axonal tracts in the spinal cord and suggest that they may be a functionally different subpopulation. However, once the ventral or dorsal OPCs are specifically ablated by expression of DTA, another source of OPCs will repopulate and form functional myelin in the areas which are occupied by the ablated OPCs (Kessaris et al., 2006). These results indicate the ventral and dorsal OPCs are functionally redundant and are able to replace each other if one source is ablated during development. Importantly, after demyelination is induced in the ventrolateral funiculus of the adult spinal cord by gliotoxin, the dorsal OPCs are able to differentiate into mature myelinating OLs and contribute to remyelination (Zhu et al., 2011). Further work is needed to quantify and compare the relative roles of the two populations in the remyelination. It is also interesting to determine whether the ventral and dorsal OPCs are equally affected by diseases and aging. Direct comparison of the gene expression profiles of ventral and dorsal OPCs will also be helpful to elucidate whether they are functionally distinct.

2.4 SIGNALS REGULATED OPC DEVELOPMENT

2.4.1 Intrinsic Factors Control the Development of OPC

The OPC specification and its subsequent differentiation, maturation, and myelination are precisely regulated in time and space by both intrinsic and extrinsic factors (Zuchero and Barres, 2013). The ventral midline Shh signaling from the floor plate induces the oligodendrogenesis from the ventral neuroepithelial progenitor cells during the development. Transcriptional regulation by these morphogens orchestrates oligodendrogenesis during CNS development (Miller, 2002; Ross et al., 2003; Rowitch, 2004). Olig1 and Olig2, two sonic hedgehog–induced basic helix-loop-helix (bHLH) transcription factors, play important roles in generating OLs during embryogenesis (Lu et al., 2000; Zhou et al., 2000). In *Olig1/2* double-mutant mice, there is a complete failure of OL development in all areas of the brain along with an apparent increase in astrocyte genesis in the spinal cord (Lu et al., 2002; Zhou and Anderson, 2002). This indicates that *Olig1/2* expression is essential for oligodendrogenesis and suggests that repression of OL development may be sufficient to cause astrogliogenesis. Additional studies in single null mutants suggest that Olig2 is required for initiation of OLs in the spinal cord, and Olig1 contributes more to OL differentiation and myelination during development (Lu et al., 2001; Lu et al., 2002; Takebayashi et al., 2002; Zhou and Anderson, 2002; Xin et al., 2005). Olig1 is also essential for remyelination (Arnett et al., 2004). Furthermore, induction of Olig2 expression in neuroepithelium leads to an increased generation of OPCs in the CNS (Maire et al., 2010). These studies suggest that Olig2 is necessary and in some brain area sufficient for OPC generation. The bHLH factor Ascl1/Mash1 also plays an important role in OL development. In Ascl1 knockout mice, the number of OPCs is significantly decreased in both the brain and spinal cord at the early development (Parras et al., 2007; Sugimori et al., 2008). However, the relatively normal number of OPCs is ultimately achieved in these mutation mice likely due to the redundant function of Ascl2 and Ascl3 in OL lineages during CNS development (Sugimori et al., 2008). Ascl1 also plays important roles in the OPC generation from SVZ neural stem cells after birth. Ascl1 is expressed in SVZ neural progenitor cells and SVZ-derived OPCs. Its expression is dramatically decreased once OPCs differentiate into OLs.

Deletion of Ascl1 in SVZ neural progenitor cells significantly decreases its OPC generation with the concurrent increase of astrocyte differentiation in the postnatal brain (Nakatani et al., 2013). Importantly, Ascl1 is required for the repopulation of OPCs and subsequent remyelination after local demyelination in the adult mice brain (Nakatani et al., 2013). In addition, other transcription factors in the homeobox and sox family also play important roles in oligodendrogenesis. For example, generation of motor neurons and OPCs from the pMN domain in the ventral spinal cord is eliminated in the Nkx6.1/Nkx6.2 double mutants demonstrating an essential role for these factors in the development of OLs (Cai et al., 2005; Vallstedt et al., 2005). The numbers of both OLs and astrocytes are significantly decreased in the spinal cord of Sox9 knockout mice and further reduced in sox 9 and sox10 double-knockout mice, suggesting a redundant function of these factors in the general glial specification (Stolt et al., 2003). There are numerous transcription factors such as Sox10, Nkx2.2, which are expressed in OPCs and OL lineages during CNS development. These factors may facilitate the development of OLs. However, they are not necessarily required for the specification of OPCs or oligodendrogenesis.

Chromatin histone modification, an important epigenetic mechanism, plays an important role in regulating oligodendrogenesis (Liu and Casaccia, 2010). For example, expression of HDACs is dramatically increased when neural stem cells exit the cell cycle and differentiate into OLs. Blocking HDAC's activity will inhibit the OL differentiation of neural stem cells (Marin-Husstege et al., 2002). Genetic deletion of HDAC1 and HDAC2 in OPCs leads to severe defects in OL differentiation and maturation (Ye et al., 2009). These results suggest that deacetylation in histones by histone deacetylases (HDACs) is critical for OL's differentiation. One likely mechanism is that deletion of both HDAC1 and HDAC2 results in stabilization and nuclear translocation of beta-catenin, which negatively regulates OL development by repressing Olig2 expression (Ye et al., 2009). Interestingly, extrinsic factors also regulate OL development at least partly by modifying histone acetylation. Shh, which promote OL differentiation, induces histone acetylation and the inhibitory BMP4, however, blocks deacetylation (Wu et al., 2012). More interestingly, Olig2 promotes OL differentiation through ATP-dependent histone remodeling. Olig2 binds to enhancers of OL specific genes and recruits the SWI/SNF chromatin remodeling protein Smarca4/Brg1, which controls the position of nucleosomes and thus increases the accessibility of OL specific genes for transcription (Yu et al., 2013b). Future work is needed to test whether chromatin remodeling by HDACs and Smarca4/Brg1 is important for remyelination after neurological diseases including SCI.

Micro-RNAs (miRNAs), another epigenetic mechanism, have also been shown to play important roles in OL development, myelination, or remyelination (Li and Yao, 2012). Some miRNAs are specifically expressed in OL lineages (Lau et al., 2008; Jovicic et al., 2013). Importantly, when Dicer1, an enzyme necessary for processing miRNAs, is specifically deleted in oligodendrocyte progenitor cells, the OL differentiation and myelination are significantly inhibited (Dugas et al., 2010; Zhao et al., 2010). Specifically, MiR-219, one of the highly enriched miRNAs in OL, plays a critical role in OL differentiation and maturation (Dugas et al., 2010). Knockdown of miR-219 in OPCs increases its proliferation but inhibit the transition from OPC to mature OLs. Overexpression of miR-219 in OPCs promotes OL differentiation.

Thus, miR-219 plays a critical role in coupling differentiation to proliferation arrest in the OL lineage, enabling the rapid transition from proliferating OPCs to myelinating OLs. MicroRNAs play an important role in regulating OL maturation and thus could be a therapeutic target to promote remyelination and functional recovery after neurological diseases.

2.4.2 Extrinsic Signaling Regulating the Development of OPC

Multiple extrinsic factors including morphogens, neurotrophin factors, cytokines, and neuronal activities play important roles in regulating OL development. It is out of the scope of this chapter to discuss all these factors. A few recent excellent reviews provide details on this topic (Boulanger and Messier, 2014; El et al., 2014). We focus on the factors that have shown important effects not only on OL development but also on remyelination, especially after spinal cord injury.

2.4.2.1 Morphogens

It is well established that a gradient of sonic hedgehog (Shh) released from the notochord and floor plate is primarily responsible for the initial specification and production of OLs in the ventral spinal cord during early development (Orentas et al., 1999). Shh signaling is present in the adult brain (Traiffort et al., 2010) and required for the establishment of adult stem cell niches (Lai et al., 2003) and migration of neuroblasts (Angot et al., 2008). Delivery of exogenous Shh increases the number of OPCs and premyelinating OLs in the normal adult cerebral cortex and corpus callosum (Loulier et al., 2006). Importantly, Shh signaling plays important roles in remyelination after local demyelination (Ferent et al., 2013). Shh treatment increases the number of OPCs and mature OLs after demyelination likely due to an enhanced proliferation, survival, and differentiation of the endogenous OPCs. Conversely, blocking Shh activity in the lesion, using its physiological antagonist, hedgehog interacting protein, results in a decrease of OPC proliferation, and differentiation as well as remyelination (Ferent et al., 2013). Increasing Shh expression in grafted OPCs also promotes its OL differentiation and functional recovery after SCI (Bambakidis et al., 2003). During early development, the dorsal BMP signaling is inhibitory for oligodendrogenesis. For example, implantation of noggin-producing cells into the early developing chicken spinal cord or anti-BMP antibody coated beads into developing *Xenopus* promoted the subsequent appearance of OL progenitors in the dorsal neural tube (Mekki-Dauriac et al., 2002; Miller et al., 2004). Conversely, elevated BMP expression inhibited the appearance of ventral OL progenitors (Mekki-Dauriac et al., 2002; Miller et al., 2004). Furthermore, transgenic overexpression of BMP4 enhances astrocyte lineage commitment in vivo and significantly inhibits the generation of OLs (Gomes et al., 2003). In vitro, BMP signaling promotes astrocyte differentiation at the expense of OLs from multipotential NSCs (Gross et al., 1996; Mabie et al., 1999; Zhu et al., 1999; Nakashima et al., 2001; Samanta and Kessler, 2004). The BMPs also inhibit postnatal OPCs to differentiate into OLs (Mabie et al., 1997; Grinspan et al., 2000) and to mature into myelinating mature OLs (See et al., 2004). Taken together, BMPs appear to inhibit the development of several stages of OLs including initiation, differentiation, and maturation. The expression of BMP 2 and 4 is dramatically increased after SCI (Ara et al., 2008; Wang et al., 2011). The upregulation of BMPs in the injured spinal cord may limit the OL differentiation of endogenous OPCs (Cheng et al., 2007) or promote the astrocyte differentiation of the engrafted NSCs (Chow et al., 2000; Cao et al., 2001; Yamamoto et al., 2001; Wu et al., 2002; Talbott et al., 2005). Similar to BMP signaling, Wnt signaling also inhibits the OL development and remyelination after demyelination (Fancy et al., 2009).

2.4.2.2 Neurotrophin Factors

Both neurotrophin 3 (NT3) and brain-derived neurotrophic factor (BDNF) regulate neuronal development and axonal regeneration (Xu et al., 1995). They are also important mediators of myelination. Mice lacking functional trkC or NT3 are deficient in both mature OLs and OPCs (Kumar et al., 1998). The NT3 enhances the survival and proliferation of OPCs in vitro (Barres and Raff, 1994; Kumar et al., 1998; Yan and Wood, 2000; Franklin et al., 2001) and in vivo (Barres et al., 1994). Myelination by OLs is also enhanced by NT3 in both cultures of neurons and the injured CNS (McTigue et al., 1998; Yan and Wood, 2000; Jean et al., 2003). The BDNF is important for myelin formation in peripheral nerve during development as inactivation of BDNF signaling by deleting trkB receptors causes myelin deficits both in vivo and in vitro (Cosgaya et al., 2002). Importantly, delivery of NT3 and BDNF increases the proliferation of endogenous OPCs and OL remyelination after spinal cord injury (McTigue et al., 1998). Overexpression of D15A, a multineurotrophin with both NT3 and BDNF activities, will promote the survival and OL differentiation of grafted glial progenitor cells, and importantly, functional recovery after spinal cord injury (Cao et al., 2005a). These studies suggest that delivery of neurotrophin factors NT3 and BDNF is an effective approach to promote oligodendrogenesis and remyelination after spinal cord injury.

2.4.2.3 Cytokines

Ciliary neurotrophic factor (CNTF) has neuroprotective effects on a variety of CNS and PNS neurons (Barbin et al., 1984; Hagg and Varon, 1993; Naumann et al., 2003). In addition, CNTF promotes OL differentiation and maturation from OPCs in vitro (Barres et al., 1996; Marmur et al., 1998; Stankoff et al., 2002). It also increases the survival of mature OLs (Barres et al., 1993; Louis et al., 1993). Importantly, CNTF and LIF decrease the severity of experimental autoimmune encephalomyelitis (EAE) by decreasing the OL apoptosis and enhancing OPC proliferation and differentiation (Butzkueven et al., 2002; Linker et al., 2002). Expression of CNTF is significantly increased around the demyelinated lesion (Vernerey et al., 2013). The upregulated CNTF acts as a chemoattractant on endogenous neural stem cells and OPCs promoting their migration into the demyelinated areas. Decreased CNTF expression by minocycline treatment leads to defects of OPC maturation after cuprizone-induced demyelination (Tanaka et al., 2013). Combinatorial approaches including CNTF delivery and OPC transplantation lead to better survival of grafted OPCs and greater remyelination and functional recovery after SCI (Cao et al., 2010). These studies suggest a potential therapeutic role for CNTF in promoting remyelination from grafted or endogenous OPCs after injury.

2.4.2.4 Astrocytes

Astrocytes promote myelination during development. The increase in GFAP-positive astrocytes during development correlates with the normal developmental period of myelination in the spinal cord (Dziewulska and Rafalowska, 2000), and experimental studies in cell culture show that astrocytes induce OLs to align their processes with axons (Meyer-Franke et al., 1999). Astrocytes from the postnatal brain promote myelination of cultured OLs by releasing LIF in response to ATP liberated from axons firing action potentials (Ishibashi et al., 2006). Type I astrocytes in the optic nerve express noggin which keeps OPC maturation along OL lineages and prevents their differentiation into type II astrocytes during development (Kondo and Raff, 2004). But reactive astrocytes in the injured, especially the chronically injured CNS, inhibit remyelination (Franklin, 2002). The reactive astrocytes in the chronic demyelinated CNS of taiep rats inhibit OL differentiation of endogenous and grafted OPCs (Foote and Blakemore, 2005). The mechanisms by which reactive astrocytes inhibit remyelination are not well defined. Re-expression of inhibitory factors for myelination, such as Jagged 1 (John et al., 2002) or hyaluronan (Back et al., 2005), in the reactive astrocytes, may contribute to remyelination failure after chronic demyelination in multiple sclerosis patients. Re-expression of BMPs by reactive astrocytes may also contribute its inhibition for OL differentiation after SCI (Wang et al., 2011). The CM of astrocytes from the injured spinal cord inhibits OL differentiation of adult OPCs and promotes their astrocytic differentiation. Future work is needed to study whether blocking the inhibitory signaling from reactive astrocytes will further promote remyelination and functional recovery after neurological diseases including SCI.

2.4.2.5 Neuronal Activity

It has been appreciated that neuronal activity plays an important role in regulating OL development. For example, blockade of activity in the developing optic nerve by axotomy or tetrodotoxin decreases OPC proliferation (Barres and Raff, 1993). Neuronal activity promotes OL differentiation and myelination in vitro (Stevens et al., 1998; Ishibashi et al., 2006; Wake et al., 2011). In vivo electrical stimulation of CNS neurons of corticospinal tract with high-frequency stimulation (333 Hz) increases proliferation of OPCs in the adult rat (Li et al., 2010). The pathological perturbations of electrical activity in a neonate with immediate postnatal onset of seizures and hemimegalencephaly also promote OL differentiation and myelination in the motor system (Goldsberry et al., 2011). One recent study using an optogenetic approach shows that the stimulation at the normal physiological firing rate increases the oligodendrogenesis and motor function (Gibson et al., 2014). Optogenetic stimulation of the projection neurons in the premotor cortex in awake mice elicits the proliferation of OPCs, promotes oligodendrogenesis, and increases myelination with the deep layers of the premotor cortex and subcortical white matter. Importantly, the increased oligodendrogenesis and myelination by optogenetic stimulation are associated with improved motor function of the corresponding limb. Blockade of OL differentiation prevents the optogenetic stimulation-elicited behavioral improvement. This study indicates that the neuronal activity plays important roles regulating oligodendrogenesis and myelination. Further studies are needed to elucidate the mechanisms by which neuronal activity regulates oligodendrogenesis and myelination. One mechanism by which neuronal activity regulates OL differentiation and myelination is to release the secreted signals. Using the co-culture system of OPC and dorsal ganglion root neurons (DRGNs), one study shows that action potentials in neurons initiate release of ATP and glutamate, which promote OL differentiation of co-cultured OPCs and drive translation of myelin basic protein mediated by Fyn kinase signaling (Fields, 2006). Another in vitro study shows that neurons secrete ATP in response to action potential. The secreted ATP combines the purinergic receptors in astrocytes to promote its release of LIF, which then enhances the differentiation and myelination of co-cultured OPCs (Ishibashi et al., 2006). Further studies are needed to confirm whether these in vitro findings occur in vivo. One alternative attractive possibility is that neuronal activity regulates the proliferation, differentiation, and myelination of OPCs via neuron-OPC synapses. Previous studies from multiple groups have demonstrated that OPCs receive functional synapses from neurons (reviewed by Bergles et al., 2010; Sun and Dietrich, 2013). Neuron-OPC synapses appear as early as OPCs are generated (Kukley et al., 2008; Velez-Fort et al., 2010) and are maintained when OPCs divide and are transferred to the daughter OPCs (Kukley et al., 2008; Ge et al., 2009). However, once OPCs differentiate into OLs, the neuronal-OPC synapses are disassembled (De Biase et al., 2010; Kukley et al., 2010). Importantly, adult-born OPCs receive synaptic input after migrating from the SVZ to the site of injury following demyelination, suggesting the neuron-OPC synapses may be involved in the regulation of remyelination after injury (Etxeberria et al., 2010). However, the precise role of these neuron-OPC synapses in vivo has yet to be determined. One in vivo study shows that OPCs receive glutamatergic synapses from thalamocortical fibers and preferentially accumulate along septa separating the barrels during the formation of the mouse barrel cortex (Mangin et al., 2012). Sensory deprivation reduces thalamocortical inputs on OPCs and increases their proliferation, leading to a more uniform distribution in the deprived barrel cortex. Another in vivo study demonstrates that early social isolation irreversibly impairs myelin development in the prefrontal cortex, causing defects in OL morphology, myelin thickness, and cognitive function (Makinodan et al., 2012). Long-term social isolation of adult mice also leads to defects in myelination, behavior, and difference in chromatin organization (Liu et al., 2012). These in vivo studies indicate that the sensory input–mediated neuronal activity regulates OL development and myelination.

2.5 FUNCTIONS OF OPCs IN NORMAL CNS

The OPCs undergo a defined series of steps to differentiate into mature OLs which extend processes to wrap and compact around axons to form the functional myelin during early postnatal development. Given that the bulk of myelination occurs early after birth, it is perhaps surprising that OPCs persist and remain abundant in the adult brain. The OPCs are distributed in both gray and white matter areas and compose around 5%–8% of all the cells in the adult brain (Pringle et al., 1992; Dawson et al., 2000). Although adult OPCs divide slowly in the normal CNS and the labeling index for cells within the adult cerebellar cortex is 0.2%–0.3%, they are the principal dividing cell population within the adult brain parenchyma (Levine et al., 1993; Horner et al., 2000). When bromodeoxyuridin (BrdU) is used for labeling the proliferative cells in the adult CNS, as many as 70% of the BrdU-labeled cells in the spinal cord and 75% of those in the cortex are NG2 + OPCs (Horner et al., 2000). Importantly, OPCs continue to differentiate into mature OLs which form new myelin in the adult CNS (Young et al., 2013). It remains unclear what the precise functions of OPCs are in the adult CNS. Recent studies suggest that one potential function of adult OPCs is to mediate white matter plasticity (Wang and Young, 2014). Many white matter regions of the brain, especially the corpus callosum, contain many unmyelinated axons in adulthood (Sturrock, 1980). The adult-born OLs from OPCs could engage in de novo myelination of previously naked axons. Previous studies show that the proportion of myelinated axons in the corpus callosum increases across the life span in rats (Nunez et al., 2000; Yates and Juraska, 2007; Peiffer et al., 2010). Myelinated axons continue to increase in the genu and splenium of the rat corpus callosum between young adulthood and middle age, after which time the level of myelination is maintained into old age (Yates and Juraska, 2007). Diffusion tensor imaging (DTI) is a very useful tool to study the myelination in white matter areas, especially in the human brain. DTI measures the diffusion anisotropy of water molecules. Fractional anisotropy (FA) in white matter will be high because axonal fasciculation and myelination allow water to move quickly in the direction of the axon but impede water movement perpendicular to the tract. In contrast, in gray matter, the diffusion of water is highly isotropic and moves more equally in all directions, producing a very low FA value. Thus, FA in DTI reflects the status of myelination in white matter. DTI studies in the human brain show that myelination continues to increase in the first decades of life (Giedd, 2004), peaks in early middle age, and declines in the elderly (Hasan et al., 2008b; Hasan et al., 2010). In contrast, no myelin loss is observed in rodents likely due to their shorter life span compared to humans. These studies suggest that myelin plasticity including the addition of new OLs and myelin is ongoing throughout adulthood in both rodents and human. In addition to de novo myelination of previously naked axons, adult OPCs may also be involved in myelin replacement. In the optic nerve, which is fully myelinated between 4 and 6 months of age in mice, OPCs continue to divide and form new myelinating OLs with progressively more and shorter myelin internodes (Young et al., 2013). As many as 30% of all myelin internodes in the optic nerve are shorter in an 8-month-old mouse. Similarly, the myelin internodes continue to shorten with age in the mouse spinal cord white matter (Lasiene et al., 2009) and rhesus monkey brain (reviewed by Peters and Kemper, 2012), likely due to addition of new myelin from adult OPC-derived myelinating OLs. The continual addition of new myelin in the fully myelinated white matter areas such as the optic nerve suggests that, in contrast to long held views, there is normally much more myelin turnover in the adult brain than has been thought. It remains to be further determined whether the continual addition of myelin internodes from newly generated OLs replaces old myelin or rather intercalates between existing internodes in the adult brain.

What is the function of myelin remodeling, such as de novo myelination or myelin internode elaboration, in the adult brain? A recent study suggests that myelin modeling in the adult brain may be involved in learning processes (McKenzie et al., 2014). Motor training such as running on a "complex wheel" with irregular spaced rungs increases the proliferation of OPCs and the subsequent generation of mature OLs in the corpus callosum of adult mice (McKenzie et al., 2014). To further study the function of OLs born later in the adult brain and the myelin that they produce, transcription factor, myelin regulatory factor (MRF) is specifically deleted in adult PDGFRα + OPCs using the inducible cre-lox system. Because MRF is necessary for OL maturation and myelination (Emery et al., 2009), the conditional deletion of MRF in adult OPCs blocks production of new OLs and myelin in the adult mice brain without affecting preexisting OLs or myelin. Importantly, the capacity to master the complex wheel is significantly compromised in the MRF conditioned knockout mice (McKenzie et al., 2014), indicating that generation of new OLs and myelin is important for learning motor skills. Studies using magnetic resonance imaging (MRI) also show changes in the structure of white matter in people trained in complex sensorimotor tasks such as playing piano, juggling, or abacus use (Bengtsson et al., 2005; Scholz et al., 2009), suggesting white matter modeling may play important roles in motor learning in humans. In addition, myelin remodeling may play important roles in cognitive function. Social isolation in adult mice decreases the differentiation of OLs, and it reduces adult myelination and myelin sheath thickness in the prefrontal cortex (Liu et al., 2012). This study suggests that adult myelination is involved in the cognitive function in the prefrontal cortex, such as socialization or working memory. In the future, it is interesting to further study whether white matter plasticity including myelin remodeling play roles in the cognitive function decline in aging or neurological diseases, e.g., Alzheimer diseases. Previous studies show that proliferation of OPCs and generation of mature OLs decreases with aging (Clarke et al., 2012; Young et al., 2013). MRI studies in humans also show myelination decreases after middle age (Hasan et al., 2008a, 2010). Future studies are needed to determine whether decreases of adult myelination are associated with or involved in cognitive function deficits associated with aging. Increasing evidence shows changes in OPCs or OLs in neurological diseases such as ALS, depression, or autism (Kang et al., 2013; Philips et al., 2013), suggesting potential roles of OPCs or OLs in the development of these neurological diseases. It is important to investigate mechanisms to regulate the generation of new OLs and myelin in the adult brain, especially in aging or neurological diseases. These studies could provide insight to develop novel therapies targeting adult myelination to slow down or reverse the neurological functional loss after aging or diseases.

In addition to serving as precursor cells for OLs and participating in myelin plasticity in the adult CNS, endogenous OPCs may have other functions that are not well defined. For example, not all endogenous OPCs differentiate into OLs (Nishiyama et al., 2014). Although myelin is mainly produced in white matter, OPCs are evenly distributed to cover the entire mature CNS parenchyma (Dawson et al., 2003). In neocortex, adjacent OPCs occupy nonoverlapping territories, and their processes seem to be contact inhibited (Hughes et al., 2013). These studies suggest that OPCs may have a yet uncovered homeostatic role in the CNS.

2.6 ENDOGENOUS OPCs AFTER SCI

OPCs are ubiquitously distributed throughout the parenchyma of the spinal cord (Horner et al., 2000). After SCI, OPCs become activated, increase their proliferation, and migrate to regions adjacent to the injury (McTigue et al., 2001; Zai et al., 2005; Lytle et al., 2009). Because OPCs, but not the differentiated OLs, are the major cells responsible for remyelination in the adult CNS following demyelination, recruitment of OPCs to the injury area is a critical first step for successful remyelination after SCI. The mechanisms to increase the proliferation and migration of OPCs are not well defined. Upregulation of several growth factors after SCI may contribute to the increased proliferation of endogenous OPCs. For example, basic fibroblast growth factor, a well-known mitogen for OPCs in vitro, is dramatically up-regulated after SCI (Tripathi and McTigue, 2008). Other growth factors, such as insulin growth factor 1, ciliary neurotrophin factor (CNTF), or glial growth factor (GGF), are also upregulated after SCI (Zai et al., 2005; Tripathi and McTigue, 2008). These factors promote the proliferation of OPCs when used alone or in combination with other factors. Delivery of exogenous growth factors, such as GGF (Whittaker et al., 2012) or neurotrophin 3 (NT3) and brain-derived neurotrophin factor (BDNF) (McTigue et al., 1998) significantly increase the proliferation of endogenous OPCs and generation of OLs and promote functional recovery after SCI. These studies suggest that these factors play important roles regulating the proliferation and differentiation of endogenous OPCs, and approaches to increase the expression of these factors could be effective therapy to increase endogenous remyelination for better functional recovery after SCI. Importantly, oligodendrogenesis is also increased in the areas around the injury likely due to OL differentiation of the enhanced proliferation of OPCs after SCI (Tripathi and McTigue, 2007). Furthermore, newly differentiated OLs remyelinate the demyelinated axons (Smith and Jeffery, 2006; Powers et al., 2013). Remyelination of the intact demyelinated axons by endogenous OPCs and invading peripheral Schwann cells may contribute to spontaneous functional recovery after SCI. However, conduction deficits persist chronically in humans (Alexeeva et al., 1997, 1998) and rodents (Blight, 1983a; Talbott et al., 2005), suggesting spontaneous remyelination is limited and incomplete after SCI. There is also ample anatomical evidence showing the presence of demyelinated axons in both the acute and chronic phases of SCI in both human (Bunge et al., 1993; Guest et al., 2005) and experimental animals (Blight, 1983a, 1993; Totoiu and Keirstead, 2005; Cao et al., 2005b). A limited number of precursor cells, the presence of inhibitory factors for OL differentiation, and/or the lack of growth factors for OL survival and myelination in the injured CNS may all contribute to this failure of endogenous remyelination. Therefore, strategies to promote significant functional recovery after SCI by remyelination from endogenous NSCs or OPCs may prove feasible in the future once mechanisms that regulate NSC and/or OPC proliferation, differentiation, and maturation are better understood. Stem cell transplantation is still considered to be a more effective approach to enhance the remyelination and locomotion functional recovery after SCI.

In addition to functioning as precursor cells to OLs and participating in the remyelination, activated OPCs contribute to the formation of the glial scar after SCI. SCI results in the complex morphological and physiological changes in glial cells, including astrocytes, microglia, and OPCs. The proliferating OPCs, together with reactive astrocytes, activated microglia, macrophages and other myeloid cells, meningeal cells, as well as a dense extracellular matrix, form the glial scar, a physical and biochemical barrier to axonal regeneration (Silver and Miller, 2004). OPCs express NG2, a member of the CSPG family which is inhibitory for axonal regrowth (Petrosyan et al., 2013). Delivery of antibodies neutralizing NG2 decreases the inhibition of the glial scar and promotes axonal regeneration after SCI (Petrosyan et al., 2013). Recent study shows that specifically, deletion of β-catenin signaling in OPCs significantly decreases its proliferation after SCI (Rodriguez et al., 2014). Importantly, decreasing OPC proliferation also reduces the accumulation of activated microglia and macrophages as well as astrocyte hypertrophy after SCI. Thus, abrogation of β-catenin signaling in OPCs reduces the glial scar and promotes axonal regeneration following injury. This study suggests that OPCs contribute to the formation of the glial scar not only by its activation and expression of NG2 but also by its role in promoting activation of microglia and astrocytes after SCI. Proliferation of OPCs is critical for both remyelination and glial scar formation following SCI. It is important to investigate whether signaling promoting the endogenous remyelination will be different or the same to ones enhancing its roles in glial scar formation. These studies will help us to develop therapeutic strategies to promote the beneficial function of endogenous OPCs in remyelination but limit its detrimental role in glial scar formation.

2.7 TRANSPLANTATION OF OPCs AFTER SCI

2.7.1 Therapeutic Potential of OPC Transplantation after SCI

Because one OL can myelinate as many as 30–80 distinct axons (Chong et al., 2012; Young et al., 2013), extensive death of OLs after SCI could elicit widespread demyelination of spared axons as shown in both anatomical (Totoiu and Keirstead, 2005; Lasiene et al., 2008) and electrophysiological studies (Blight, 1983a; James et al., 2011). Myelination results in axons with high resistance and low capacitance, enabling efficient and rapid salutatory conduction of nerve electrical impulses, which may be necessary for high-level cognitive and motor functions in mammals, especially in humans. In addition, increasing evidence demonstrates that myelin also plays important roles in supporting and protecting axons. For example, deletion of single myelin genes such as proteolipid protein (PLP) (Griffiths et al., 1998), 2′,3′-cyclic-nucleotide 3′-phosphodiesterase (CNP) (Lappe-Siefke et al., 2003), or myelin-associated glycoprotein (MAG) (Nguyen et al., 2009) does not result in overt myelin defects. However, the integrity of axons is affected, and extensive chronic axonal degeneration is evident in all mutation mice. Similarly, overexpression of suicide genes (e.g., diphtheria toxin fragment A, caspase 9) in OLs elicits primary OL death and demyelination without dramatic immune response (Caprariello et al., 2012; Oluich et al., 2012). Not surprisingly, primary demyelination in these animals results in robust axonal damage. These studies suggest that myelin is important in maintaining axonal integrity and function. Without the support and protection of myelin, demyelinated axons are more vulnerable to axonal degeneration in the hostile injury environment following SCI (Irvine and Blakemore, 2008). Thus, enhancing remyelination not only potentially restores axonal conduction but also most likely prevents the secondary degeneration of demyelinated axons. These two mechanisms may work synergistically to mediate functional recovery by remyelination after SCI.

As discussed above, spontaneous remyelination from endogenous OPCs occurs after SCI. However, endogenous remyelination is incomplete. Cell transplantation is an alternative therapeutic strategy to promote remyelination and functional recovery after SCI.

Previous studies show that transplantation of OPCs from different resources promotes functional recovery after SCI (McDonald et al., 1999; Keirstead et al., 2005; Lee et al., 2005; Cao et al., 2010). These studies show that grafted OPCs differentiate into mature OLs and increase OL remyelination. Importantly, OPC transplantation partially restores the axonal conduction after SCI (Cao et al., 2010). Furthermore, the electrophysiological and behavioral functional recovery is correlated with the numbers of OL remyelinated axons in the injured spinal cord. These studies suggest that OL replacement and remyelination is one of the important mechanisms for OPC graft-mediated functional recovery. A combination of delivery of growth factors with OPC transplantation further increases OL remyelination and locomotion recovery after SCI (Bambakidis and Miller, 2004; Cao et al., 2010). For example, delivery of Shh along with OPC grafting enhances the survival, proliferation, and migration of grafted OPCs and, importantly, promotes greater functional recovery compared to OPC grafting or Shh delivery alone (Bambakidis and Miller, 2004). Overexpression of multineurotrophin D15A by genetic modification promotes OL differentiation and survival of grafted OPCs (Cao et al., 2005a). Importantly, animals receiving D15A-expressing OPCs show greater electrophysiological and locomotion recovery. D15A is a multineurotrophin that binds trkB and trkC and has both NT3 and BDNF activities (Urfer et al., 1994). These results are consistent with a previous study showing NT3 and BDNF promote OL differentiation and myelination of endogenous OPCs after SCI (McTigue et al., 1998). Similarly, expression of CNTF significantly enhances the survival of grafted OPCs in the injured spinal cord, and the survival of CNTF-OPCs is more than threefold of control GFP-expressed OPCs (Cao et al., 2010). Transplantation of GFP- or CNTF-OPCs significantly increases the number of OL remyelinated axons in the injured epicenter and enhances functional recovery (Figure 2.3). The CNTF-OPC grafts result in a greater number of remyelinated axons and much better functional recovery than GFP-OPC grafts. The functional recovery is closely related to the number of remyelinated axons in the injured spinal cord (Figure 2.4). In contrast, transplantation of neural progenitor cells (NPCs) isolated from MBP mutation shiverer mice fails to provide the functional benefits of its counterpart NPCs from wild-type animals (Yasuda et al., 2011; Hawryluk et al., 2014). The NPCs from shiverer mice are able to differentiate into OLs but do not form functional compact myelin. These studies further highlight the importance of remyelination for functional recovery after transplantation of OPCs or NPCs. It is worth noting that remyelination is likely one of several mechanisms for functional benefits after OPC transplantation following SCI. Other mechanisms may also be involved. For example, OPCs express a variety of growth factors and neurotrophin factors that could decrease the injury and increase the spared tissue after transplantation into the injured spinal cord (Zhang et al., 2006; Sharp et al., 2010; Hawryluk et al., 2012). Grafted OPCs may also modulate the immune responses to decrease the injury and promote functional recovery (Pluchino et al., 2005).

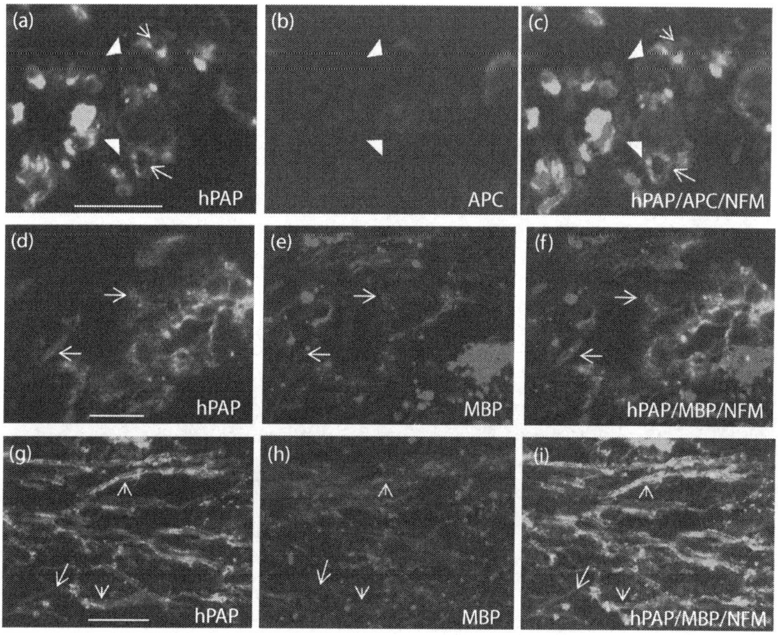

FIGURE 2.3 Oligodendrocyte remyelination by grafted OPCs after SCI. The grafted OPC differentiated into mature APC^+ OLs (a–c, arrowheads), which formed myelin rings around NFM^+ axons (a–c, arrows). Double-staining for hPAP and MBP in cross sections further confirmed that hPAP immunoreactive rings around NFM^+ axons were MBP^+ myelin (d–f, arrows). In longitudinal section, MBP^+ myelin sheaths from the grafted OPCs were more clearly shown along the NFM^+ axons (g–i, arrows). Scale bars = 20 μm (a–i). (Reproduced from Cao Q et al., 2010. *J Neurosci* 30:2989–3001.)

Besides OPCs, other types of cells have also been shown to promote remyelination and functional recovery after SCI. Multipotent NSCs, isolated from the fetal or adult CNS, are able to differentiate into OLs as well as neurons and astrocytes in vitro (Shihabuddin et al., 2000; Cao et al., 2001; Karimi-Abdolrezaee et al., 2006). They also successfully differentiate into mature myelin-forming OLs after being transplanted into dysmyelinated shiverer mice (Eftekharpour et al., 2007). However, the majority of grafted NSCs differentiate into astrocytes with very few grafted NSCs becoming mature OLs after SCI, especially after contusion SCI (Chow et al., 2000; Shihabuddin et al., 2000; Cao et al., 2001; Wu et al., 2002; Vroemen et al., 2003; Hofstetter et al., 2005). These studies suggest that the microenvironment in the injured spinal cord favors the astrocyte differentiation but inhibits the neuronal and OL differentiation from the grafted NSCs. Importantly, the astrocyte differentiation from grafted NSCs enhance the plasticity of pain fibers and promote allodynia, a condition in which a pain response is observed from a stimulus that normally does not elicit pain, after SCI (Hofstetter et al., 2005). Thus, it may be necessary to modify NSCs prior to transplantation to inhibit astrocyte differentiation and enhance OL differentiation. For example, genetically, overexpression of transcription factor neurogenin-2 in grafted NSCs not only inhibits astrocytes differentiation and allodynia

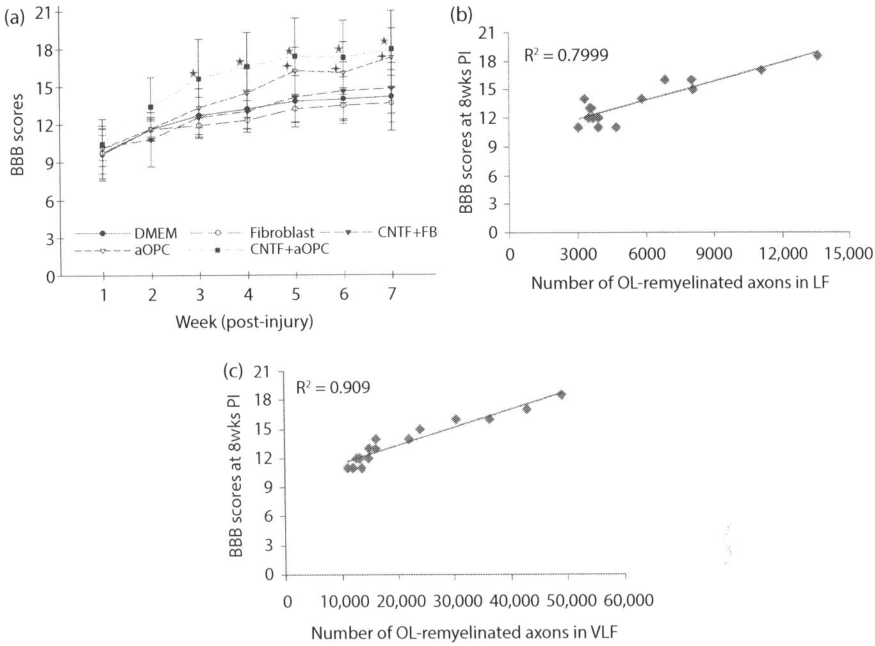

FIGURE 2.4 Hindlimb locomotor recovery after transplantation of adult OPCs. Locomotor function, determined using the BBB locomotor scores, was significantly recovered in CNTF-OPC-grafted animals from weeks 3–7 after injury, and also in EGFP-OPC-grafted animals from weeks 5–7 after injury, compared to FB- or DMEM-grafted animals (a) (data are the mean ± SD, $n = 10$; $*p < .05$). BBB scores were closely correlated with the number of OL-remyelinated axons in both the LF (b) and VLF (c) at the injury epicenter of spinal cord. (Reproduced from Cao Q et al., 2010. *J Neurosci* 30:2989–3001.)

but also promotes remyelination and locomotor recovery (Hofstetter et al., 2005). Priming NSCs into OL lineage before transplantation proved effective to increase OL differentiation and remyelination (Zhang et al., 1999; Smith and Blakemore, 2000; Keirstead et al., 2005). Delivery of growth factors promoting OL differentiation and survival also enhances OL remyelination of grafted NSCs and promotes functional recovery after SCI (Karimi-Abdolrezaee et al., 2006). OPCs more efficiently differentiate into OLs after transplantation into the injured spinal cord. Importantly, OPCs may not have the potential to differentiate into astrocytes in vivo and could bypass the astrocyte differentiating signaling in the injured spinal cord. Schwann cells (SCs) and olfactory ensheathing cells (OECs) have also shown potential to remyelinate the demyelinated axons and promote functional recovery after transplantation following SCI (Pearse et al., 2004; Pearse et al., 2007; Richter and Roskams, 2008). Compared to OPCs, SCs or OECs are less efficient for remyelination. Each SC or OEC myelinates one axon while one OL is able to myelinate as many as 30–50 axons. Furthermore, the integration and migration are limited in both SCs and OECs after transplantation following SCI. Thus, they often remyelinate axons within or immediately adjacent to the lesion site only. Importantly, other benefits such as enhancing axonal regeneration or sprouting play more important roles than remyelination in functional recovery resulting from SC or OEC transplantation.

Although OPC transplantation promotes remyelination and functional recovery after acute or subacute SCI, this approach is not effective after chronic injury (Keirstead et al., 2005; Karimi-Abdolrezaee et al., 2006). It is probably due to a hostile microenvironment in the chronic injury that does not support the survival and differentiation of grafted OPCs. For example, the glial scar in the chronic SCI may inhibit the OL differentiation and maturation by increasing the expression or the inhibiting signaling, such as bone morphogenetic proteins (Cheng et al., 2007). Our previous study shows that activated astrocytes or their conditioned mediums significantly block OL differentiation of cocultured OPCs with a concurrent increase of astrocyte differentiation (Wang et al., 2011) (Figure 2.5). Importantly, activated astrocytes after SCI significantly increase the expression of BMP2 and 4, and blocking BMP signaling partially reverse the inhibition of activated astrocytes on OL differentiation of OPCs (Wang et al., 2011). These results indicate that activated astrocytes after SCI inhibit OL differentiation by upregulation of BMP signaling. It is also possible that there are few spared demyelinated axons available for remyelination after chronic SCI (Lasiene et al., 2008; Powers et al., 2012). It remains debatable whether demyelination persists after chronic SCI. Previous ultrastructure study using electron microscopy shows that demyelination is robust after acute SCI and then decreases over time during 10 weeks after injury due to endogenous remyelination (Totoiu and Keirstead, 2005). Surprisingly, there is a secondary wave of demyelination that continues until 64 weeks after SCI, when little remyelination occurs (Totoiu and Keirstead, 2005). This study suggests that demyelination persists in chronic SCI, and remyelination is incomplete. However, recent studies show that spared rubrospinal tract axons labeled by actively transporting fluoro-Ruby are mostly remyelinated at 3 months after SCI, although the remyelinated axons have thinner myelin with shorter myelin internodes (Lasiene et al., 2008; Powers et al., 2012). These studies show that there is no, or very limited, demyelination in the spared rubrospinal tracts after chronic SCI. Further studies are needed to determine whether this is the case in other tracts after chronic SCI. It is also unknown whether demyelinated axons degenerate at the chronic SCI because they are vulnerable to the hostile environment after injury.

2.7.2 Resources of GPCs for Potential Clinical Translation

2.7.2.1 Human CNS-Derived OPCs

Human OPCs have been successfully purified from subcortical white matter of adult human brain tissue dissected from surgical procedures using fluorescence-activated cell sorting (FACS) (Nunes et al., 2003; Windrem et al., 2004). These cells can differentiate into mature OLs to form myelin in the MBP mutation shiverer mice (Windrem et al., 2008) or remyelinate the demyelinated axons following the local chemical demyelination (Windrem et al., 2002). Human GRPs are successfully isolated from fetal cadaver brain tissue of gestational age 18–24 weeks by Advanced Bioscience Resources (ABR; Alameda, California) and have been directly tested in different SCI models by several laboratories (Davies et al., 2011; Jin et al., 2011).

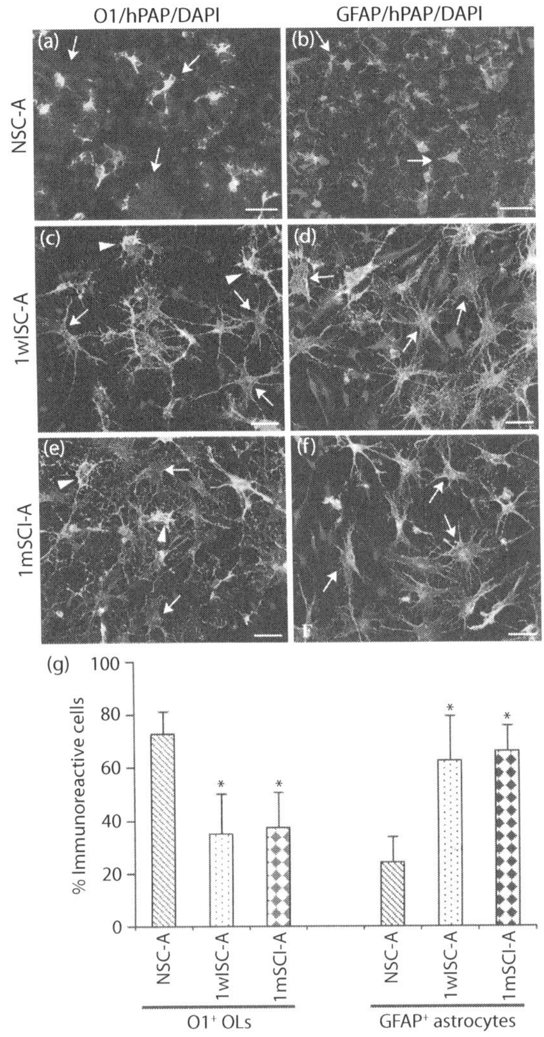

FIGURE 2.5 Active astrocytes from the injured spinal cord inhibit OL differentiation of co-cultured OPCs. Three days after being co-cultured with astrocytes from normal spinal cord, the majority of OPCs labeled by hPAP differentiate into $O1^+$ OLs with membrane sheets, the typical morphology of mature OLs (a, arrows). Most $hPAP^+$ OPCs do not differentiate into $GFAP^+$ astrocytes (b, arrowheads). However, in the co-culture with astrocytes from 1w (c, arrows) or 1m (e, arrows) postinjury spinal cord, most OPCs labeled by hPAP fail to differentiate into $O1^+$ OLs. OPCs, which differentiate into $O1^+$ OLs, lack the complex membrane sheets (c,e, arrowheads). Most OPCs labeled by hPAP differentiate into $GFAP^+$ astrocytes (d,f, arrows). Quantitative data confirm that reactive astrocytes from the injured spinal cord significantly decrease OL differentiation with a concurrent increase of astrocyte differentiation of co-cultured OPCs (g) (Data represent the mean ± SD from four repeated experiments from separately generated cultures; $*p < .05$). Scale bar = 50 μm. (Reproduced from Wang Y et al., 2011. *J Neurosci* 31:6053–6058.)

Human NSCs are also purified from the fetal brain and spinal cord of gestational age of 18–24 weeks by FACS using CD133 antibody (Uchida et al., 2000). Human NSCs differentiate into neurons, astrocytes, and OLs and promote locomotion recovery after transplantation following SCI (Cummings et al., 2005; Salazar et al., 2010). Such human NSCs have been commercialized by StemCell, Inc. (http://www.Stemcells.com) and are being studied in a clinical trial for thoracic spinal cord injury (https://clinicaltrials.gov). These studies show that the fetal or adult CNS is a valuable resource for OPCs or NSCs to potentially treat SCI. The advantages of CNS-derived stem cells or progenitor cells include (1) the cells are derived from the CNS and are well characterized. Overall, they are safe and (2) the cells could be "off-the-shelf" products and are available for the treatment of acute and chronic SCI. Their disadvantages are also obvious: (1) limited available tissue for cell purification; (2) the ethical controversy of using fetal tissue from abortion; (3) the need to use immunosuppression after transplantation.

2.7.2.2 Human ESC-Derived OPCs

Embryonic stem cells (ESCs) are pluripotent cells that have the capability to differentiate into nearly all cell types in the body (Zhang et al., 2001). The protocols to differentiate and purify OPCs from human ESCs have been established (Keirstead et al., 2005; Hu et al., 2009; All et al., 2012). Importantly, transplantation of OPCs from one of the differentiation protocols has shown locomotor functional recovery after acute but not chronic SCI (Keirstead et al., 2005; All et al., 2012). The clinical trial to use these human ESC-derived OPCs in acute SCI patients was approved by the U.S. Food and Drug Administration (FDA), and a few SCI patients received the OPC grafts. This clinical trial was suspended and then restarted. Regardless, the approval of this clinical trial by the FDA indicates that OPCs or NSCs from human ESCs are a valuable cell resource for the treatment of neurological diseases including SCI. The advantages of human ESCs included the following: (1) ESCs can replicate indefinitely without aging; (2) they are pluripotent, i.e., can give rise to all the different types of cells in the body; (3) they give rise to genetically normal cells; (4) they can be easily manipulated genetically; and (5) OPCs or NSCs from human ESCs could also be "off-the-shelf" products and, therefore, easier to commercialize and attract the investment of the private sector. The disadvantages of human ESCs include (1) their controversiality; (2) their long-term safety, especially the potential of teratoma formation; and (3) the need for lifelong immunosuppression.

2.7.2.3 Human-Induced Pluripotent Stem Cells-Derived OPCs

The human-induced pluripotent stem (iPS) cells are ESC-like cells reprogrammed from embryonic or adult somatic cells by overexpression of four pluripotency transcription factors, SRY (sex determining region Y)-box 2 (Sox2), POU class 5 homeobox 1 (Pou5f1/Oct4), Kruppel-like factor 4 (Klf4), and c-Myc that successfully reprogrammed mouse somatic cells into pluripotent stem cells (Takahashi et al., 2007; Yu et al., 2007). Human iPSCs have tremendous potential for individualized patient- and disease-specific therapy. However, there are a few hurdles to overcome before their application in patients. Original protocols for producing iPSCs relied on conventional retroviral or lentiviral vectors to induce the transcription factors into the genome of host somatic cells, a process associated with risks including mutation, dysregulation of gene expression, and the development of cancers after iPS-cell transplantation (Maherali and Hochedlinger, 2008; Soldner et al., 2009). Generation of integration-free human iPSCs is a prerequisite for its translation. Progress has been made to solve this hurdle, such as using nonintegrating adenoviruses (Stadtfeld et al., 2008), nonviral episomal transfected plasmids (Okita et al., 2008; Yu et al., 2009), or direct delivery of reprogramming proteins (Kim et al., 2009; Zhou et al., 2009) or RNAs (Warren et al., 2010). The second hurdle for application of human iPSCs is to direct their cell-specific differentiation in vitro and in vivo. Directing iPSCs to differentiation along desired cell type(s) is critical not only for successful cell replacement therapy but also for avoiding side effects such as tumor formation from the undifferentiated iPSCs or the undesired differentiation. The protocol to differentiate human iPSCs to OPCs is also established, although the efficiency needs further improvement (Wang et al., 2013). However, the therapeutic efficacy and long-term safety of iPSCs-derived OPCs is not well tested after SCI. Human iPSCs may be more tumorigenic than hESCs due to genetic and epigenetic aberrations (Ben-David and Benvenisty, 2011). Each iPSC line may have to be pre-evaluated to assess the teratoma formation after cell transplantation in animal models, as different iPSC lines vary in differentiation capacity and teratoma formation (Tsuji et al., 2010). Thus, more studies need to be done before iPSC-derived NSCs or OPCs can be eventually transferred into clinical application.

Human iPSCs share many advantages of human ESCs, such as the unlimited proliferative capacity and pluripotency to differentiate into any specific type of cell in the human body. They also have additional benefits of bypassing the ethical controversy of human ESCs, and for autologous transplantation avoiding immune rejection. However, human iPSCs still share the major risk of human ESCs: formation of teratoma.

2.7.2.4 Directly Induced OPCs

Recent studies show that using a similar approach but different transcription factors can also directly induce mice and rat fibroblasts into OPCs, which can differentiate into mature OLs myelinating axons both in vitro and in vivo (Najm et al., 2013; Yang et al., 2013). The directly induced NSCs or GPCs could be better resources for cell therapies in SCI than their counterparts from fetal NSCs, hESCs, or hiPSCs. They share the major benefits of hiPSCs, such as no ethical concerns and potentially autologous transplantation, the two big advantages of hiPSCs compared to hESCs or fetal NSCs. In addition, directly induced NSCs or GPCs can be produced much easier and faster than iPSCs. The derivation of patient-specific iPSCs still is a lengthy and cumbersome procedure, although substantial progress has been made in terms of improving the efficiency and robustness of reprogramming. Furthermore, the directly induced NSCs or OPCs are much safer than human iPSCs because directly induced OPCs are lineage-specific stem cells or precursor cells without passing through the stem cell stage and, thus, the risk of tumor formation. However, it remains unknown whether a similar approach works on human cells. There is no report for successful directly induced OPCs from human fibroblasts or other somatic cells. Importantly, the directly induced NSCs or OPCs are relatively new, and their efficacy and safety have not been extensively tested in animal models, such as SCI and other neurological disease models.

2.7.3 Challenges of Translating Stem Cell Therapies to Clinic

Remarkable progress has been made in the fields of stem cells and regenerative medicine in the last decade, including the generation of new types of stem cells, such as iPSCs. There is increasing enthusiasm for using stem cell therapies for many human diseases including SCI. However, there are several challenges that need to be addressed before the stem cell therapies can be clinically translated to benefit patients with SCI.

2.7.3.1 Safety of Stem Cell Therapy after SCI

Safety is the first priority for any successful clinical translation of any stem cell therapy in SCI and other diseases. Stem cell therapy must be safe and do no harm to SCI patients who receive it. The major concern with iPSCs and ESCs is their potential tumorigenicity as the current protocols cannot completely eliminate undifferentiated iPSCs or ESCs from the purified NSCs or OPCs. The most generally used approach is to differentiate ESCs or iPSCs into a specific cell type, such as OPCs, and then purify these cells using cell lineage–specific markers by fluorescence-activated cell sorting. Currently, there is little evidence from animal studies that purified NSCs or progenitor cells from hESCs or hiPSCs form tumors after transplantation. However, these studies are relatively short lived, and SCI patients may live many decades following stem cell transplantation. Further studies are needed to study the long-term safety of hESC- or hiPSC-derived NSCs or OPCs.

Another safety concern for stem cell therapy is the potential complication due to undesired cell differentiation of grafted stem cells. For example, astrocyte differentiation from NSC grafting increases allodynia, and such devastating side effects will seriously compromise the use of these cells for SCI repair. Given the early stage at which the field of stem cell therapy is at for SCI, further studies are needed to determine whether there are other complications. The safeguards, such as the elimination of the grafted cells once severe complications appear, will significantly decrease the risk of complications of stem cell therapy. Such approaches are available for experimental animals. The transplanted cells could be modified to express the "suicide gene," the herpes simplex virus thymidine kinase in the undifferentiated stem cell, which could be eliminated after administration of ganciclovir (Jung et al., 2007). However, it may be difficult to get this approach approved in clinical use. Development of inducible suicide mechanisms in the grafted stem cells that will be suitable for clinical application may be an important advance.

2.7.3.2 Mechanisms Mediating the Therapeutic Effects of Stem Cells after SCI

It is well documented that transplantation of OPCs can promote functional recovery. However, the mechanisms mediating the therapeutic effects of these cells remain poorly understood. Oligodendrocyte differentiation and remyelination of transplanted stem cells are well demonstrated. However, it remains undetermined whether mechanisms other than remyelination also contribute to functional recovery and how much each mechanism contributes to functional recovery. In addition, it remains unknown how the signals in the injured environment govern the proliferation and differentiation of transplanted stem cells. It is also poorly understood how signals derived from transplanted stem cells affect the functions of local host tissue and endogenous NSCs. Rigorous mechanistic studies need to be performed in order to elucidate signaling pathways ongoing in the lesion in addition to those activated by transplanted cells. Once these critical signaling pathways are identified, the therapeutic capacity of stem cells may be enhanced through increasing their expression of the relevant molecules. Deep understanding of the mechanisms of transplanted stem cell–mediated functional recovery will lead to increased reproducibility and safety of cell therapy in humans.

2.7.3.3 Reproducibility of Stem Cell Therapy after SCI

Before any particular therapy is able to translate into clinical trials, sufficient reproducibility is required by other laboratories that are independent of the laboratory that originally developed a particular therapy. However, it is challenging to reproduce work between one laboratory and another, a far more difficult issue than one might have anticipated. In 2008, the U.S. National Institutes of Health (NIH) made the laudatory decision of funding three centers focused on reproducing reported successes in the treatment of SCI (referred to as "Facilities of Research-Spinal Cord Injury" grants). Thus far, attempts have been made to reproduce the previous published studies on transplantation of olfactory ensheathing cells (Steward et al., 2006), a Nogo-66 receptor antagonist (Steward et al., 2008), Schwann cell transplantation combined with delivery of cyclic AMP (Sharp et al., 2012), EGFR inhibitor (209), minocycline (Pinzon et al., 2008b), or erythropoietin (Pinzon et al., 2008a). It is very disappointing that not a single one of these studies reproduced the observations of the original report. The reasons for not reproducing these studies are not completely understood. But the difficulties in reproducibility could slow the translation of successful therapies in experimental animals into clinic application for SCI patients. Compared to pharmacological agents and bioactive proteins, stem cell therapies are even more difficult to reproduce because stem cells are inherently more complex than single agents. For example, relatively minor changes in their growth conditions may alter their properties to provide benefits. Furthermore, stem cell therapies for SCI are mostly tested on rodents. Although rodents provide a useful laboratory model to investigate the pathology of SCI and to evaluate therapeutic interventions after SCI, rodent models should not be the sole predictors of therapeutic success in humans. Reproducibility of the therapeutic efficacy of any particular stem cell therapy should also be performed in large animals, especially primates, which are evolutionally closer to humans. The reproducibility in different laboratories and different species will increase the success of future clinical trials.

CONCLUSION

Myelin plays important roles in the functions of the CNS. Myelin enables efficient and rapid conduction of axons and optimizes space and energy utilization in the CNS.

Importantly, myelin provides support and protection for the axons. Demyelination due to extensive OL cell death following SCI causes not only the dysfunction of axonal conduction but also subsequent degeneration of demyelinated axons. Without support and protection from myelin, the demyelinated axons are more vulnerable to degeneration in the hostile injury environment. Thus, enhancing remyelination is an effective therapeutic approach to promote functional recovery after SCI. Endogenous OPCs become activated to proliferate after SCI and later differentiate into OLs to contribute to remyelination and spontaneous functional recovery. A better understanding of the mechanisms that regulate proliferation, differentiation, and maturation of endogenous OPC could further enhance the remyelination and locomotion functional recovery after SCI. In addition, transplantation of exogenous OPCs has been shown to be an effective approach to promote functional recovery after SCI by increasing remyelination as well as other mechanisms, e.g., release of trophic factors. The clinical trials to use hESC-derived OPCs or fetal brain–derived NSCs have been approved for SCI. More OPC resources such as hiPSC-derived OPCs or directly induced OPCs could be available in the near future. Although there are a number of clinical and technical obstacles to be resolved before its clinical application, stem cell grafting will ultimately prove effective as one aspect of a combinatorial therapy to treat SCI. The rapid progress in the field will make its use to treat SCI as well as other neurological disorders more feasible than ever.

REFERENCES

Alexeeva N, Broton JG, Calancie B. 1998. Latency of changes in spinal motoneuron excitability evoked by transcranial magnetic brain stimulation in spinal cord injured individuals. *Electroencephalogr Clin Neurophysiol* 109:297–303.

Alexeeva N, Broton JG, Suys S, Calancie B. 1997. Central cord syndrome of cervical spinal cord injury: Widespread changes in muscle recruitment studied by voluntary contractions and transcranial magnetic stimulation. *Exp Neurol* 148:399–406.

All AH, Bazley FA, Gupta S, Pashai N, Hu C, Pourmorteza A, Kerr C. 2012. Human embryonic stem cell-derived oligodendrocyte progenitors aid in functional recovery of sensory pathways following contusive spinal cord injury. *PLoS One* 7:e47645.

Angot E, Loulier K, Nguyen-Ba-Charvet KT, Gadeau AP, Ruat M, Traiffort E. 2008. Chemoattractive activity of sonic hedgehog in the adult subventricular zone modulates the number of neural precursors reaching the olfactory bulb. *Stem Cells* 26:2311–2320.

Ara J, See J, Mamontov P, Hahn A, Bannerman P, Pleasure D, Grinspan JB. 2008. Bone morphogenetic proteins 4, 6, and 7 are up-regulated in mouse spinal cord during experimental autoimmune encephalomyelitis. *J Neurosci Res* 86:125–135.

Arnett HA, Fancy SP, Alberta JA, Zhao C, Plant SR, Kaing S, Raine CS, Rowitch DH, Franklin RJ, Stiles CD. 2004. bHLH transcription factor Olig1 is required to repair demyelinated lesions in the CNS. *Science* 306:2111–2115.

Azim K, Raineteau O, Butt AM. 2012. Intraventricular injection of FGF-2 promotes generation of oligodendrocyte-lineage cells in the postnatal and adult forebrain. *Glia* 60:1977–1990.

Back SA, Tuohy TM, Chen H, Wallingford N, Craig A, Struve J, Luo NL et al. 2005. Hyaluronan accumulates in demyelinated lesions and inhibits oligodendrocyte progenitor maturation. *Nat Med* 11:966–972.

Bambakidis NC, Miller RH. 2004. Transplantation of oligodendrocyte precursors and sonic hedgehog results in improved function and white matter sparing in the spinal cords of adult rats after contusion. *Spine J* 4:16–26.

Bambakidis NC, Wang RZ, Franic L, Miller RH. 2003. Sonic hedgehog-induced neural precursor proliferation after adult rodent spinal cord injury. *J Neurosurg* 99:70–75.

Barbin G, Manthorpe M, Varon S. 1984. Purification of the chick eye ciliary neuronotrophic factor. *J Neurochem* 43:1468–1478.

Barres BA, Burne JF, Holtmann B, Thoenen H, Sendtner M, Raff MC. 1996. Ciliary neurotrophic factor enhances the rate of oligodendrocyte generation. *Mol Cell Neurosci* 8:146–156.

Barres BA, Raff MC. 1993. Proliferation of oligodendrocyte precursor cells depends on electrical activity in axons. *Nature* 361:258–260.

Barres BA, Raff MC. 1994. Control of oligodendrocyte number in the developing rat optic nerve. *Neuron* 12:935–942.

Barres BA, Raff MC, Gaese F, Bartke I, Dechant G, Barde YA. 1994. A crucial role for neurotrophin-3 in oligodendrocyte development. *Nature* 367:371–375.

Barres BA, Schmid R, Sendnter M, Raff MC. 1993. Multiple extracellular signals are required for long-term oligodendrocyte survival. *Development* 118:283–295.

Beattie MS, Hermann GE, Rogers RC, Bresnahan JC. 2002. Cell death in models of spinal cord injury. *Prog Brain Res* 137:37–47.

Ben-David U, Benvenisty N. 2011. The tumorigenicity of human embryonic and induced pluripotent stem cells. *Nat Rev Cancer* 11:268–277.

Bengtsson SL, Nagy Z, Skare S, Forsman L, Forssberg H, Ullen F. 2005. Extensive piano practicing has regionally specific effects on white matter development. *Nat Neurosci* 8:1148–1150.

Bergles DE, Jabs R, Steinhauser C. 2010. Neuron-glia synapses in the brain. *Brain Res Rev* 63:130–137.

Blight AR. 1983a. Axonal physiology of chronic spinal cord injury in the cat: Intracellular recording in vitro. *Neuroscience* 10:1471–1486.

Blight AR. 1983b. Cellular morphology of chronic spinal cord injury in the cat: Analysis of myelinated axons by line-sampling. *Neuroscience* 10:521–543.

Blight AR. 1993. Remyelination, revascularization, and recovery of function in experimental spinal cord injury. *Adv Neurol* 59:91–104.

Blight AR, DeCrescito V. 1986. Morphometric analysis of experimental spinal cord injury in the cat: The relation of injury intensity to survival of myelinated axons. *Neuroscience* 19:321–341.

Boulanger JJ, Messier C. 2014. From precursors to myelinating oligodendrocytes: Contribution of intrinsic and extrinsic factors to white matter plasticity in the adult brain. *Neuroscience* 269:343–366.

Bunge RP, Puckett WR, Becerra JL, Marcillo A, Quencer RM. 1993. Observations on the pathology of human spinal cord injury. A review and classification of 22 new cases with details from a case of chronic cord compression with extensive focal demyelination. *Adv Neurol* 59:75–89.

Butzkueven H, Zhang JG, Soilu-Hanninen M, Hochrein H, Chionh F, Shipham KA, Emery B et al. 2002. LIF receptor signaling limits immune-mediated demyelination by enhancing oligodendrocyte survival. *Nat Med* 8:613–619.

Cai J, Qi Y, Hu X, Tan M, Liu Z, Zhang J, Li Q, Sander M, Qiu M. 2005. Generation of oligodendrocyte precursor cells from mouse dorsal spinal cord independent of Nkx6 regulation and Shh signaling. *Neuron* 45:41–53.

Cao Q, He Q, Wang Y, Cheng X, Howard RM, Zhang Y, DeVries WH et al. 2010. Transplantation of ciliary neurotrophic factor-expressing adult oligodendrocyte precursor cells promotes remyelination and functional recovery after spinal cord injury. *J Neurosci* 30:2989–3001.

Cao Q, Xu XM, DeVries WH, Enzmann GU, Ping P, Tsoulfas P, Wood PM, Bunge MB, Whittemore SR. 2005a. Functional recovery in traumatic spinal cord injury after transplantation of multineurotrophin-expressing glial-restricted precursor cells. *J Neurosci* 25:6947–6957.

Cao Q, Zhang YP, Iannotti C, DeVries WH, Xu XM, Shields CB, Whittemore SR. 2005b. Functional and electrophysiological changes after graded traumatic spinal cord injury in adult rat. *Exp Neurol* 191 (Suppl 1):S3–S16.

Cao QL, Zhang YP, Howard RM, Walters WM, Tsoulfas P, Whittemore SR. 2001. Pluripotent stem cells engrafted into the normal or lesioned adult rat spinal cord are restricted to a glial lineage. *Exp Neurol* 167:48–58.

Caprariello AV, Mangla S, Miller RH, Selkirk SM. 2012. Apoptosis of oligodendrocytes in the central nervous system results in rapid focal demyelination. *Ann Neurol* 72:395–405.

Casha S, Yu WR, Fehlings MG. 2001. Oligodendroglial apoptosis occurs along degenerating axons and is associated with FAS and p75 expression following spinal cord injury in the rat. *Neuroscience* 103:203–218.

Cheng X, Wang Y, He Q, Qiu M, Whittemore SR, Cao Q. 2007. Bone morphogenetic protein signaling and olig1/2 interact to regulate the differentiation and maturation of adult oligodendrocyte precursor cells. *Stem Cells* 25:3204–3214.

Chong SY, Rosenberg SS, Fancy SP, Zhao C, Shen YA, Hahn AT, McGee AW et al. 2012. Neurite outgrowth inhibitor Nogo-A establishes spatial segregation and extent of oligodendrocyte myelination. *Proc Natl Acad Sci USA* 109:1299–1304.

Chow SY, Moul J, Tobias CA, Himes BT, Liu Y, Obrocka M, Hodge L, Tessler A, Fischer I. 2000. Characterization and intraspinal grafting of EGF/bFGF-dependent neurospheres derived from embryonic rat spinal cord. *Brain Res* 874:87–106.

Clarke LE, Young KM, Hamilton NB, Li H, Richardson WD, Attwell D. 2012. Properties and fate of oligodendrocyte progenitor cells in the corpus callosum, motor cortex, and piriform cortex of the mouse. *J Neurosci* 32:8173–8185.

Compston A. 2006. The basis for treatment in multiple sclerosis. *Acta Neurol Scand Suppl* 183:41–47.

Cosgaya JM, Chan JR, Shooter EM. 2002. The neurotrophin receptor p75NTR as a positive modulator of myelination. *Science* 298:1245–1248.

Crowe MJ, Bresnahan JC, Shuman SL, Masters JN, Beattie MS. 1997. Apoptosis and delayed degeneration after spinal cord injury in rats and monkeys. *Nat Med* 3:73–76.

Cummings BJ, Uchida N, Tamaki SJ, Salazar DL, Hooshmand M, Summers R, Gage FH, Anderson AJ. 2005. Human neural stem cells differentiate and promote locomotor recovery in spinal cord-injured mice. *Proc Natl Acad Sci USA* 102:14069–14074.

Davies SJ, Shih CH, Noble M, Mayer-Proschel M, Davies JE, Proschel C. 2011. Transplantation of specific human astrocytes promotes functional recovery after spinal cord injury. *PLoS One* 6:e17328.

Dawson MR, Levine JM, Reynolds R. 2000. NG2-expressing cells in the central nervous system: Are they oligodendroglial progenitors? *J Neurosci Res* 61:471–479.

Dawson MR, Polito A, Levine JM, Reynolds R. 2003. NG2-expressing glial progenitor cells: An abundant and widespread population of cycling cells in the adult rat CNS. *Mol Cell Neurosci* 24:476–488.

De Biase LM, Nishiyama A, Bergles DE. 2010. Excitability and synaptic communication within the oligodendrocyte lineage. *J Neurosci* 30:3600–3611.

Dugas JC, Cuellar TL, Scholze A, Ason B, Ibrahim A, Emery B, Zamanian JL, Foo LC, McManus MT, Barres BA. 2010. Dicer1 and miR-219 are required for normal oligodendrocyte differentiation and myelination. *Neuron* 65:597–611.

Dusart I, Marty S, Peschanski M. 1992. Demyelination, and remyelination by Schwann cells and oligodendrocytes after kainate-induced neuronal depletion in the central nervous system. *Neuroscience* 51:137–148.

Dziewulska D, Rafalowska J. 2000. Astrogliosis and blood vessel development during human spinal cord myelination. *Folia Neuropathol* 38:61–67.

Eftekharpour E, Karimi-Abdolrezaee S, Wang J, El BH, Morshead C, Fehlings MG. 2007. Myelination of congenitally dysmyelinated spinal cord axons by adult neural precursor cells results in formation of nodes of Ranvier and improved axonal conduction. *J Neurosci* 27:3416–3428.

El WB, Macchi M, Cayre M, Durbec P. 2014. Oligodendrogenesis in the normal and pathological central nervous system. *Front Neurosci* 8:145.

Emery B, Agalliu D, Cahoy JD, Watkins TA, Dugas JC, Mulinyawe SB, Ibrahim A, Ligon KL, Rowitch DH, Barres BA. 2009. Myelin gene regulatory factor is a critical transcriptional regulator required for CNS myelination. *Cell* 138:172–185.

Emery E, Aldana P, Bunge MB, Puckett W, Srinivasan A, Keane RW, Bethea J, Levi AD. 1998. Apoptosis after traumatic human spinal cord injury. *J Neurosurg* 89:911–920.

Etxeberria A, Mangin JM, Aguirre A, Gallo V. 2010. Adult-born SVZ progenitors receive transient synapses during remyelination in corpus callosum. *Nat Neurosci* 13:287–289.

Fancy SP, Baranzini SE, Zhao C, Yuk DI, Irvine KA, Kaing S, Sanai N, Franklin RJ, Rowitch DH. 2009. Dysregulation of the Wnt pathway inhibits timely myelination and remyelination in the mammalian CNS. *Genes Dev* 23:1571–1585.

Fancy SP, Zhao C, Franklin RJ. 2004. Increased expression of Nkx2.2 and Olig2 identifies reactive oligodendrocyte progenitor cells responding to demyelination in the adult CNS. *Mol Cell Neurosci* 27:247–254.

Ferent J, Zimmer C, Durbec P, Ruat M, Traiffort E. 2013. Sonic Hedgehog signaling is a positive oligodendrocyte regulator during demyelination. *J Neurosci* 33:1759–1772.

Ffrench-Constant C, Raff MC. 1986. Proliferating bipotential glial progenitor cells in adult rat optic nerve. *Nature* 319:499–502.

Fields RD. 2006. Nerve impulses regulate myelination through purinergic signalling. *Novartis Found Symp* 276:148–158.

Foote AK, Blakemore WF. 2005. Inflammation stimulates remyelination in areas of chronic demyelination. *Brain* 128:528–539.

Franklin RJ. 2002. Why does remyelination fail in multiple sclerosis? *Nat Rev Neurosci* 3:705–714.

Franklin RJ, Hinks GL, Woodruff RH, O'Leary MT. 2001. What roles do growth factors play in CNS remyelination? *Prog Brain Res* 132:185–193.

Furusho M, Dupree JL, Nave KA, Bansal R. 2012. Fibroblast growth factor receptor signaling in oligodendrocytes regulates myelin sheath thickness. *J Neurosci* 32:6631–6641.

Furusho M, Kaga Y, Ishii A, Hebert JM, Bansal R. 2011. Fibroblast growth factor signaling is required for the generation of oligodendrocyte progenitors from the embryonic forebrain. *J Neurosci* 31:5055–5066.

Ge WP, Zhou W, Luo Q, Jan LY, Jan YN. 2009. Dividing glial cells maintain differentiated properties including complex morphology and functional synapses. *Proc Natl Acad Sci USA* 106:328–333.

Gibson EM, Purger D, Mount CW, Goldstein AK, Lin GL, Wood LS, Inema I. et al. 2014. Neuronal activity promotes oligodendrogenesis and adaptive myelination in the mammalian brain. *Science* 344:1252304.

Giedd JN. 2004. Structural magnetic resonance imaging of the adolescent brain. *Ann NY Acad Sci* 1021:77–85.

Gledhill RF, Harrison BM, McDonald WI. 1973. Demyelination and remyelination after acute spinal cord compression. *Exp Neurol* 38:472–487.

Goldsberry G, Mitra D, MacDonald D, Patay Z. 2011. Accelerated myelination with motor system involvement in a neonate with immediate postnatal onset of seizures and hemimegalencephaly. *Epilepsy Behav* 22:391–394.

Gomes WA, Mehler MF, Kessler JA. 2003. Transgenic overexpression of BMP4 increases astroglial and decreases oligodendroglial lineage commitment. *Dev Biol* 255:164–177.

Gonzalez-Perez O, Alvarez-Buylla A. 2011. Oligodendrogenesis in the subventricular zone and the role of epidermal growth factor. *Brain Res Rev* 67:147–156.

Griffiths I, Klugmann M, Anderson T, Yool D, Thomson C, Schwab MH, Schneider A. et al. 1998. Axonal swellings and degeneration in mice lacking the major proteolipid of myelin. *Science* 280:1610–1613.

Grinspan JB, Edell E, Carpio DF, Beesley JS, Lavy L, Pleasure D, Golden JA. 2000. Stage-specific effects of bone morphogenetic proteins on the oligodendrocyte lineage. *J Neurobiol* 43:1–17.

Gross RE, Mehler MF, Mabie PC, Zang Z, Santschi L, Kessler JA. 1996. Bone morphogenetic proteins promote astroglial lineage commitment by mammalian subventricular zone progenitor cells. *Neuron* 17:595–606.

Guest JD, Hiester ED, Bunge RP. 2005. Demyelination and Schwann cell responses adjacent to injury epicenter cavities following chronic human spinal cord injury. *Exp Neurol* 192:384–393.

Hagg T, Varon S. 1993. Ciliary neurotrophic factor prevents degeneration of adult rat substantia nigra dopaminergic neurons in vivo. *Proc Natl Acad Sci USA* 90:6315–6319.

Hasan KM, Ewing-Cobbs L, Kramer LA, Fletcher JM, Narayana PA. 2008a. Diffusion tensor quantification of the macrostructure and microstructure of human midsagittal corpus callosum across the lifespan. *NMR Biomed* 21:1094–1101.

Hasan KM, Kamali A, Abid H, Kramer LA, Fletcher JM, Ewing-Cobbs L. 2010. Quantification of the spatiotemporal microstructural organization of the human brain association, projection and commissural pathways across the lifespan using diffusion tensor tractography. *Brain Struct Funct* 214:361–373.

Hasan KM, Kamali A, Kramer LA, Papnicolaou AC, Fletcher JM, Ewing-Cobbs L. 2008b. Diffusion tensor quantification of the human midsagittal corpus callosum subdivisions across the lifespan. *Brain Res* 1227:52–67.

Hawryluk GW, Mothe A, Wang J, Wang S, Tator C, Fehlings MG. 2012. An in vivo characterization of trophic factor production following neural precursor cell or bone marrow stromal cell transplantation for spinal cord injury. *Stem Cells Dev* 21:2222–2238.

Hawryluk GW, Spano S, Chew D, Wang S, Erwin M, Chamankhah M, Forgione N, Fehlings MG. 2014. An examination of the mechanisms by which neural precursors augment recovery following spinal cord injury: A key role for remyelination. *Cell Transplant* 23:365–380.

Hofstetter CP, Holmstrom NA, Lilja JA, Schweinhardt P, Hao J, Spenger C, Wiesenfeld-Hallin Z, Kurpad SN, Frisen J, Olson L. 2005. Allodynia limits the usefulness of intraspinal neural stem cell grafts; directed differentiation improves outcome. *Nat Neurosci* 8:346–353.

Horner PJ, Power AE, Kempermann G, Kuhn HG, Palmer TD, Winkler J, Thal LJ, Gage FH. 2000. Proliferation and differentiation of progenitor cells throughout the intact adult rat spinal cord. *J Neurosci* 20:2218–2228.

Hu BY, Du ZW, Zhang SC. 2009. Differentiation of human oligodendrocytes from pluripotent stem cells. *Nat Protoc* 4:1614–1622.

Hughes EG, Kang SH, Fukaya M, Bergles DE. 2013. Oligodendrocyte progenitors balance growth with self-repulsion to achieve homeostasis in the adult brain. *Nat Neurosci* 16:668–676.

Hulsebosch CE. 2002. Recent advances in pathophysiology and treatment of spinal cord injury. *Adv Physiol Educ* 26:238–255.

Irvine KA, Blakemore WF. 2008. Remyelination protects axons from demyelination-associated axon degeneration. *Brain* 131:1464–1477.

Ishibashi T, Dakin KA, Stevens B, Lee PR, Kozlov SV, Stewart CL, Fields RD. 2006. Astrocytes promote myelination in response to electrical impulses. *Neuron* 49:823–832.

Itoyama Y, Webster HD, Richardson EP, Jr., Trapp BD. 1983. Schwann cell remyelination of demyelinated axons in spinal cord multiple sclerosis lesions. *Ann Neurol* 14:339–346.

James ND, Bartus K, Grist J, Bennett DL, McMahon SB, Bradbury EJ. 2011. Conduction failure following spinal cord injury: Functional and anatomical changes from acute to chronic stages. *J Neurosci* 31:18543–18555.

Jean I, Lavialle C, Barthelaix-Pouplard A, Fressinaud C. 2003. Neurotrophin-3 specifically increases mature oligodendrocyte population and enhances remyelination after chemical demyelination of adult rat CNS. *Brain Res* 972:110–118.

Jin Y, Neuhuber B, Singh A, Bouyer J, Lepore A, Bonner J, Himes T, Campanelli JT, Fischer I. 2011. Transplantation of human glial restricted progenitors and derived astrocytes into a contusion model of spinal cord injury. *J Neurotrauma* 28:579–594.

John GR, Shankar SL, Shafit-Zagardo B, Massimi A, Lee SC, Raine CS, Brosnan CF. 2002. Multiple sclerosis: Re-expression of a developmental pathway that restricts oligodendrocyte maturation. *Nat Med* 8:1115–1121.

Jovicic A, Roshan R, Moisoi N, Pradervand S, Moser R, Pillai B, Luthi-Carter R. 2013. Comprehensive expression analyses of neural cell-type-specific miRNAs identify new determinants of the specification and maintenance of neuronal phenotypes. *J Neurosci* 33:5127–5137.

Jung J, Hackett NR, Pergolizzi RG, Pierre-Destine L, Krause A, Crystal RG. 2007. Ablation of tumor-derived stem cells transplanted to the central nervous system by genetic modification of embryonic stem cells with a suicide gene. *Hum Gene Ther* 18:1182–1192.

Kang SH, Li Y, Fukaya M, Lorenzini I, Cleveland DW, Ostrow LW, Rothstein JD, Bergles DE. 2013. Degeneration and impaired regeneration of gray matter oligodendrocytes in amyotrophic lateral sclerosis. *Nat Neurosci* 16:571–579.

Karimi-Abdolrezaee S, Eftekharpour E, Wang J, Morshead CM, Fehlings MG. 2006. Delayed transplantation of adult neural precursor cells promotes remyelination and functional neurological recovery after spinal cord injury. *J Neurosci* 26:3377–3389.

Keirstead HS, Nistor G, Bernal G, Totoiu M, Cloutier F, Sharp K, Steward O. 2005. Human embryonic stem cell-derived oligodendrocyte progenitor cell transplants remyelinate and restore locomotion after spinal cord injury. *J Neurosci* 25:4694–4705.

Kessaris N, Fogarty M, Iannarelli P, Grist M, Wegner M, Richardson WD. 2006. Competing waves of oligodendrocytes in the forebrain and postnatal elimination of an embryonic lineage. *Nat Neurosci* 9:173–179.

Kim D, Kim CH, Moon JI, Chung YG, Chang MY, Han BS, Ko S et al. 2009. Generation of human induced pluripotent stem cells by direct delivery of reprogramming proteins. *Cell Stem Cell* 4:472–476.

Kondo T, Raff MC. 2004. A role for Noggin in the development of oligodendrocyte precursor cells. *Dev Biol* 267:242–251.

Kukley M, Kiladze M, Tognatta R, Hans M, Swandulla D, Schramm J, Dietrich D. 2008. Glial cells are born with synapses. *FASEB J* 22:2957–2969.

Kukley M, Nishiyama A, Dietrich D. 2010. The fate of synaptic input to NG2 glial cells: Neurons specifically downregulate transmitter release onto differentiating oligodendroglial cells. *J Neurosci* 30:8320–8331.

Kumar S, Kahn MA, Dinh L, De Vellis J. 1998. NT-3-mediated TrkC receptor activation promotes proliferation and cell survival of rodent progenitor oligodendrocyte cells in vitro and in vivo. *J Neurosci Res* 54:754–765.

Lai K, Kaspar BK, Gage FH, Schaffer DV. 2003. Sonic hedgehog regulates adult neural progenitor proliferation in vitro and in vivo. *Nat Neurosci* 6:21–27.

Lappe-Siefke C, Goebbels S, Gravel M, Nicksch E, Lee J, Braun PE, Griffiths IR, Nave KA. 2003. Disruption of Cnp1 uncouples oligodendroglial functions in axonal support and myelination. *Nat Genet* 33:366–374.

Lasiene J, Matsui A, Sawa Y, Wong F, Horner PJ. 2009. Age-related myelin dynamics revealed by increased oligodendrogenesis and short internodes. *Aging Cell* 8:201–213.

Lasiene J, Shupe L, Perlmutter S, Horner P. 2008. No evidence for chronic demyelination in spared axons after spinal cord injury in a mouse. *J Neurosci* 28:3887–3896.

Lau P, Verrier JD, Nielsen JA, Johnson KR, Notterpek L, Hudson LD. 2008. Identification of dynamically regulated microRNA and mRNA networks in developing oligodendrocytes. *J Neurosci* 28:11720–11730.

Lee KH, Yoon DH, Park YG, Lee BH. 2005. Effects of glial transplantation on functional recovery following acute spinal cord injury. *J Neurotrauma* 22:575–589.

Levine JM, Stincone F, Lee YS. 1993. Development and differentiation of glial precursor cells in the rat cerebellum. *Glia* 7:307–321.

Li GL, Farooque M, Holtz A, Olsson Y. 1999. Apoptosis of oligodendrocytes occurs for long distances away from the primary injury after compression trauma to rat spinal cord. *Acta Neuropathol (Berl)* 98:473–480.

Li JS, Yao ZX. 2012. MicroRNAs: Novel regulators of oligodendrocyte differentiation and potential therapeutic targets in demyelination-related diseases. *Mol Neurobiol* 45:200–212.

Li Q, Brus-Ramer M, Martin JH, McDonald JW. 2010. Electrical stimulation of the medullary pyramid promotes proliferation and differentiation of oligodendrocyte progenitor cells in the corticospinal tract of the adult rat. *Neurosci Lett* 479:128–133.

Linker RA, Maurer M, Gaupp S, Martini R, Holtmann B, Giess R, Rieckmann P et al. 2002. CNTF is a major protective factor in demyelinating CNS disease: A neurotrophic cytokine as modulator in neuroinflammation. *Nat Med* 8:620–624.

Liu J, Casaccia P. 2010. Epigenetic regulation of oligodendrocyte identity. *Trends Neurosci* 33:193–201.

Liu J, Dietz K, DeLoyht JM, Pedre X, Kelkar D, Kaur J, Vialou V et al. 2012. Impaired adult myelination in the prefrontal cortex of socially isolated mice. *Nat Neurosci* 15:1621–1623.

Liu XZ, Xu XM, Hu R, Du C, Zhang SX, McDonald JW, Dong HX et al. 1997. Neuronal and glial apoptosis after traumatic spinal cord injury. *J Neurosci* 17:5395–5406.

Lois C, Alvarez-Buylla A. 1993. Proliferating subventricular zone cells in the adult mammalian forebrain can differentiate into neurons and glia. *Proc Natl Acad Sci USA* 90:2074–2077.

Louis JC, Magal E, Takayama S, Varon S. 1993. CNTF protection of oligodendrocytes against natural and tumor necrosis factor-induced death. *Science* 259:689–692.

Loulier K, Ruat M, Traiffort E. 2006. Increase of proliferating oligodendroglial progenitors in the adult mouse brain upon Sonic hedgehog delivery in the lateral ventricle. *J Neurochem* 98:530–542.

Lu QR, Cai L, Rowitch D, Cepko CL, Stiles CD. 2001. Ectopic expression of Olig1 promotes oligodendrocyte formation and reduces neuronal survival in developing mouse cortex. *Nat Neurosci* 4:973–974.

Lu QR, Sun T, Zhu Z, Ma N, Garcia M, Stiles CD, Rowitch DH. 2002. Common developmental requirement for Olig function indicates a motor neuron/oligodendrocyte connection. *Cell* 109:75–86.

Lu QR, Yuk D, Alberta JA, Zhu Z, Pawlitzky I, Chan J, McMahon AP, Stiles CD, Rowitch DH. 2000. Sonic hedgehog—Regulated oligodendrocyte lineage genes encoding bHLH proteins in the mammalian central nervous system. *Neuron* 25:317–329.

Lytle JM, Chittajallu R, Wrathall JR, Gallo V. 2009. NG2 cell response in the CNP-EGFP mouse after contusive spinal cord injury. *Glia* 57:270–285.

Mabie PC, Mehler MF, Kessler JA. 1999. Multiple roles of bone morphogenetic protein signaling in the regulation of cortical cell number and phenotype. *J Neurosci* 19:7077–7088.

Mabie PC, Mehler MF, Marmur R, Papavasiliou A, Song Q, Kessler JA. 1997. Bone morphogenetic proteins induce astroglial differentiation of oligodendroglial-astroglial progenitor cells. *J Neurosci* 17:4112–4120.

Maherali N, Hochedlinger K. 2008. Guidelines and techniques for the generation of induced pluripotent stem cells. *Cell Stem Cell* 3:595–605.

Maire CL, Wegener A, Kerninon C, Nait OB. 2010. Gain-of-function of Olig transcription factors enhances oligodendrogenesis and myelination. *Stem Cells* 28:1611–1622.

Makinodan M, Rosen KM, Ito S, Corfas G. 2012. A critical period for social experience-dependent oligodendrocyte maturation and myelination. *Science* 337:1357–1360.

Mangin JM, Li P, Scafidi J, Gallo V. 2012. Experience-dependent regulation of NG2 progenitors in the developing barrel cortex. *Nat Neurosci* 15:1192–1194.

Marin-Husstege M, Muggironi M, Liu A, Casaccia-Bonnefil P. 2002. Histone deacetylase activity is necessary for oligodendrocyte lineage progression. *J Neurosci* 22:10333–10345.

Marmur R, Kessler JA, Zhu G, Gokhan S, Mehler MF. 1998. Differentiation of oligodendroglial progenitors derived from cortical multipotent cells requires extrinsic signals including activation of gp130/LIFbeta receptors. *J Neurosci* 18:9800–9811.

McDonald JW, Liu XZ, Qu Y, Liu S, Mickey SK, Turetsky D, Gottlieb DI, Choi DW. 1999. Transplanted embryonic stem cells survive, differentiate and promote recovery in injured rat spinal cord. *Nat Med* 5:1410–1412.

McKenzie IA, Ohayon D, Li H, de Faria JP, Emery B, Tohyama K, Richardson WD. 2014. Motor skill learning requires active central myelination. *Science* 346:318–322.

McTigue DM, Horner PJ, Stokes BT, Gage FH. 1998. Neurotrophin-3 and brain-derived neurotrophic factor induce oligodendrocyte proliferation and myelination of regenerating axons in the contused adult rat spinal cord. *J Neurosci* 18:5354–5365.

McTigue DM, Wei P, Stokes BT. 2001. Proliferation of NG2-positive cells and altered oligodendrocyte numbers in the contused rat spinal cord. *J Neurosci* 21:3392–3400.

Mekki-Dauriac S, Agius E, Kan P, Cochard P. 2002. Bone morphogenetic proteins negatively control oligodendrocyte precursor specification in the chick spinal cord. *Development* 129:5117–5130.

Menn B, Garcia-Verdugo JM, Yaschine C, Gonzalez-Perez O, Rowitch D, Alvarez-Buylla A. 2006. Origin of oligodendrocytes in the subventricular zone of the adult brain. *J Neurosci* 26:7907–7918.

Meyer-Franke A, Shen S, Barres BA. 1999. Astrocytes induce oligodendrocyte processes to align with and adhere to axons. *Mol Cell Neurosci* 14:385–397.

Miller RH. 1996. Oligodendrocyte origins. *Trends Neurosci* 19:92–96.

Miller RH. 2002. Regulation of oligodendrocyte development in the vertebrate CNS. *Prog Neurobiol* 67:451–467.

Miller RH, Dinsio K, Wang R, Geertman R, Maier CE, Hall AK. 2004. Patterning of spinal cord oligodendrocyte development by dorsally derived BMP4. *J Neurosci Res* 76:9–19.

Najm FJ, Lager AM, Zaremba A, Wyatt K, Caprariello AV, Factor DC, Karl RT, Maeda T, Miller RH, Tesar PJ. 2013. Transcription factor-mediated reprogramming of fibroblasts to expandable, myelinogenic oligodendrocyte progenitor cells. *Nat Biotechnol* 31:426–433.

Nakashima K, Takizawa T, Ochiai W, Yanagisawa M, Hisatsune T, Nakafuku M, Miyazono K, Kishimoto T, Kageyama R, Taga T. 2001. BMP2-mediated alteration in the developmental pathway of fetal mouse brain cells from neurogenesis to astrocytogenesis. *Proc Natl Acad Sci USA* 98:5868–5873.

Nakatani H, Martin E, Hassani H, Clavairoly A, Maire CL, Viadieu A, Kerninon C et al. 2013. Ascl1/Mash1 promotes brain oligodendrogenesis during myelination and remyelination. *J Neurosci* 33:9752–9768.

Naumann T, Schnell O, Zhi Q, Kirsch M, Schubert KO, Sendtner M, Hofmann HD. 2003. Endogenous ciliary neurotrophic factor protects GABAergic, but not cholinergic, septohippocampal neurons following fimbria-fornix transection. *Brain Pathol* 13:309–321.

Nguyen T, Mehta NR, Conant K, Kim KJ, Jones M, Calabresi PA, Melli G et al. 2009. Axonal protective effects of the myelin-associated glycoprotein. *J Neurosci* 29:630–637.

Nishiyama A, Chang A, Trapp BD. 1999. NG2+ glial cells: A novel glial cell population in the adult brain. *J Neuropathol Exp Neurol* 58:1113–1124.

Nishiyama A, Komitova M, Suzuki R, Zhu X. 2009. Polydendrocytes (NG2 cells): Multifunctional cells with lineage plasticity. *Nat Rev Neurosci* 10:9–22.

Nishiyama A, Suzuki R, Zhu X. 2014. NG2 cells (polydendrocytes) in brain physiology and repair. *Front Neurosci* 8:133.

Nottingham S, Knapp P, Springer J. 2002. FK506 treatment inhibits caspase-3 activation and promotes oligodendroglial survival following traumatic spinal cord injury. *Exp Neurol* 177:242–251.

Nunes MC, Roy NS, Keyoung HM, Goodman RR, McKhann G, Jiang L, Kang J, Nedergaard M, Goldman SA. 2003. Identification and isolation of multipotential neural progenitor cells from the subcortical white matter of the adult human brain. *Nat Med* 9:439–447.

Nunez JL, Nelson J, Pych JC, Kim JH, Juraska JM. 2000. Myelination in the splenium of the corpus callosum in adult male and female rats. *Brain Res Dev Brain Res* 120:87–90.

Okita K, Nakagawa M, Hyenjong H, Ichisaka T, Yamanaka S. 2008. Generation of mouse induced pluripotent stem cells without viral vectors. *Science* 322:949–953.

Oluich LJ, Stratton JA, Xing YL, Ng SW, Cate HS, Sah P, Windels F, Kilpatrick TJ, Merson TD. 2012. Targeted ablation of oligodendrocytes induces axonal pathology independent of overt demyelination. *J Neurosci* 32:8317–8330.

Orentas DM, Hayes JE, Dyer KL, Miller RH. 1999. Sonic hedgehog signaling is required during the appearance of spinal cord oligodendrocyte precursors. *Development* 126:2419–2429.

Ortega F, Gascon S, Masserdotti G, Deshpande A, Simon C, Fischer J, Dimou L, Chichung LD, Schroeder T, Berninger B. 2013. Oligodendrogliogenic and neurogenic adult subependymal zone neural stem cells constitute distinct lineages and exhibit differential responsiveness to Wnt signalling. *Nat Cell Biol* 15:602–613.

Park E, Velumian AA, Fehlings MG. 2004. The role of excitotoxicity in secondary mechanisms of spinal cord injury: A review with an emphasis on the implications for white matter degeneration. *J Neurotrauma* 21:754–774.

Parras CM, Hunt C, Sugimori M, Nakafuku M, Rowitch D, Guillemot F. 2007. The proneural gene Mash1 specifies an early population of telencephalic oligodendrocytes. *J Neurosci* 27:4233–4242.

Pearse DD, Pereira FC, Marcillo AE, Bates ML, Berrocal YA, Filbin MT, Bunge MB. 2004. cAMP and Schwann cells promote axonal growth and functional recovery after spinal cord injury. *Nat Med* 10:610–616.

Pearse DD, Sanchez AR, Pereira FC, Andrade CM, Puzis R, Pressman Y, Golden K et al. 2007. Transplantation of Schwann cells and/or olfactory ensheathing glia into the contused spinal cord: Survival, migration, axon association, and functional recovery. *Glia* 55:976–1000.

Peiffer AM, Shi L, Olson J, Brunso-Bechtold JK. 2010. Differential effects of radiation and age on diffusion tensor imaging in rats. *Brain Res* 1351:23–31.

Peters A, Kemper T. 2012. A review of the structural alterations in the cerebral hemispheres of the aging rhesus monkey. *Neurobiol Aging* 33:2357–2372.

Petrosyan HA, Hunanyan AS, Alessi V, Schnell L, Levine J, Arvanian VL. 2013. Neutralization of inhibitory molecule NG2 improves synaptic transmission, retrograde transport, and locomotor function after spinal cord injury in adult rats. *J Neurosci* 33:4032–4043.

Philips T, Bento-Abreu A, Nonneman A, Haeck W, Staats K, Geelen V, Hersmus N et al. 2013. Oligodendrocyte dysfunction in the pathogenesis of amyotrophic lateral sclerosis. *Brain* 136:471–482.

Pinzon A, Marcillo A, Pabon D, Bramlett HM, Bunge MB, Dietrich WD. 2008a. A re-assessment of erythropoietin as a neuroprotective agent following rat spinal cord compression or contusion injury. *Exp Neurol* 213:129–136.

Pinzon A, Marcillo A, Quintana A, Stamler S, Bunge MB, Bramlett HM, Dietrich WD. 2008b. A re-assessment of minocycline as a neuroprotective agent in a rat spinal cord contusion model. *Brain Res* 1243:146–151.

Pluchino S, Zanotti L, Rossi B, Brambilla E, Ottoboni L, Salani G, Martinello M et al. 2005. Neurosphere-derived multipotent precursors promote neuroprotection by an immunomodulatory mechanism. *Nature* 436:266–271.

Powers BE, Lasiene J, Plemel JR, Shupe L, Perlmutter SI, Tetzlaff W, Horner PJ. 2012. Axonal thinning and extensive remyelination without chronic demyelination in spinal injured rats. *J Neurosci* 32:5120–5125.

Powers BE, Sellers DL, Lovelett EA, Cheung W, Aalami SP, Zapertov N, Maris DO, Horner PJ. 2013. Remyelination reporter reveals prolonged refinement of spontaneously regenerated myelin. *Proc Natl Acad Sci USA* 110:4075–4080.

Pringle NP, Mudhar HS, Collarini EJ, Richardson WD. 1992. PDGF receptors in the rat CNS: During late neurogenesis, PDGF alpha-receptor expression appears to be restricted to glial cells of the oligodendrocyte lineage. *Development* 115:535–551.

Profyris C, Cheema SS, Zang D, Azari MF, Boyle K, Petratos S. 2004. Degenerative and regenerative mechanisms governing spinal cord injury. *Neurobiol Dis* 15:415–436.

Psachoulia K, Jamen F, Young KM, Richardson WD. 2009. Cell cycle dynamics of NG2 cells in the postnatal and ageing brain. *Neuron Glia Biol* 5:57–67.

Raff M. 2003. Adult stem cell plasticity: Fact or artifact? *Annu Rev Cell Dev Biol* 19:1–22.

Raff MC, Miller RH, Noble M. 1983. A glial progenitor cell that develops in vitro into an astrocyte or an oligodendrocyte depending on culture medium. *Nature* 303:390–396.

Richardson WD, Kessaris N, Pringle N. 2006. Oligodendrocyte wars. *Nat Rev Neurosci* 7:11–18.

Richter MW, Roskams AJ. 2008. Olfactory ensheathing cell transplantation following spinal cord injury: Hype or hope? *Exp Neurol* 209:353–367.

Rodriguez JP, Coulter M, Miotke J, Meyer RL, Takemaru K, Levine JM. 2014. Abrogation of beta-catenin signaling in oligodendrocyte precursor cells reduces glial scarring and promotes axon regeneration after CNS injury. *J Neurosci* 34:10285–10297.

Ross SE, Greenberg ME, Stiles CD. 2003. Basic helix-loop-helix factors in cortical development. *Neuron* 39:13–25.

Rowitch DH. 2004. Glial specification in the vertebrate neural tube. *Nat Rev Neurosci* 5:409–419.

Salazar DL, Uchida N, Hamers FP, Cummings BJ, Anderson AJ. 2010. Human neural stem cells differentiate and promote locomotor recovery in an early chronic spinal cord injury NOD-scid mouse model. *PLoS One* 5:e12272.

Samanta J, Kessler JA. 2004. Interactions between ID and OLIG proteins mediate the inhibitory effects of BMP4 on oligodendroglial differentiation. *Development* 131:4131–4142.

Scholz J, Klein MC, Behrens TE, Johansen-Berg H. 2009. Training induces changes in white-matter architecture. *Nat Neurosci* 12:1370–1371.

See J, Zhang X, Eraydin N, Mun SB, Mamontov P, Golden JA, Grinspan JB. 2004. Oligodendrocyte maturation is inhibited by bone morphogenetic protein. *Mol Cell Neurosci* 26:481–492.

Sekhon LH, Fehlings MG. 2001. Epidemiology, demographics, and pathophysiology of acute spinal cord injury. *Spine* 26:S2–S12.

Sharp J, Frame J, Siegenthaler M, Nistor G, Keirstead HS. 2010. Human embryonic stem cell-derived oligodendrocyte progenitor cell transplants improve recovery after cervical spinal cord injury. *Stem Cells* 28:152–163.

Sharp KG, Flanagan LA, Yee KM, Steward O. 2012. A re-assessment of a combinatorial treatment involving Schwann cell transplants and elevation of cyclic AMP on recovery of motor function following thoracic spinal cord injury in rats. *Exp Neurol* 233:625–644.

Shi J, Marinovich A, Barres BA. 1998. Purification and characterization of adult oligodendrocyte precursor cells from the rat optic nerve. *J Neurosci* 18:4627–4636.

Shihabuddin LS, Horner PJ, Ray J, Gage FH. 2000. Adult spinal cord stem cells generate neurons after transplantation in the adult dentate gyrus. *J Neurosci* 20:8727–8735.

Silver J, Miller JH. 2004. Regeneration beyond the glial scar. *Nat Rev Neurosci* 5:146–156.

Smith PM, Blakemore WF. 2000. Porcine neural progenitors require commitment to the oligodendrocyte lineage prior to transplantation in order to achieve significant remyelination of demyelinated lesions in the adult CNS. *Eur J Neurosci* 12:2414–2424.

Smith PM, Jeffery ND. 2006. Histological and ultrastructural analysis of white matter damage after naturally-occurring spinal cord injury. *Brain Pathol* 16:99–109.

Sohn J, Selvaraj V, Wakayama K, Orosco L, Lee E, Crawford SE, Guo F et al. 2012. PEDF is a novel oligodendrogenic morphogen acting on the adult SVZ and corpus callosum. *J Neurosci* 32:12152–12164.

Soldner F, Hockemeyer D, Beard C, Gao Q, Bell GW, Cook EG, Hargus G et al. 2009. Parkinson's disease patient-derived induced pluripotent stem cells free of viral reprogramming factors. *Cell* 136:964–977.

Springer JE, Azbill RD, Knapp PE. 1999. Activation of the caspase-3 apoptotic cascade in traumatic spinal cord injury. *Nat Med* 5:943–946.

Stadtfeld M, Nagaya M, Utikal J, Weir G, Hochedlinger K. 2008. Induced pluripotent stem cells generated without viral integration. *Science* 322:945–949.

Stankoff B, Aigrot MS, Noel F, Wattilliaux A, Zalc B, Lubetzki C. 2002. Ciliary neurotrophic factor (CNTF) enhances myelin formation: A novel role for CNTF and CNTF-related molecules. *J Neurosci* 22:9221–9227.

Stevens B, Tanner S, Fields RD. 1998. Control of myelination by specific patterns of neural impulses. *J Neurosci* 18:9303–9311.

Steward O, Sharp K, Selvan G, Hadden A, Hofstadter M, Au E, Roskams J. 2006. A re-assessment of the consequences of delayed transplantation of olfactory lamina propria following complete spinal cord transection in rats. *Exp Neurol* 198:483–499.

Steward O, Sharp K, Yee KM, Hofstadter M. 2008. A re-assessment of the effects of a Nogo-66 receptor antagonist on regenerative growth of axons and locomotor recovery after spinal cord injury in mice. *Exp Neurol* 209:446–468.

Stolt CC, Lommes P, Sock E, Chaboissier MC, Schedl A, Wegner M. 2003. The Sox9 transcription factor determines glial fate choice in the developing spinal cord. *Genes Dev* 17:1677–1689.

Sturrock RR. 1980. Myelination of the mouse corpus callosum. *Neuropathol Appl Neurobiol* 6:415–420.

Sugimori M, Nagao M, Parras CM, Nakatani H, Lebel M, Guillemot F, Nakafuku M. 2008. Ascl1 is required for oligodendrocyte development in the spinal cord. *Development* 135:1271–1281.

Sun W, Dietrich D. 2013. Synaptic integration by NG2 cells. *Front Cell Neurosci* 7:255.

Suzuki SO, Goldman JE. 2003. Multiple cell populations in the early postnatal subventricular zone take distinct migratory pathways: A dynamic study of glial and neuronal progenitor migration. *J Neurosci* 23:4240–4250.

Takahashi K, Tanabe K, Ohnuki M, Narita M, Ichisaka T, Tomoda K, Yamanaka S. 2007. Induction of pluripotent stem cells from adult human fibroblasts by defined factors. *Cell* 131:861–872.

Takebayashi H, Nabeshima Y, Yoshida S, Chisaka O, Ikenaka K, Nabeshima Y. 2002. The basic helix-loop-helix factor olig2 is essential for the development of motoneuron and oligodendrocyte lineages. *Curr Biol* 12:1157–1163.

Talbott JF, Loy DN, Liu Y, Qiu MS, Bunge MB, Rao MS, Whittemore SR. 2005. Endogenous Nkx2.2+/Olig2+ oligodendrocyte precursor cells fail to remyelinate the demyelinated adult rat spinal cord in the absence of astrocytes. *Exp Neurol* 192:11–24.

Tanaka T, Murakami K, Bando Y, Yoshida S. 2013. Minocycline reduces remyelination by suppressing ciliary neurotrophic factor expression after cuprizone-induced demyelination. *J Neurochem* 127:259–270.

Tator CH, Fehlings MG. 1991. Review of the secondary injury theory of acute spinal cord trauma with emphasis on vascular mechanisms. *J Neurosurg* 75:15–26.

Totoiu MO, Keirstead HS. 2005. Spinal cord injury is accompanied by chronic progressive demyelination. *J Comp Neurol* 486:373–383.

Traiffort E, Angot E, Ruat M. 2010. Sonic hedgehog signaling in the mammalian brain. *J Neurochem* 113:576–590.

Tripathi R, McTigue DM. 2007. Prominent oligodendrocyte genesis along the border of spinal contusion lesions. *Glia* 55:698–711.

Tripathi RB, Clarke LE, Burzomato V, Kessaris N, Anderson PN, Attwell D, Richardson WD. 2011. Dorsally and ventrally derived oligodendrocytes have similar electrical properties but myelinate preferred tracts. *J Neurosci* 31:6809–6819.

Tripathi RB, McTigue DM. 2008. Chronically increased ciliary neurotrophic factor and fibroblast growth factor-2 expression after spinal contusion in rats. *J Comp Neurol* 510:129–144.

Tsuji O, Miura K, Okada Y, Fujiyoshi K, Mukaino M, Nagoshi N, Kitamura K et al. 2010. Therapeutic potential of appropriately evaluated safe-induced pluripotent stem cells for spinal cord injury. *Proc Natl Acad Sci USA* 107:12704–12709.

Tsunoda I, Fujinami RS. 2002. Inside-Out versus Outside-In models for virus induced demyelination: Axonal damage triggering demyelination. *Springer Semin Immunopathol* 24:105–125.

Uchida N, Buck DW, He D, Reitsma MJ, Masek M, Phan TV, Tsukamoto AS, Gage FH, Weissman IL. 2000. Direct isolation of human central nervous system stem cells. *Proc Natl Acad Sci USA* 97:14720–14725.

Urfer R, Tsoulfas P, Soppet D, Escandon E, Parada LF, Presta LG. 1994. The binding epitopes of neurotrophin-3 to its receptors trkC and gp75 and the design of a multifunctional human neurotrophin. *EMBO J* 13:5896–5909.

Vallstedt A, Klos JM, Ericson J. 2005. Multiple dorsoventral origins of oligodendrocyte generation in the spinal cord and hindbrain. *Neuron* 45:55–67.

Velez-Fort M, Maldonado PP, Butt AM, Audinat E, Angulo MC. 2010. Postnatal switch from synaptic to extrasynaptic transmission between interneurons and NG2 cells. *J Neurosci* 30:6921–6929.

Vernerey J, Macchi M, Magalon K, Cayre M, Durbec P. 2013. Ciliary neurotrophic factor controls progenitor migration during remyelination in the adult rodent brain. *J Neurosci* 33:3240–3250.

Vroemen M, Aigner L, Winkler J, Weidner N. 2003. Adult neural progenitor cell grafts survive after acute spinal cord injury and integrate along axonal pathways. *Eur J Neurosci* 18:743–751.

Wake H, Lee PR, Fields RD. 2011. Control of local protein synthesis and initial events in myelination by action potentials. *Science* 333:1647–1651.

Wang S, Bates J, Li X, Schanz S, Chandler-Militello D, Levine C, Maherali N et al. 2013. Human iPSC-derived oligodendrocyte progenitor cells can myelinate and rescue a mouse model of congenital hypomyelination. *Cell Stem Cell* 12:252–264.

Wang S, Young KM. 2014. White matter plasticity in adulthood. *Neuroscience* 276:148–160.

Wang Y, Cheng X, He Q, Zheng Y, Kim DH, Whittemore SR, Cao QL. 2011. Astrocytes from the contused spinal cord inhibit oligodendrocyte differentiation of adult oligodendrocyte precursor cells by increasing the expression of bone morphogenetic proteins. *J Neurosci* 31:6053–6058.

Warren L, Manos PD, Ahfeldt T, Loh YH, Li H, Lau F, Ebina W et al. 2010. Highly efficient reprogramming to pluripotency and directed differentiation of human cells with synthetic modified mRNA. *Cell Stem Cell* 7:618–630.

Whittaker MT, Zai LJ, Lee HJ, Pajoohesh-Ganji A, Wu J, Sharp A, Wyse R, Wrathall JR. 2012. GGF2 (Nrg1-beta3. treatment enhances NG2+ cell response and improves functional recovery after spinal cord injury. *Glia* 60:281–294.

Wilkins A, Chandran S, Compston A. 2001. A role for oligodendrocyte-derived IGF-1 in trophic support of cortical neurons. *Glia* 36:48–57.

Wilkins A, Majed H, Layfield R, Compston A, Chandran S. 2003. Oligodendrocytes promote neuronal survival and axonal length by distinct intracellular mechanisms: A novel role for oligodendrocyte-derived glial cell line-derived neurotrophic factor. *J Neurosci* 23:4967–4974.

Windrem MS, Nunes MC, Rashbaum WK, Schwartz TH, Goodman RA, McKhann G, Roy NS, Goldman SA. 2004. Fetal and adult human oligodendrocyte progenitor cell isolates myelinate the congenitally dysmyelinated brain. *Nat Med* 10:93–97.

Windrem MS, Roy NS, Wang J, Nunes M, Benraiss A, Goodman R, McKhann GM, Goldman SA. 2002. Progenitor cells derived from the adult human subcortical white matter disperse and differentiate as oligodendrocytes within demyelinated lesions of the rat brain. *J Neurosci Res* 69:966–975.

Windrem MS, Schanz SJ, Guo M, Tian GF, Washco V, Stanwood N, Rasband M et al. 2008. Neonatal chimerization with human glial progenitor cells can both remyelinate and rescue the otherwise lethally hypomyelinated shiverer mouse. *Cell Stem Cell* 2:553–565.

Wolswijk G, Noble M. 1989. Identification of an adult-specific glial progenitor cell. *Development* 105:387–400.

Wren D, Wolswijk G, Noble M. 1992. In vitro analysis of the origin and maintenance of O-2Aadult progenitor cells. *J Cell Biol* 116:167–176.

Wu M, Hernandez M, Shen S, Sabo JK, Kelkar D, Wang J, O'Leary R, Phillips GR, Cate HS, Casaccia P. 2012. Differential modulation of the oligodendrocyte transcriptome by sonic hedgehog and bone morphogenetic protein 4 via opposing effects on histone acetylation. *J Neurosci* 32:6651–6664.

Wu S, Suzuki Y, Noda T, Bai H, Kitada M, Kataoka K, Nishimura Y, Ide C. 2002. Immunohistochemical and electron microscopic study of invasion and differentiation in spinal cord lesion of neural stem cells grafted through cerebrospinal fluid in rat. *J Neurosci Res* 69:940–945.

Xin M, Yue T, Ma Z, Wu FF, Gow A, Lu QR. 2005. Myelinogenesis and axonal recognition by oligodendrocytes in brain are uncoupled in Olig1-null mice. *J Neurosci* 25:1354–1365.

Xu XM, Guenard V, Kleitman N, Aebischer P, Bunge MB. 1995. A combination of BDNF and NT-3 promotes supraspinal axonal regeneration into Schwann cell grafts in adult rat thoracic spinal cord. *Exp Neurol* 134:261–272.

Yamamoto S, Yamamoto N, Kitamura T, Nakamura K, Nakafuku M. 2001. Proliferation of parenchymal neural progenitors in response to injury in the adult rat spinal cord. *Exp Neurol* 172:115–127.

Yan H, Wood PM. 2000. NT-3 weakly stimulates proliferation of adult rat O1(–)O4(+) oligodendrocyte-lineage cells and increases oligodendrocyte myelination in vitro. *J Neurosci Res* 62:329–335.

Yang N, Zuchero JB, Ahlenius H, Marro S, Ng YH, Vierbuchen T, Hawkins JS, Geissler R, Barres BA, Wernig M. 2013. Generation of oligodendroglial cells by direct lineage conversion. *Nat Biotechnol* 31:434–439.

Yasuda A, Tsuji O, Shibata S, Nori S, Takano M, Kobayashi Y, Takahashi Y et al. 2011. Significance of remyelination by neural stem/progenitor cells transplanted into the injured spinal cord. *Stem Cells* 29:1983–1994.

Yates MA, Juraska JM. 2007. Increases in size and myelination of the rat corpus callosum during adulthood are maintained into old age. *Brain Res* 1142:13–18.

Ye F, Chen Y, Hoang T, Montgomery RL, Zhao XH, Bu H, Hu T et al. 2009. HDAC1 and HDAC2 regulate oligodendrocyte differentiation by disrupting the beta-catenin-TCF interaction. *Nat Neurosci* 12:829–838.

Yong C, Arnold PM, Zoubine MN, Citron BA, Watanabe I, Berman NE, Festoff BW. 1998. Apoptosis in cellular compartments of rat spinal cord after severe contusion injury. *J Neurotrauma* 15:459–472.

Young KM, Psachoulia K, Tripathi RB, Dunn SJ, Cossell L, Attwell D, Tohyama K, Richardson WD. 2013. Oligodendrocyte dynamics in the healthy adult CNS: Evidence for myelin remodeling. *Neuron* 77:873–885.

Yu J, Hu K, Smuga-Otto K, Tian S, Stewart R, Slukvin II, Thomson JA. 2009. Human induced pluripotent stem cells free of vector and transgene sequences. *Science* 324:797–801.

Yu J, Vodyanik MA, Smuga-Otto K, Antosiewicz-Bourget J, Frane JL, Tian S, Nie J et al. 2007. Induced pluripotent stem cell lines derived from human somatic cells. *Science* 318:1917–1920.

Yu K, McGlynn S, Matise MP. 2013a. Floor plate-derived sonic hedgehog regulates glial and ependymal cell fates in the developing spinal cord. *Development* 140:1594–1604.

Yu Y, Chen Y, Kim B, Wang H, Zhao C, He X, Liu L et al. 2013b. Olig2 targets chromatin remodelers to enhancers to initiate oligodendrocyte differentiation. *Cell* 152:248–261.

Zai LJ, Yoo S, Wrathall JR. 2005. Increased growth factor expression and cell proliferation after contusive spinal cord injury. *Brain Res* 1052:147–155.

Zhang SC, Ge B, Duncan ID. 1999. Adult brain retains the potential to generate oligodendroglial progenitors with extensive myelination capacity. *Proc Natl Acad Sci USA* 96:4089–4094.

Zhang SC, Wernig M, Duncan ID, Brustle O, Thomson JA. 2001. In vitro differentiation of transplantable neural precursors from human embryonic stem cells. *Nat Biotechnol* 19:1129–1133.

Zhang YW, Denham J, Thies RS. 2006. Oligodendrocyte progenitor cells derived from human embryonic stem cells express neurotrophic factors. *Stem Cells Dev* 15:943–952.

Zhao X, He X, Han X, Yu Y, Ye F, Chen Y, Hoang T et al. 2010. MicroRNA-mediated control of oligodendrocyte differentiation. *Neuron* 65:612–626.

Zhou H, Wu S, Joo JY, Zhu S, Han DW, Lin T, Trauger S et al. 2009. Generation of induced pluripotent stem cells using recombinant proteins. *Cell Stem Cell* 4:381–384.

Zhou Q, Anderson DJ. 2002. The bHLH transcription factors OLIG2 and OLIG1 couple neuronal and glial subtype specification. *Cell* 109:61–73.

Zhou Q, Wang S, Anderson DJ. 2000. Identification of a novel family of oligodendrocyte lineage-specific basic helix-loop-helix transcription factors. *Neuron* 25:331–343.

Zhu G, Mehler MF, Zhao J, Yu YS, Kessler JA. 1999. Sonic hedgehog and BMP2 exert opposing actions on proliferation and differentiation of embryonic neural progenitor cells. *Dev Biol* 215:118–129.

Zhu Q, Whittemore SR, DeVries WH, Zhao X, Kuypers NJ, Qiu M. 2011. Dorsally-derived oligodendrocytes in the spinal cord contribute to axonal myelination during development and remyelination following focal demyelination. *Glia* 59:1612–1621.

Zuchero JB, Barres BA. 2013. Intrinsic and extrinsic control of oligodendrocyte development. *Curr Opin Neurobiol* 23:914–920.

3 Cell-Based Therapy for Stroke

Vivek Misra, Bing Yang, Sushil Sharma, and Sean I. Savitz

CONTENTS

3.1 INTRODUCTION

Stroke is the leading cause of serious long-term disability and ranks fifth among all causes of mortality in the United States. The only U.S. Food and Drug Administration (FDA)–approved treatment for acute ischemic stroke is intravenous thrombolysis using tissue plasminogen activator (tPA) within 3 hours from symptom onset. However, tPA has limited applicability largely due to a short therapeutic time window. Such limitations have prompted researchers to investigate neuroprotective agents that could potentially minimize neuronal injury from stroke and perhaps prolong therapeutic time windows for currently approved treatments. To date, none of these investigational agents have been effective in pivotal phase III clinical trials. Given the tremendous socioeconomic burden from stroke, it is critical to expand research to identify novel therapeutic strategies in wider time windows that could promote brain repair and improve outcome. Cellular therapy has emerged as a promising therapeutic modality with positive results from various basic science studies in animal models of ischemic stroke.

We present a comprehensive review of current knowledge from preclinical studies on cellular therapy for stroke. We also discuss some possible mechanisms of cellular therapy for ischemic stroke. Finally, we highlight the current challenges involved in translating this new therapeutic approach from the bench to the bedside, and briefly discuss the first preliminary clinical trials testing the safety of administering cells to stroke patients.

3.2 CELL TYPES AND SOURCES

There are two main types of categories of cell therapies under investigation for stroke: neural stem cells (derived from embryonic/fetal cells or cell lines) and adult non-neural, cell types.

1. *Neural stem cells:* Neural stem cells (NSCs) have the potential capacity to differentiate into neurons, astrocytes, and oligodendrocytes (Arnhold et al. 2000; Brustle et al. 1999; Flax et al. 1998). They can be derived from several sources:
   a. *Human embryonic stem cell–derived NSCs:* NSCs can be differentiated from human embryonic stem cells (hESCs) (Kim 2004; Koch et al. 2009; Reubinoff et al. 2001). Many investigators have shown cell integration into the host stroke brain and enhanced recovery after hESC transplantation (Theus et al. 2008; Wei et al. 2005). The capacity of unlimited propagation in culture for hESC7 makes hESC-derived NSCs potentially feasible and scalable for a clinical trial. However, batch-to-batch variations and the tumorigenic potential of these cells are serious concerns that need to be addressed before their clinical application.
   b. *Human fetal-derived NSCs:* Fetal human NSCs have similar features of hESC-derived NSCs (Kelly et al. 2004; Messina et al. 2003). The former have limited capacity for expansion in vitro, and this renders scaling to sufficient quantity for clinical studies a major challenge.
   c. *Cell lines:* Several human neural cell lines have been cloned and developed with genetic engineering techniques to possess progenitor-cell like abilities of self-renewal, integration into the host brain, and differentiation into neuronal and glial lineages (Bani-Yaghoub et al. 1999; Hara et al. 2008; Kim 2007; Kim et al. 2008c). They are reported to elicit functional recovery after stroke (Borlongan et al. 1998; Hara et al. 2007; Lee et al. 2007a). Examples include:
      i. HB1.F3 cell has a normal human karyotype 46XX (Kim et al. 2008c; Lee et al. 2007a).
      ii. CTX0E03 cell is also derived from human somatic cells with karyotype 46XY (Pollock et al. 2006).
      iii. hNT cell which originates from the teratocarcinoma NT2 cell line (Borlongan et al. 1998; Hara et al. 2008; Nelson et al. 2002). The hNT cells were some of the first to be taken forward to clinical trials for transplantation in stroke patients.

iv. The human fetal NSC line ReN001 (Stroemer et al. 2008). These cells have been taken forward to clinical trials.

2. *Non-neural cell types:* Owing to the ethical concerns associated with embryonic- and fetal-derived cells, much work has been devoted to developing progenitor cells derived from other body tissues (Brenneman et al. 2010; Kang et al. 2003; Koh et al. 2008; Kranz et al. 2010). Some of these cells may show the capacity for differentiation into cells with neural phenotypes and show other activities such as survival, migration to the area of injury, and improvement in functional recovery in animal models of stroke (Gerschat et al. 2008; Jiang et al. 2006; Keene et al. 2003; Song et al. 2007). However, the predominant mechanism is not differentiation into neurons but rather secretion of various bioactive factors that may promote endogenous repair and attenuate inflammation (Brenneman et al. 2010; Giraldi-Guimardes et al. 2009; Kim et al. 2007, 2008a; Pluchino et al. 2005; Vendrame et al. 2005; Wang et al. 2008; Wei et al. 2009; Yashuhara et al. 2010). Below is a listing of some of the non-neural cell therapies:
   a. *Bone marrow–derived mononuclear cells (BMMNCs):* BMMNCs can be isolated rapidly from autologous bone marrow without culture (Brenneman et al. 2010). This feature makes them very convenient for preparation for clinical use. MNCs are enriched with hematopoietic, mesenchymal, and endothelial progenitor cells but also contain many other types of cells such as lymphocytes and monocytes.
   b. *Bone marrow mesenchymal stem cells (BMMSCs):* BMMSCs are isolated from bone marrow and purified by culture in vitro (Hayase et al. 2009; Honma et al. 2006; Vu et al. 2014). Another type of stem cell purified and cultured from bone marrow is the multipotent adult progenitor cell (MAPC) (Keene et al. 2003).
   c. *Umbilical cord blood cells (UCBC):* UCBCs contain a mixture of various types of stem cells including hematopoietic stem cells, endothelial stem cells, and mesenchymal stem cells (Gerschat et al. 2008; Liao et al. 2009)
   d. *Adipose tissue mesenchymal progenitor cells (ASCs):* ASCs are isolated from adipose stromal tissue (Kang et al. 2003).
   e. *Peripheral blood progenitor cells and peripheral-derived mononuclear cells:* Peripheral blood progenitor cells and peripheral-derived mononuclear cells are reported to have benefits similar to those of bone marrow–derived cells (Moore et al. 2005; Ukai et al. 2007; Willing et al. 2003b). Many of these adult cell preparations are composed of numerous cell types including hematopoietic and endothelial stem/progenitor cells, immature lymphocytes, and monocytes (Parr et al. 2007; Vendrame et al. 2004).
   f. *Induced pluripotent stem cells (IPSCs):* Mature adult cells can be reprogrammed into pluripotent progenitor cells using various transcription factors. These IPSCs are currently being investigated in preclinical studies of ischemic stroke (Oki et al. 2012).

3.3 POSSIBLE MECHANISMS OF CELLULAR THERAPY

There are many potential mechanisms of cellular therapy depending on the cell type, timing, and route of cell administration. Studies also suggest that neuronal replacement is not an important mechanism underlying the recovery effects of adult cells, but rather adult cells induce modifications to the host microenvironment (Chen et al. 2003a,b; Copone et al. 2007; Lee et al. 2008).

1. *Migration and integration into the host brain:* NSCs have the potential to differentiate into neurons, astrocytes, and oligodendrocytes. NSCs are potently activated by stroke and exhibit directional migration to injured brain areas (Guzman et al. 2008a,b; Jiang et al. 2006). It has been reported that NSCs express synaptic proteins and elaborate functional synaptic connections with host neural circuits in rodent ischemic stroke (Daadi et al. 2009; Ishibashi et al. 2004). NSCs have been reported to have electrophysiological properties characteristic of functional neurons (Bühnemann et al. 2006). However, functional recovery after stroke reported in most studies occurs earlier than the formation of new neural cells. Because a small number of surviving donor cells are found in the host brain (Roh et al. 2008), it is believed that cellular therapy even with NSCs is unlikely to be related to replacing lost brain tissue.
2. *Neuroprotection:* Both NSCs and non-NSCs show neuroprotective effects in vitro and in vivo (Arien-Zekav et al. 2009; Chen et al. 2003a,b; Chu et al. 2004; Copone et al. 2007; Hau et al. 2008; Yamashita et al. 2006). Transplanted cells can directly secrete various trophic-growth factors, such as vascular endothelial growth factor (VEGF), fibroblast growth factor (FGF), glial cell–derived neurotrophic factor (GDNF), insulin-like growth factor (IGF), nerve growth factor (NGF), brain-derived neurotrophic factor (BDNF), etc. (Li et al. 2002). Transplanted cells can also stimulate the release of neurotrophins within the host parenchymal cells (Hara et al. 2007; Kim 2004; Kurozumi et al. 2004; Yamashita et al. 2006). BMMNCs also secrete trophic survival factors and inhibit neuronal toxicity induced by activated microglia in vitro (Sharma et al. 2010).
3. *Immunomodulation:* There is increasing evidence to suggest that cell-induced immunomodulation exists not only in the microenvironment of the injured brain but also in the macroenvironment of the entire body (Chu et al. 2005; Copone et al. 2007; Hogduijn et al. 2007; Vendrame et al. 2005). UCBC, for example, can affect not only the release of inflammatory cytokines but also the release and biology of immune cells from the spleen (Copeland et al. 2009; Vendrame et al. 2005). Cells administered during the acute phase of stroke can attenuate the injury caused by stroke-induced inflammation (Lindvall and Kokaia 2006; Park et al. 2009; Vendrame et al. 2005). Intravenous injection of human UCBC or human NSCs reduces the number of leukocytes in the brain (Lee et al. 2008; Vendrame et al. 2005). MSC and NSC transplantation also suppresses T cell proliferation (Nasef et al. 2007). MSCs or MNCs can modify the proliferation of microglia in the peri-infarct region in animal stroke models (Daadi et al. 2010; Denes et al. 2007; Jiang et al. 2010). Cellular therapy

also induces the upregulation of gene expression for anti-inflammatory cytokines and the downregulation of genes related to inflammatory and immune response in the brain. This shifts the balance from pro- to anti-inflammatory cytokines (Ohtaki et al. 2008; Pluchino et al. 2005). Cells interact with distant organ systems to generate an alteration in the systemic inflammatory and immunologic response. Intravenous injection of human UCBC changes the splenic mass, which correlates with decreased $CD8^+$ T cell release (Vendrame et al. 2005). The increased anti-inflammatory cytokine (IL-10) level in the spleen has been postulated as one source of increased IL-10 levels in the serum and the brain (Vendrame et al. 2005).

4. *Enhancing endogenous progenitors:* After stroke, endogenous progenitors recruited from local brain tissue are involved in natural repair processes. Cell transplantation may enhance this phenomenon (Chen et al. 2003a; Zheng et al. 2007). In animal models with stem cell transplantation, endogenous neural progenitor cells proliferate and migrate to the injured brain (Park et al. 2010; Zhang et al. 2011). Oligodendrocyte precursors are also elevated, which could be beneficial in remyelinating injured axons (Li et al. 2006; Shen et al. 2006).
5. *Reduction in the blood-brain barrier (BBB) permeability and enhancing angiogenesis:* Adult stem cells such as MSCs can reduce BBB permeability by increasing endothelial tight junction proteins (Zhang et al. 2006). Angiogenesis and vasculogenesis may be important to the repair process after stroke. Both neural stem cells and non-neural stem cells can directly secrete or induce host expression of angiogenic factors that stimulate proliferation of existing vascular endothelial cells and mobilization of endogenous endothelial progenitors leading to revascularization (Ding et al. 2008; Kim 2007; Taguchi et al. 2004; Tang et al. 2014).
6. *Improvement in neural plasticity:* NSCs are also reported to increase the sprouting of nerve fibers. Axonal projections emanate from neurons in the ipsilateral striatum and extrude into the peri-infarct area or white matter (Ishibashi et al. 2004; Jiang et al. 2006). This has been reported even when cells are grafted at sites contralateral to the ischemic hemisphere (Xiao et al. 2005). They are also observed to enhance dendrite branching and synaptogenesis (Ishibashi et al. 2004; Jiang et al. 2006). These features have also been reported in non-neural stem cells (Andrews et al. 2008; Brustle et al. 1999; Ding et al. 2007). Cell-induced changes in neurotrophin and growth factors may also influence neuroblast migration, synaptogenesis, and axonal reorganization to promote functional recovery (Crigler et al. 2006; Daadi et al. 2010; Ding et al. 2008; Li et al. 2006; Shen et al. 2006; Sohur et al. 2006; Zhang et al. 2006).

3.4 KEY TRANSLATIONAL BARRIERS TO ADMINISTERING CELLS AS A THERAPY FOR STROKE

1. *Cell tracking and imaging:* Monitoring transplanted cells and tracking their biodistribution is very important to study the mechanisms and fate of injected cells in clinical trials. Many noninvasive imaging techniques

have been used in cell tracking in vivo (Daadi et al. 2009; Gera et al. 2010; Guzman et al. 2008a,b; Kiessling 2008); however, none of these techniques are ready for deployment in stroke patients.

a. Magnetic resonance imaging (MRI) with iron oxide nanoparticles is the most studied technique (Beaulieu 2002; Callera and Demela 2007; Hauger et al. 2006; Rice et al. 2007) but is limited by the problem that iron is released by dead cells into the environment or digested by phagocytes. MRI may therefore not be able to assess graft survival (Ferreira 2009). In addition, there is no commercially available iron labeling reagent approved by the FDA.
b. Bioluminescence imaging (BLI) offers the possibility to monitor cell migration and assess graft cell viability; however, it currently has no direct application to humans. A luciferase reporter gene is transfected into cells prior to transplantation, which allows for the trafficking of cells by photon emission after the substrate D-luciferin is injected (Kim et al. 2004; Reumers et al. 2008). Accuracy may be disturbed by the downregulation of luciferase expression.
c. Positron emission tomography (PET) uses a PET reporter gene, allowing for high sensitivity. However, PET is limited by a lack of fine anatomical information and has limited availability.
d. Single photon emission computed tomography (SPECT) similarly uses radionucleotide imaging and is more available, but concerns over the use of radiation need to be considered as well as the viability of nucleotide labels on the cellular product.

2. *Potential risks of stem cell therapy:* Although stem cell therapy could potentially provide hope to patients affected by neurologic diseases such as stroke, there are potential risks including malignant transformation of transplanted cells (Lindvall and Kokaia 2006). The injection of pluripotent precursor cells in rodents leads to the development of teratomas or teratocarcinomas (Erdo et al. 2003; Reubinoff et al. 2001). Earlier studies by Erdo et al. (2003) have provided evidence that embryonic stem cells seem more prone to generate tumors when implanted into the same species from which they are derived (Erdo et al. 2003). A patient with ataxia telangiectasia who was treated by fetal neural stem cell transplantation reportedly developed a multifocal brain tumor (Amariglio et al. 2009). The tumorigenic potential of hESCs seems to be greatly reduced when they are predifferentiated in vitro before implantation (Arnhold et al. 2000; Brustle et al. 1999; Reubinoff et al. 2001). These findings suggest that extensive basic as well as clinical research are required to understand the molecular biology of stem cells and investigate long-term safety to maximize the potential benefits of regenerative medicine while minimizing the risks.
3. *Routes of stem cell delivery:* Basic science studies in animal models of ischemic stroke have used various routes of delivery including intracerebral (IC), intracerebroventricular/intracisternal (ICV), intravenous (IV), and intra-arterial (IA) injections.

a. *Intracerebral delivery (IC):* Direct intracerebral injection as a route of stem cell delivery has been investigated extensively in animal studies. Intracerebral injection involves insertion of a needle placed stereotactically in the brain parenchyma for infusion of cells. After IC injection into the ipsilateral (Kim et al. 2008a) or contralateral side (Kim et al. 2008a; Veizovic et al. 2001), stem cells exhibit pathotropism and migrate to the area of infarction (Daadi et al. 2009). Intracerebral stem cell engraftment has been shown to promote neuronal survival and improvement in functional deficits (Andrews et al. 2008; Borlongan et al. 1998; Chen et al. 2001b; Daadi et al. 2009; Ikeda et al. 2005; Kim et al. 2008a,b,c; Koh et al. 2008; Li et al. 2000; Modo et al. 2009; Theus et al. 2008; Veizovic et al. 2001; Wei et al. 2005) and reduce infarct volumes (Koh et al. 2008; Modo et al. 2009). This method provides a precise method of targeted stem cell delivery but is invasive and results in poor cell distribution throughout the area of ischemia (Li et al. 2000). Multiple injections into the ischemic lesions could also result in further tissue injury and nonuniform distribution of cells in the target lesion (Olanow et al. 2003).
b. *Intracerebroventricular/intracisternal (ICV):* Injection of stem cells in the cerebrospinal fluid pathways ensures stem cell access to a larger surface area of the brain (Jiang et al. 2005). Studies have demonstrated variable cell migration after ICV delivery (Jiang et al. 2005; Modo et al. 2002; Zhang et al. 2003). ICV injection in one study demonstrated reduction in neuronal injury with functional improvement (Copone et al. 2007). Another study demonstrated improvement in cognitive deficits with ICV injections compared to sensorimotor improvement with IC injections (Modo et al. 2002). However, this route of cell delivery is invasive, and uniform distribution of cells at the site of ischemia is still questionable.
c. *Intravenous (IV):* Intravenous administration of stem cells carries the advantage of being easy and least invasive. It also allows wide systemic distribution of cells with the potential for exposure to chemoattractant signals resulting in selective accumulation in the target lesion. Several authors have also demonstrated improvement in neurologic deficit with functional recovery (Chen et al. 2001a; Chu et al. 2004, 2005; Honma et al. 2006; Iihoshi et al. 2004; Li et al. 2002; Onda et al. 2008; Pavlichenko et al. 2008; Sykova and Jendelova 2007; Taguchi et al. 2004; Ukai et al. 2007; Vendrame et al. 2004; Wang et al. 2008; Willing et al. 2003a,b) as well as reduction in infarct volume (Honma et al. 2006; Iihoshi et al. 2004; Onda et al. 2008; Pavlichenko et al. 2008; Ukai et al. 2007; Vendrame et al. 2004) after IV cell delivery. A study also demonstrated greater long-term improvement in functional deficits with IV cell delivery compared to IC transplantation (Willing et al. 2003a,b). However, systemic stem cell injection results in poor cell engraftment at the target ischemic lesion in the brain (Borlongan et al. 2004; Chu et al. 2005; Li et al. 2002; Mäkinen et al. 2006; Walczak et al. 2008). This

is largely due to the cells being trapped peripherally by filtering organs as the lungs, liver, spleen, and kidneys (Borlongan et al. 2004; Hauger et al. 2006; Kraitchman et al. 2005; Mäkinen et al. 2006; Pluchino et al. 2005). Ischemic stroke is often associated with nephropathy and myocardial injury which could also serve as concomitant target organs and further reduce stem cell engraftment in the ischemic brain lesion (Hauger et al. 2006; Kraitchman et al. 2005). Despite these findings, the entry of cells in the infarcted tissue is not a prerequisite for enhancing neurologic recovery (Borlongan et al. 2004).

d. *Intra-arterial (IA):* Intra-arterial stem cell delivery involves injecting the cells directly into the carotid artery ipsilateral to the side of the ischemic lesion. It has the potential to bypass the peripheral filtering organs, thereby increasing cell delivery with uniform distribution in the ischemic brain lesion. Authors have demonstrated increased cell engraftment in the ischemic brain (Guzman et al. 2008a,b; Li et al. 2001; Shen et al. 2006; Walczak et al. 2008; Zhang et al. 2006) as well as functional recovery (Guzman et al. 2008a,b; Li et al. 2001; Shen et al. 2006; Zhang et al. 2006) after IA injections on the affected side. A study demonstrated 21% of injected cells engrafted in the ischemic lesion (Li et al. 2001) after intracarotid injection compared to 0.75%–18.5% reported after IV administration in another study (Liu et al. 2006), which also demonstrated stem cell concentrations in the infarct area after IA delivery to be significantly higher than those reported after IV injection (Lee et al. 2007b). However, intracarotid injections are invasive and have the potential for inducing microvascular occlusion and worsening cerebral blood flow (Walczak et al. 2008). The resulting higher engraftment rates could be offset by impaired flow and worsened ischemia (Walczak et al. 2008). Subsequent studies attributed these adverse findings to faulty IA infusion technique which caused flow arrest during cell delivery (Chua et al. 2011), high infusion rates, and larger cell sizes (Ge et al. 2014; Janowski et al. 2013).

Clearly, the optimal route of stem cell delivery is not yet determined, and it is doubtful whether IA delivery of BMMNCs results in better neurologic outcomes compared to IV infusion (Vasconcelos-dos-Santos et al. 2012; Yang et al. 2013).

4. *Dose of stem cells:* There is a great variability in animal studies with regard to stem cell doses. Studies that investigated IC injections utilized doses ranging from 1×10^5 to 3×10^6 cells with multiple injections to deliver the entire dose. Similar doses have been delivered using ICV/Intracisternal injections. Studies that investigated IA injections into the carotid arteries on the side of the ischemic lesions delivered doses ranging from 3×10^5 to 2×10^6 cells completed during one phase of injection. However, studies that investigated stem cells via IV injection used larger doses ranging from 1×10^6 to 5×10^7 cells delivered within a single infusion. Larger doses used with IV stem cell delivery might presumably be needed to ensure better cerebral engraftment after filtering by peripheral parenchymal organs;

however, it remains unclear if brain engraftment is necessary for cells to promote recovery. Two separate animal studies have demonstrated a dose-response relationship with IV stem cell delivery after ischemic stroke. In one study, 10^6 or more human cord blood cells were required to produce behavioral improvement at 4 weeks when delivered intravenously 24 hours after middle cerebral artery occlusion (MCAO) in a rat model (Vendrame et al. 2004). There was also an inverse relationship between cell dose and infarct volume, with doses of 10^7 cells or above causing significant reduction in infarct volumes (Vendrame et al. 2004). These results were also reproduced in a separate study that investigated IV cell delivery 12 hours after induction of transient MCAO in rats (Honma et al. 2006). There was a progressive reduction in infarct volume and improved functional outcome with increasing number of infused cells. In one study, IV injection of 10^6 cells produced greater long-term improvement in functional deficits compared to striatal IC implantation of 2.5×10^5 cells 24 hours after MCAO in a rat model (Willing et al. 2003a,b). However, multiple IV infusions of UCBCs did not result in improved outcomes in a preclinical study compared to a single dose alone (Shehadah et al. 2013). One study demonstrated that IC injection of NSCs over a maximal threshold dose resulted in impaired graft survival likely from oversaturation of the recipient tissue with insufficient availability of nutrients at the transplanted site (Darsalia et al. 2011).

5. *Timing of stem cell delivery:* There is no clearly defined therapeutic window for stem cell therapy in stroke. Animal studies have investigated the effects of administering cells using various methods of delivery from time windows ranging from 30 minutes to 1 month after stroke onset. Intravenous infusion of BMMNCs in preclinical studies has been demonstrated to be efficacious up to 72 hours from stroke onset (Iihoshi et al. 2004; Yang et al. 2011). Intravenous infusion of UCBCs has also been reported to result in maximal improvement within a time window of 72 hours, though some improvement was observed with cell delivery up to 120 hours from stroke onset (Boltze et al. 2012). A meta-analysis of animal studies investigating BMMSCs reported improvement in behavioral outcomes with cell delivery using various routes at time windows ranging from over 24 hours to up to 1 month from stroke onset (Vu et al. 2014). Another study investigating IA delivery of NSCs demonstrated maximal cell survival and engraftment at a time window of 3 days, though neuronal differentiation was highest at 7–14 days from stroke onset (Rosenblum et al. 2012). Intracerebral transplantation of NSCs in the infarcted tissue has been reported to have maximal survival when performed within 48 hours of the onset of cerebral ischemia (Darsalia et al. 2011). Cells, unlike pharmacologic agents, likely do not follow well-defined pharmacokinetics, and the window for their therapeutic effect is likely influenced by the intended mechanism of action. A cellular product with a principal action to induce neuroplasticity might theoretically have a longer window up to months after stroke, and a mechanism involving immunomodulation might be confined to the first few days after stroke while active inflammatory processes are at play.

3.5 CLINICAL TRIALS

The early clinical trials investigated stem cell therapy in patients with chronic stroke with the presumption of neuronal replacement as the potential mechanism and utilized cells primarily from fetal sources. Subsequently, early phase human studies emerged worldwide where various cell types were investigated in acute as well as chronic stroke patients.

Bone marrow–derived cells from autologous sources are an attractive choice for clinical trials due to lack of ethical concerns and also because their application would not require concomitant immunosuppression. The BMMNCs are most feasible for investigation in the acute setting, because the mononuclear fraction can be harvested rapidly over a few hours for subsequent infusion within the first few days after symptom onset. The BMMSCs from autologous sources would, however, require a longer time to separate and culture, making it less practical to investigate in shorter therapeutic time windows. The BMMSCs from allogeneic sources are, therefore, more feasible to use in clinical trials in acute/subacute stroke.

3.5.1 Chronic Stroke

3.5.1.1 Intracerebral Implantation

Kondziolka et al. conducted one of the first studies investigating the safety and feasibility of human neuronal cell (human carcinoma–derived cell line) transplantation in 12 patients with chronic basal ganglia infarcts (Kondziolka et al. 2000). They performed stereotactic implantation of 2 to 6 million cells injected directly into the area of infarction. The patients also received methylprednisolone intraoperatively and cyclosporine-A 1 week prior and up to 8 weeks after surgery. At 24 weeks, there was improvement in six patients on the European Stroke Scale (ESS) score, no change in three, and deterioration in three patients. On the National Institutes of Health (NIH) stroke scale, eight patients showed improvement, one was unchanged, and three had deteriorated. Positron emission tomography scans at 6 months demonstrated increased fluorodeoxyglucose at the implant site in six patients. One patient died 27 months after transplantation from acute myocardial infarction. Autopsy revealed the presence of neural cells derived from the injection at the transplant site. There were no procedure-related deaths or cell-related adverse events or radiologic abnormalities. They subsequently conducted a phase 2 randomized observer blinded trial of 18 patients (nine ischemic and nine hemorrhagic strokes) with stable motor deficits (Kondziolka et al. 2005). Patients in the surgical group received either 5 or 10 million cells implanted at the site of infarction. They reported adverse events in three patients (one had a single seizure, one syncope, and one an asymptomatic subdural hematoma). There was measurable improvement in some patients on the ESS score; however, there was no improvement in motor function compared to controls. They also demonstrated sustained cell viability with the surgical technique (Kondziolka et al. 2004).

Savitz et al. evaluated the safety and feasibility of porcine fetal cells (obtained from the porcine primordial striatum) transplanted stereotactically in five patients with stable chronic basal ganglia infarcts (Savitz et al. 2005). They injected doses of 10×10^6 cells per needle tract with four patients receiving up to five needle tracts. The fifth patient received 800 μL of cell suspension (20×10^6 cells/mL) injected at the site in less than 1 minute. One patient developed temporary worsening of motor deficits, and one developed seizures. Magnetic resonance imaging in these two patients demonstrated enhancement in areas remote from the site of transplantation. These MRI findings, however, fully resolved on subsequent repeat imaging. The FDA terminated the study following these events. Three patients reported functional improvement in motor and/or language deficits, and two had no change from their baseline deficits.

3.5.1.2 Intravenous Infusion

Bang et al. conducted a phase I/II clinical trial investigating the safety and feasibility of IV infusion of MSCs in five patients with disabling MCA distribution ischemic strokes (Bang et al. 2005). They calculated the human dose based on mean body mass at a total of 1×10^8 cells. These patients received 5×10^7 cells each at 4–5 weeks and then at 7–9 weeks from symptom onset. Patients who received cells had a reported better outcome at 3, 6, and 12 months with no adverse events compared to a control group (25 patients) who received standard of care. They expanded the study to include a larger patient population; 16 randomized to MSCs and 36 controls for a longer follow-up duration of 5 years. Patients receiving IV MSCs had reportedly lesser mortality and improved long-term outcomes (Lee et al. 2010).

Honmou et al. administered $0.6–1.6 \times 10^8$ MSCs intravenously within 36–133 days of symptom onset in 12 patients with ischemic stroke and reported safety, feasibility, as well as reduction in infarct volumes (Honmou et al. 2011). Bhasin et al. also conducted two separate pilot studies reporting safety and feasibility of IV infusion of autologous BMMNC and BMMSCs in 12 and 6 chronic stroke patients, respectively (Bhasin et al. 2011, 2012).

3.5.1.3 Intra-Arterial Delivery

Battistella et al. reported safety of IA infusion of $1–5 \times 10^8$ BMMNCs in the middle cerebral artery (MCA) of six patients within 90 days of symptom onset. Their group had earlier reported their anecdotal experience of safety of IA infusion of Tc-99m labeled BMMNCs in the MCA of patients with chronic stroke (Barbosa da Fonseca et al. 2009, 2010).

3.5.2 Acute Stroke

3.5.2.1 Intravenous Infusion

Savitz et al. investigated IV infusion of autologous BMMNCs in patients with acute ischemic stroke within 72 hours of symptom onset (Savitz et al. 2011). In the initial phase, 10 patients were included, eight of these receiving 10 million cells/kg and two infused with ≥7 million cells/kg intravenously. The study demonstrated feasibility and safety of the trial regimen, and no adverse events related to bone marrow harvest and cell delivery in the acute setting of ischemic stroke were reported. There was a trend toward improved outcome in most patients, and the trial was subsequently expanded to include 25 patients with results pending publication.

Prasad et al. also reported safety and feasibility of IV BMMNC infusion in 11 patients with subacute ischemic stroke (Prasad et al. 2012). Patients in this study received a mean dose of 80×10^6 MNCs within a time window of 7–30 days of stroke onset.

3.5.2.2 Intra-Arterial Delivery

Mendonça et al. performed an IA injection of BMMNCs in a patient 3 days after left MCA territory infarction (Mendonça et al. 2006). They injected 30×10^7 cells into the left MCA and demonstrated increased metabolism in the left parietal lobe on PET imaging as well as some neurological improvement. The same group also injected 3.0×10^7 cells in the left MCA of another patient 9 days after symptom onset (Correa et al. 2007). They labeled approximately 1% of these cells with 150 MBq Tc-99m by incubation with hexamethylpropylene amine oxime (HMPAO). They performed brain SPECT scans 8 hours after cell delivery and reported intense accumulation of the labeled cells in the left hemisphere. A whole-body scan also demonstrated uptake in the liver and spleen, but none in the lungs. They also reported good clinical recovery in this patient with NIHSS score 0 at 4 months after stroke onset. This anecdotal report claimed that SPECT imaging after Tc-99m HMPAO cell labeling was a feasible noninvasive method of in vivo cell tracking after IA delivery. Their group subsequently reported safety and feasibility of IA infusion of BMMNCs in the MCA of 20 patients with moderate to severe ischemic strokes within 3–7 days of symptom onset (Friedrich et al. 2012). They conducted another study in which patients with subacute stroke received IV or IA delivery of Tc-99m labeled BMMNCs and reported safety of this method of cell tracking (Rosado-de-Castro et al. 2013).

Moniche et al. also reported safety and feasibility of IA infusion of mean doses of 1.59×10^8 BMMNCs within 5–9 days of onset in patients with MCA ischemic strokes (Moniche et al. 2012). Banerjee et al. also safely performed IA infusion of bone marrow–derived immunoselected $CD34^+$ progenitor cells in the MCA of five stroke patients within 7 days of onset (Banerjee et al. 2014).

3.6 STEPS GUIDELINES

There are several challenges to translating cell therapy in stroke from the bench to the bedside. Although clinical trials have begun worldwide investigating various cell types, many questions require extensive study, including the optimal cell types, dose, route, and timing of delivery. The Stem Cell Therapies as an Emerging Paradigm in Stroke (STEPS) consortium, which includes academic investigators, leaders from industry, and the NIH, have provided guidelines that focus on key issues for the development of cell-based therapies for stroke (Savitz et al. 2014).

REFERENCES

Amariglio N, Hirshberg A, Scheithauer BW et al. 2009. Donor-derived brain tumor following neural stem cell transplantation in an ataxia telangiectasia patient. *PLoS Medicine* 6:e1000029.

Andrews EM, Tsai SY, Johnson SC et al. 2008. Human adult bone marrow-derived somatic cell therapy results in functional recovery and axonal plasticity following stroke in the rat. *Exp Neurol* 211:588–592.

Arien-Zekav H, Lecht S, Bercu MM et al. 2009. Neuroprotection by cord blood neural progenitors involves antioxidants, neurotrophic and angiogenic factors. *Exp Neurol* 216:83–94.

Arnhold S, Lenartz D, Kruttwig K et al. 2000. Differentiation of green fluorescent protein-labeled embryonic stem cell-derived neural precursor cells into Thy-1-positive neurons and glia after transplantation into adult rat striatum. *J Neurosurg* 93:1026–1032.

Banerjee S, Bentley P, Chataway J et al. 2014. Intra-arterial immunoselected CD34+ stem cells for acute ischemic stroke. *Stem Cells Transl Med* 3:1322–1330.

Bang OY, Lee JS, Lee PH et al. 2005. Autologous mesenchymal stem cell transplantation in stroke patients. *Ann Neurol* 57:874–882.

Bani-Yaghoub M, Felker JM, Naus CC. 1999. Human NT2/D1 cells differentiate into functional astrocytes. *Neuroreport* 10:3843–3846.

Barbosa da Fonseca LM, Battistella V, Andre C et al. 2009. Early tissue distribution of bone marrow mononuclear cells after intra-arterial delivery in a patient with chronic stroke. *Circulation* 120:539–541.

Barbosa da Fonseca LM, Gutfilen B, de Freitas GR et al. 2010. Migration and homing of bone-marrow mononuclear cells in chronic ischemic stroke after intra-arterial injection. *Exp Neurol* 221:122–128.

Beaulieu C 2002. The basis of anisotropic water diffusion in the nervous system—A technical review. *NMR Biomed* 15:435–455.

Bhasin A, Srivastava MV, Airan B et al. 2011. Autologous mesenchymal stem cells in chronic stroke. *Cerebrovasc Dis Extra* 1:93–104.

Bhasin A, Srivastava M, Bose S et al. 2012. Autologous intravenous mononuclear stem cell therapy in chronic ischemic stroke. *J Stem Cells Regen Med* 8:181–189.

Boltze J, Schmidt UR, Schäbitz WR et al. 2012. Determination of the therapeutic time window for human umbilical cord blood mononuclear cell transplantation following experimental stroke in rats. *Cell Transplant* 21:1199–1211.

Borlongan CV, Hadman M, Sanberg CD et al. 2004. Central nervous system entry of peripherally injected umbilical cord blood cells is not required for neuroprotection in stroke. *Stroke* 35:2385–2389.

Borlongan CV, Tajima Y, Trojanowski JQ et al. 1998. Transplantation of cryopreserved human embryonal carcinoma-derived neurons (NT2N cells) promotes functional recovery in ischemic rats. *Exp Neurol* 149:310–321.

Brenneman M, Sharma S, Savitz SI et al. 2010. Autologous bone marrow mononuclear cells enhance recovery after acute ischemic stroke in young and middle-aged rats. *J Cereb Blood Flow Metab* 30:140–149.

Brustle O, Jones KN, Learish RD et al. 1999. Embryonic stem cell-derived glial precursors: A source of myelinating transplants. *Science* 285:754–756.

Bühnemann C, Scholz A, Dihné M et al. 2006. Neuronal differentiation of transplanted embryonic stem cell-derived precursors in stroke lesions of adult rats. *Brain* 129:3238–3248.

Callera F, Demela CM. 2007. Magnetic resonance tracking of magnetically labeled autologous bone marrow CD34+ cells transplanted into the spinal cord via lumber puncture technique in patients with chronic spinal cord injury: CD34+ cells migration into the injured site. *Stem Cell Dev* 16:461–466.

Chen J, Li Y, Wang L et al. 2001a. Therapeutic benefits of intravenous administration of bone marrow stromal cells after cerebral ischemia in rats. *Stroke* 32:1005–1011.

Chen J, Li Y, Wang L et al. 2001b. Therapeutic benefit of intracerebral transplantation of bone marrow stromal cells after cerebral ischemia in rats. *J Neurol Sci* 189:49–57.

Chen J, Li Y, Katakowski M et al. 2003a. Intravenous bone marrow stromal cell therapy reduces apoptosis and promotes endogenous cell proliferation after stroke in female rat. *J Neurosci Res* 73:778–786.

Chen J, Zhang ZG, Li Y et al. 2003b. Intravenous administration of human bone marrow stromal cells induces angiogenesis in the ischemic boundary zone after stroke in rats. *Circ Res* 92:692–699.

Chu K, Kim M, Park KI et al. 2004. Human neural stem cells improve sensorimotor deficits in the adult rat brain with experimental focal ischemia. *Brain Res* 1016:145–153.

Chu K, Park KI, Lee ST et al. 2005. Combined treatment of vascular endothelial growth factor and human neural stem cells in experimental focal cerebral ischemia. *Neurosci Res* 53:384–390.

Chua JY, Pendharkar AV, Guzman R et al. 2011. Intra-arterial injection of neural stem cells using a microneedle technique does not cause microembolic strokes. *J Cereb Blood Flow Metab* 31:1263–1271.

Copeland N, Harris D, Gaballa MA 2009. Human umbilical cord blood stem cells, myocardial infarction and stroke. *Clin Med* 9:342–345.

Copone C, Fregerio S, Fumagalli S et al. 2007. Neurosphere-derived cells exert a neuroprotective action by changing ischemic microenvironment. *PLoS One* 2:e373.

Correa PL, Mesquita CT, Felix RM et al. 2007. Assessment of intra-arterial injected autologous bone marrow mononuclear cell distribution by radioactive labeling in acute ischemic stroke. *Clin Nucl Med* 32:839–841.

Crigler L, Robey RC, Asawachaicharn A et al. 2006. Human mesenchymal stem cell subpopulation express a variety of neuro-regulatory molecules and promote neuronal cell survival and neuritogenesis. *Exp Neurol* 198:54–64.

Daadi MM, Davis AS, Steinberg GK. 2010. Human neural stem cell grafts modify microglial response and enhance axonal sprouting in neonatal hypoxic-ischemic brain injury. *Stroke* 41:516–523.

Daadi MM, Li Z, Steinberg GK. 2009. Molecular and magnetic resonance imaging of human embryonic stem cell-derived neural stem cell grafts in ischemic rat brain. *Mol Ther* 17:1282–1291.

Darsalia V, Allison SJ, Kokaia Z et al. 2011. Cell number and timing of transplantation determine survival of human neural stem cell grafts in stroke-damaged rat brain. *J Cereb Blood Flow Metab* 31:235–242.

Denes A, Vidyasagar R, Allan SM et al. 2007. Proliferating resident microglia after focal cerebral ischaemia in mice. *J Cereb Blood Flow Metab* 27:1941–1953.

Ding DC, Shyu WC, Li H et al. 2007. Enhancement of neuroplasticity through upregulation of β1-integrin in human umbilical cord-derived stromal cell implanted stroke model. *Neurobiol Dis* 27:339–353.

Ding G, Jiang Q, Chopp M et al. 2008. Magnetic resonance imaging investigation of axonal remodeling and angiogenesis after embolic stroke in sildenafil-treated rats. *J Cereb Blood Flow Metab* 28:1440–1448.

Erdo F, Buhrle C, Blunk, J et al. 2003. Host-dependent tumorigenesis of embryonic stem cell transplantation in experimental stroke. *J Cereb Blood Flow Metab* 23:780–785.

Ferreira L. 2009. Nanoparticles as tools to study and control stem cells. *J Cell Biochem* 108:746–752.

Flax JD, Aurora S, Yang C et al. 1998. Engraftable human neural stem cells respond to developmental cues, replace neurons, and express foreign genes. *Nat Biotechnol* 16:1033–1039.

Friedrich MA, Martins MP, Freitas GR et al. 2012. Intra-arterial infusion of autologous bone marrow mononuclear cells in patients with moderate to severe middle cerebral artery acute ischemic stroke. *Cell Transplant* 21 Suppl 1:S13–21.

Ge J, Guo L, Wu Y et al. 2014. The size of mesenchymal stem cells is a significant cause of vascular obstructions and stroke. *Stem Cell Rev* 10:295–303.

Gera A, Steinberg GK, Guzman R. 2010. In vivo neural stem cell imaging: Current modalities and future directions. *Regen Med* 5:73–86.

Gerschat S, Schira J, Kury P et al. 2008. Unrestricted somatic stem cells from human umbilical cord blood can be differentiated into neurons with a dopaminergic phenotype. *Stem Cells Dev* 17:221–232.

Giraldi-Guimardes A, Rezende-Lima M, Mendez-Otero R et al. 2009. Treatment with bone marrow mononuclear cells induces functional recovery and decreases neurodegeneration after sensorimotor cortical ischemia in rats. *Brain Res* 1266:108–120.

Guzman R, Bliss T, Los Angles Ade et al. 2008a. Neural progenitor cells transplanted into the uninjured brain undergo targeted migration after stroke onset. *J Neurosci Res* 86:873–882.

Guzman R, De Los Angeles A, Cheshier S et al. 2008b. Intracarotid injection of fluorescence activated cell-sorted CD49d-positive neural stem cells improves targeted cell delivery and behavior after stroke in a mouse stroke model. *Stroke* 39:1300–1306.

Hara K, Matsukawa N, Borlongan CV et al. 2007. Transplantation of post-mitotic human neuroteratocarcinoma-overexpressing Nurrl cells provides therapeutic benefits in experimental stroke: In vitro evidence of expedited neuronal differentiation and GDNF secretion. *J Neurosci Res* 85:1240–1251.

Hara K, Yasuhara T, Borlongan CV et al. 2008. Neural progenitor NT2N cell lines from teratocarcinoma for transplantation therapy in stroke. *Prog Neurobiol* 85:318–334.

Hau S, Reich DM, Scholz M et al. 2008. Evidence for neuroprotective properties of human umbilical cord blood cells after neuronal hypoxia in vitro. *BMC Neurosci* 9:30.

Hauger O, Frost EE, van Heeswijk R et al. 2006. MR evaluation of the glomerular homing of magnetically labeled mesenchymal stem cells in a rat model of nephropathy. *Radiology* 238:200–210.

Hayase M, Kitada M, Dezawa M et al. 2009. Committed neural progenitor cells derived from genetically modified bone marrow stromal cells ameliorate deficits in a rat model of stroke. *J Cereb Blood Flow Metab* 29:1409–1420.

Hogduijn MJ, Crop MJ, Peeters AM et al. 2007. Human heart, spleen, and peri-renal fat derived stem cells have immunomodulatory capacities. *Stem Cells Dev* 16:587–604.

Honma T, Honmou O, Iihoshi S et al. 2006. Intravenous infusion of immortalized human mesenchymal stem cells protects against injury in a cerebral ischemia model in adult rat. *Exp Neurol* 199:56–66.

Honmou O, Houkin K, Kocsis JD et al. 2011. Intravenous administration of auto serum-expanded autologous mesenchymal stem cells in stroke. *Brain* 134:1790–1807.

Iihoshi S, Honmou O, Houkin K et al. 2004. A therapeutic window for intravenous administration of autologous bone marrow after cerebral ischemia in adult rats. *Brain Res* 1007:1–9.

Ikeda R, Kurokawa MS, Chiba S et al. 2005. Transplantation of neural cells derived from retinoic acid-treated cynomolgus monkey embryonic stem cells successfully improved motor function of hemiplegic mice with experimental brain injury. *Neurobiol Dis* 20:38–48.

Ishibashi S, Sakaguchi M, Mizusawa H et al. 2004. Human neural stem/progenitor cells, expanded in long-term neurosphere culture, promote functional recovery after focal ischemia in Mongolian gerbils. *J Neurosci Res* 78:215–223.

Janowski M, Lyczek A, Walczak P et al. 2013. Cell size and velocity of injection are major determinants of the safety of intracarotid stem cell transplantation. *J Cereb Blood Flow Metab* 33:921–927.

Jiang Q, Zhang ZG, Ding GL et al. 2005. Investigation of neural progenitor cell induced angiogenesis after embolic stroke in rat using MRI. *Neuroimage* 28:698–707.

Jiang L, Womble T, Willing AE et al. 2010. Human umbilical cord blood cells decrease microglial survival in vitro. *Stem Cells Dev* 19:221–228.

Jiang Q, Zhang ZG, Ding GL et al. 2006. MRI detects white matter reorganization after neural progenitor cell treatment of stroke. *Neuroimage* 32:1080–1089.

Kang SK, Jun ES, Bay YC et al. 2003. Interactions between human adipose stromal cells and mouse neural stem cells in vitro. *Brain Res Dev Brain Res* 145:141–149.

Keene CD, Ortiz-Gonzalez XR, Jiang Y et al. 2003. Neural differentiation and incorporation of bone marrow-derived multipotent adult progenitor cells after single cell transplantation into blastocyst stage mouse embryos. *Cell Transplant* 12:201–213.

Kelly S, Bliss TM, Steinberg GK et al. 2004. Transplanted human fetal neural stem cells survive, migrate, and differentiate in ischemic rat cerebral cortex. *Proc Natl Acad Sci USA* 101:11839–11844.

Kiessling F. 2008. Noninvasive cell tracking. *Handb Exp Pharmacol* (185 Pt 2):305–321.

Kim D, Chun BG, Kim YK et al. 2008a. In vivo tracking of human mesenchymal stem cells in experimental stroke. *Cell Transplant* 16:1007–1012.

Kim DE, Schellingerhout D, Weissleder R et al. 2004. Imaging of stem cell recruitment to ischemic infarcts in a murine model. *Stroke* 35:952–957.

Kim JM, Le ST, Chu K et al. 2007. Systemic transplantation of human adipose stem cells attenuated cerebral inflammation and degeneration in a hemorrhagic stroke model. *Brain Res* 1183:43–50.

Kim SS, Yoo SW, Park TS et al. 2008b. Neural induction with neurogenin1 increases the therapeutic effects of mesenchymal stem cells in the ischemic brain. *Stem Cells* 26:2217–2228.

Kim SU. 2004. Human neural stem cells genetically modified for brain repair in neurological disorders. *Neuropathology* 24:154–171.

Kim SU. 2007. Genetically engineered human neural stem cells for brain repair in neurological diseases. *Brain Dev* 29:193–201.

Kim SU, Nagai A, Park IH et al. 2008c. Production and characterization of immortal human neural stem cell line with multipotent differentiation property. *Methods Mol Biol* 438:103–121.

Koch P, Opitz T, Brüstle O et al. 2009. A rosette-type, self-renewing human ES cell-derived neural stem cell with potential for in vitro instruction and synaptic integration. *Proc Natl Acad Sci USA* 106:3225–3230.

Koh SH, Kim KS, Choi MR et al. 2008. Implantation of human umbilical cord-derived mesenchymal stem cells as a neuroprotective therapy for ischemic stroke in rats. *Brain Res* 1229:233–248.

Kondziolka D, Steinberg GK, McGrogan M et al. 2004. Evaluation of surgical techniques for neuronal cell transplantation used in patients with stroke. *Cell Transplant* 13:749–754.

Kondziolka D, Steinberg GK, Wechsler L et al. 2005. Neurotransplantation for patients with subcortical motor stroke: A phase 2 randomized trial. *J Neurosurg* 103:38–45.

Kondziolka D, Wechsler L, Goldstein S et al. 2000. Transplantation of cultured human neuronal cells for patients with stroke. *Neurology* 55:565–569.

Kraitchman DL, Tatsumi M, Gilson WD et al. 2005. Dynamic imaging of allogeneic mesenchymal stem cells trafficking to myocardial infarction. *Circulation* 112:1451–1461.

Kranz A, Wagner DC, Boltze J et al. 2010. Transplantation of placenta-derived mesenchymal stromal cells upon experimental stroke in rats. *Brain Res* 1315:128–136.

Kurozumi K, Nakamura K, Hamada H et al. 2004. BDNF gene-modified mesenchymal stem cells promote functional recovery and reduce infarct size in the rat middle cerebral artery occlusion model. *Mol Ther* 9:189–197; Erratum in: *Mol Ther* 9:766.

Lee HJ, Kim KS, Kim SU et al. 2007a. Brain transplantation of immortalized human neural stem cells promotes functional recovery in mouse intracerebral hemorrhage stroke model. *Stem Cells* 25:1204–1212.

Lee HJ, Kim KS, Kim SU et al. 2007b. Human neural stem cells over-expressing VEGF provide neuroprotection, angiogenesis and functional recovery in mouse stroke model. *PLoS One* 2:e156.

Lee JS, Hong JM, Bang OY et al. 2010. A long-term follow-up study of intravenous autologous mesenchymal stem cell transplantation in patients with ischemic stroke. *Stem Cells* 28:1099–1106.

Lee ST, Chu K, Roh JK et al. 2008. Anti-inflammatory mechanism of intravascular neural stem cell transplantation in haemorrhagic stroke. *Brain* 131:616–629.

Li Y, Chen J, Chen XG et al. 2002. Human marrow stromal cell therapy for stroke in rat: Neurotrophins and functional recovery. *Neurology* 59:514–523.

Li Y, Chen J, Wang L et al. 2001. Treatment of stroke in rat with intracarotid administration of bone marrow stromal cells. *Neurology* 56:1666–1672.

Li Y, Chopp M, Chen J et al. 2000. Intrastriatal transplantation of bone marrow nonhematopoitic cells improves functional recovery after stroke in adult mice. *J Cereb Blood Flow Metab* 20:1311–1319.

Li Y, McIntosh K, Chopp M et al. 2006. Allogeneic bone marrow stromal cells promote glial-axonal remodeling without immunologic sensitization after stroke in rats. *Exp Neurol* 198:313–325.

Liao W, Xie J, Han ZC et al. 2009. Therapeutic effect of human umbilical cord multipotent mesenchymal stromal cells in a rat model of stroke. *Transplantation* 87:350–359.

Lindvall O, Kokaia Z. 2006. Stem cells for the treatment of neurological disorders. *Nature* 441:1094–1096.

Liu H, Honmou O, Harada K et al. 2006. Neuroprotection by PLGF gene-modified human mesenchymal stem cells after cerebral ischaemia. *Brain* 129:2734–2745.

Mäkinen S, Kekarainen T, Nystedt J et al. 2006. Human umbilical cord blood cells do not improve sensorimotor or cognitive outcome following transient middle cerebral artery occlusion in rats. *Brain Res* 1123:207–215.

Mendonça ML, Freitas GR, Silva SA et al. 2006. Safety of intra-arterial autologous bone marrow mononuclear cell transplantation for acute ischemic stroke. *Arq Bras Cardiol* 86:52–55.

Messina DJ, Alder L, Tresco PA. 2003. Comparison of pure and mixed populations of human fetal-derived neural progenitors transplanted into intact adult rat brain. *Exp Neurol* 184:816–829.

Modo M, Beech JS, Meade TJ et al. 2009. A chronic 1 year assessment of MRI contrast agent-labelled neural stem cell transplants in stroke. *Neuroimage* 47:T133–T142.

Modo M, Stroemer RP, Tang E et al. 2002. Effects of implantation site of stem cell grafts on behavioral recovery from stroke damage. *Stroke* 33:2270–2278.

Moniche F, Gonzalez A, Gil-Peralta A et al. 2012. Intra-arterial bone marrow mononuclear cells in ischemic stroke: A pilot clinical trial. *Stroke* 43:2242–2244.

Moore DF, Li H, Baird AE et al. 2005. Using peripheral blood mononuclear cells to determine a gene expression profile of acute ischemic stroke: A pilot investigation. *Circulation* 111:212–221.

Nasef A, Mathieu N, Fouillard L et al. 2007. Immunosuppressive effects of mesenchymal stem cells: Involvement of HLA-G. *Transplantation* 84:231–237.

Nelson PT, Kondziolka D, Trojanowski JQ et al. 2002. Clonal human (hNT) neuron grafts for stroke therapy: Neuropathology in a patient 27 months after implantation. *Am J Pathol* 160:1201–1206.

Ohtaki H, Ylostalo JH, Prockop DJ et al. 2008. Stem/progenitor cells from bone marrow decrease neuronal death in global ischemia by modulation of inflammatory/immune responses. *Proc Natl Acad Sci USA* 105:14638–14643.

Oki K, Tatarishvili J, Kokaia Z et al. 2012. Human-induced pluripotent stem cells form functional neurons and improve recovery after grafting in stroke-damaged brain. *Stem Cells* 30:1120–1133.

Olanow CW, Goetz CG, Kordower JH et al. 2003. A double-blind controlled trial of bilateral fetal nigral transplantation in Parkinson's disease. *Ann Neurol* 54:403–414.

Onda T, Honmou O, Harada K et al. 2008. Therapeutic benefits by human mesenchymal stem cells (hMSCs) and Ang-1 gene-modified hMSCs after cerebral ischemia. *J Cereb Blood Flow Metab* 28:329–340.

Park DH, Eve DJ, Gemma C et al. 2010. Increased neuronal proliferation in the dentate gyrus of aged rats following neural stem cell implantation. *Stem Cells Dev* 19:175–180.

Park DH, Eve DJ, Sanberg PR et al. 2009. Inflammation and stem cell migration to the injured brain in higher organisms. *Stem Cells Dev* 18:693–702.

Parr AM, Tator CH, Keating A. 2007. Bone marrow-derived mesenchymal stromal cells for the repair of central nervous system injury. *Bone Marrow Transplant* 40:609–619.

Pavlichenko N, Sokolova I, Vijde S et al. 2008. Mesenchymal stem cells transplantation could be beneficial for treatment of experimental ischemic stroke in rats. *Brain Res* 1233:203–213.

Pluchino S, Zanotti L, Rossi B et al. 2005. Neurosphere-derived multipotent precursors promote neuroprotection by an immunomodulatory mechanism. *Nature* 436:266–271.

Pollock K, Stroemer P, Sinden JD et al. 2006. A conditionally immortal clonal stem cell line from human cortical neuroepithelium for the treatment of ischemic stroke. *Exp Neurol* 199:143–155.

Prasad K, Mohanty S, Mishra NK et al. 2012. Autologous intravenous bone marrow mononuclear cell therapy for patients with subacute ischaemic stroke: A pilot study. *Indian J Med Res* 136:221–228.

Reubinoff BE, Itsykson P, Turetsky T et al. 2001. Neural progenitors from human embryonic stem cells. *Nat Biotechnol* 19:1134–1140.

Reumers V, Deroose CM, Baekelandt V et al. 2008. Noninvasive and quantitative monitoring of adult neuronal stem cell migration in mouse brain using bioluminescence imaging. *Stem Cells* 26:2382–2390.

Rice HE, Hsu EW, Johnson GA et al. 2007. Superparamagnetic iron oxide labeling and transplantation of adipose-derived stem cells in middle cerebral artery occlusion-injured mice. *AJR Am J Roentgenol* 188:1101–1108.

Roh JK, Jung KH, Chu K. 2008. Adult stem cell transplantation in stroke: Its limitations and prospects. *Curr Stem Cell Res Ther* 3:185–196.

Rosado-de-Castro PH, Schmidt Fda R, Barbosa da Fonseca LM et al. 2013. Biodistribution of bone marrow mononuclear cells after intra-arterial or intravenous transplantation in subacute stroke patients. *Regen Med* 8:145–155.

Rosenblum S, Wang N, Guzman R et al. 2012. Timing of intra-arterial neural stem cell transplantation after hypoxia-ischemia influences cell engraftment, survival, and differentiation. *Stroke* 43:1624–1631.

Savitz SI, Cramer SC, Wechsler L; STEPS 3 Consortium. 2014. Stem cells as an emerging paradigm in stroke 3: Enhancing the development of clinical trials. *Stroke* 45:634–639.

Savitz SI, Dinsmore J, Wu J et al. 2005. Neurotransplantation of fetal porcine cells in patients with basal ganglia infarcts: A preliminary safety and feasibility study. *Cerebrovasc Dis* 20:101–107.

Savitz SI, Misra V, Grotta JC et al. 2011. Intravenous autologous bone marrow mononuclear cells for ischemic stroke. *Ann Neurol* 70:59–69.

Sharma S, Yang B, Savitz SI et al. 2010. Bone marrow mononuclear cells protect neurons and modulate microglia in cell culture models of ischemic stroke. *J Neurosci Res* 88:2869–2876.

Shehadah A, Chen J, Chopp M et al. 2013. Efficacy of single and multiple injections of human umbilical tissue-derived cells following experimental stroke in rats. *PLoS One* 8:e54083.

Shen LH, Li Y, Chen J et al. 2006. Intracarotid transplantation of bone marrow stromal cells increases axon-myelin remodeling after stroke. *Neuroscience* 137:393–399.

Sohur US, Emsley JG, Mitchell BD et al. 2006. Adult neurogenesis and cellular brain repair with neural progenitors, precursors and stem cells. *Philos Trans R Soc Lond B Biol Sci* 361:1477–1497.

Song S, Song S, Zhang H et al. 2007. Comparison of neuron-like cells derived from bone marrow stem cells to those differentiated from adult brain neural stem cells. *Stem Cell Dev* 16: 747–756.

Stroemer P, Hope A, Sinden J et al. 2008. Development of a human neural stem cell line for use in recovery from disability after stroke. *Front Biosci* 13:2290–2292.

Sykova E, Jendelova P. 2007. In vivo tracking of stem cells in brain and spinal cord injury. *Prog Brain Res* 161:367–383.

Taguchi A, Soma T, Tanaka H et al. 2004. Administration of CD34+ cells after stroke enhances neurogenesis via angiogenesis in a mouse model. *J Clin Invest* 114:330–338.

Tang Y, Wang J, Yang GY et al. 2014. Neural stem cell protects aged rat brain from ischemia-reperfusion injury through neurogenesis and angiogenesis. *J Cereb Blood Flow Metab* 34:1138–1147.

Theus MH, Wei L, Cui L et al. 2008. In vitro hypoxic preconditioning of embryonic stem cells as a strategy of promoting cell survival and functional benefits after transplantation into the ischemic rat brain. *Exp Neurol* 210:656–670.

Ukai R, Honmou O, Harada K et al. 2007. Mesenchymal stem cells derived from peripheral blood protects against ischemia. *J Neurotrauma* 24:508–520.

Vasconcelos-dos-Santos A, Rosado-de-Castro PH, Mendez-Otero R et al. 2012. Intravenous and intra-arterial administration of bone marrow mononuclear cells after focal cerebral ischemia: Is there a difference in biodistribution and efficacy? *Stem Cell Res* 9:1–8.

Veizovic T, Beech JS, Stroemer RP et al. 2001. Resolution of stroke deficits following contralateral grafts of conditionally immortal neuroepithelial stem cells. *Stroke* 32:1012–1019.

Vendrame M, Cassady J, Newcomb J et al. 2004. Infusion of human umbilical cord blood cells in a rat model of stroke dose-dependently rescues behavioral deficits and reduces infarct volume. *Stroke* 35:2390–2395.

Vendrame M, Gemma C, Willing AE et al. 2005. Anti-inflammatory effects of human cord blood cells in a rat model of stroke. *Stem Cells Dev* 14:595–604.

Vu Q, Xie K, Cramer SC et al. 2014. Meta-analysis of preclinical studies of mesenchymal stromal cells for ischemic stroke. *Neurology* 82:1277–1286.

Walczak P, Zhang J, Gilad AA et al. 2008. Dual-modality monitoring of targeted intraarterial delivery of mesenchymal stem cells after transient ischemia. *Stroke* 39:1569–1574.

Wang Y, Deng Y, Zhou GQ. 2008. SDF-1α/CXCR4-mediated migration of systemically transplanted bone marrow stromal cells towards ischemic brain lesion in a rat model. *Brain Res* 1195:104–112.

Wei L, Cui L, Snider BJ et al. 2005. Transplantation of embryonic stem cells overexpressing Bcl-2 promotes functional recovery after transient cerebral ischemia. *Neurobiol Dis* 19:183–193.

Wei X, Zhao L, Du Y et al. 2009. Adipose stromal cells-secreted neuroprotective media against neuronal apoptosis. *Neurosci Lett* 462:76–79.

Willing AE, Lixian J, Milliken M et al. 2003a. Intravenous versus intrastriatal cord blood administration in a rodent model of stroke. *J Neurosci Res* 73:296–307.

Willing AE, Vendrame M, Mallery J et al. 2003b. Mobilized peripheral blood cells administered intravenously produce functional recovery in stroke. *Cell Transplant* 12:449–454.

Xiao J, Nan Z, Low WC et al. 2005. Transplantation of a novel cell line population of umbilical cord blood stem cells ameliorates neurological deficits associated with ischemic brain injury. *Stem Cells Dev* 14:722–733.

Yamashita T, Deguchi K, Abe K et al. 2006. Neuroprotection and neurosupplementation in ischaemic brain. *Biochem Soc Trans* 34:1310–1312.

Yang B, Migliati E, Savitz SI et al. 2013. Intra-arterial delivery is not superior to intravenous delivery of autologous bone marrow mononuclear cells in acute ischemic stroke. *Stroke* 44:3463–3472.

Yang B, Strong R, Savitz SI et al. 2011. Therapeutic time window and dose response of autologous bone marrow mononuclear cells for ischemic stroke. *J Neurosci Res* 89:833–839.

Yashuhara T, Hara K, Borlongan CV et al. 2010. Mannitol facilitates neurotrophic factor upregulation and behavioral recovery in neonatal hypoxic-ischemic rats with human umbilical cord blood grafts. *J Cell Mol Med* 14:914–921.

Zhang C, Li Y, Chen J et al. 2006. Bone marrow stromal cells upregulate expression of bone morphogenetic proteins 2 and 4, gap junction protein connexin-43 and synaptophysin after stroke in rats. *Neuroscience* 141:687–695.

Zhang P, Li J, Gao M et al. 2011. Human embryonic neural stem cell transplantation increases subventricular zone cell proliferation and promotes peri-infarct angiogenesis after focal cerebral ischemia. *Neuropathology* 31:384–391.

Zhang ZG, Jiang Q, Zhang R et al. 2003. Magnetic resonance imaging and neurosphere therapy of stroke in rat. *Ann Neurol* 53:259–263.

Zheng H, Fu G, Huang H et al. 2007. Migration of endothelial progenitor cells mediated by stromal cell-derived factor-1α/CXCR4 via PI3K/Akt/eNOS signal transduction pathway. *J Cardiovasc Pharmacol* 50:274–280.

4 Experimental Models of Traumatic Brain Injury

Clinical Relevance and Shortcomings

Edward W. Vogel III, Barclay Morrison III, Megan N. Evilsizor, Daniel R. Griffiths, Theresa C. Thomas, Jonathan Lifshitz, Richard L. Sutton, Joseph B. Long, David Ritzel, Geoffrey S.F. Ling, James Huh, Ramesh Raghupathi, and Tracy K. McIntosh

CONTENTS

4.1 INTRODUCTION

Experimental models of traumatic brain injury (TBI) have been designed to mimic closely the clinical sequelae of human TBI and play a crucial role in the process of evaluating and understanding the physiologic, behavioral, and histopathologic changes associated with TBI, with a view toward developing novel treatment strategies for this devastating disease. Because human TBI is very much a heterogenous disease, no single animal model of TBI developed, to date, can mimic the whole spectrum of clinical TBI. Rather, the concurrent use of a number of distinct yet complementary models are necessary to reliably reproduce the whole range of injury severity and characteristic features observed upon clinical and postmortem examination of TBI patients. Although imperfect, experimental TBI models have contributed enormously to our insight into the posttraumatic sequelae and have prompted the development of several novel diagnostic and treatment strategies that either are now part of clinical standard practice or are under intense preclinical and clinical investigation.

This chapter attempts to provide a succinct but broad overview of the most widely used and popular experimental models of mechanically induced TBI in whole-animal and in vitro cellular models. Because rodents are the species of choice in the vast majority of studies, due to their obvious advantages (small size, modest cost, extensive normative data available), our discussion will focus mainly on results obtained in studies with rodents except where such data are not available. Moreover, space limitations preclude an extensive review of the existing literature concerning nonmechanical models to produce brain damage (thermal, chemical, or electrical), or inanimate finite element computational model characterizations.

4.2 IMPORTANT VARIABLES IN MODELING TRAUMATIC BRAIN INJURY

Despite differing objectives, any model designed to reliably produce experimental TBI should fulfill a number of criteria. The injury severity and the exact injury location must be defined, and the injury response must be quantifiable and reproducible, not only in the same laboratory but also between different laboratories and institutions. Additionally, the damage caused from the traumatic event should be part of a continuum, increasing with increasing mechanical forces applied to the head or brain. The most widely used models in the majority of recent studies employ standardized surgical protocols and techniques including sham (uninjured) animals with identical surgical treatment to control for systemic variables. This experimental design controls for the possible influence of the operative procedure, anesthesia, changes in body temperature or brain temperature, brain damage due to head restraint or the placement of intracranial probes, etc., on posttraumatic outcome. Additionally, the majority of the trauma devices employ computer-based measurements of the applied load, such as pressure gradients, the velocity of the impactor, or the speed of acceleration/deceleration forces to measure variations of the mechanical parameters that define inflicted injury severity. This information is used to make adjustments to the device and allows for the maintenance of a narrow range of inflicted injury severity within a particular study.

4.2.1 Injury Severity

To closely mimic the range of TBI severity in the clinical situation, experimental studies have modified the existing injury models to be capable of producing brain trauma over a spectrum of injury severity. This goal is accomplished by adjusting the main mechanical parameters of the injury device (e.g., height or mass of the free-falling weight, depth of the traumatic impact or impact velocity, and height of the pressure impulse) by adjusting the pendulum of the fluid-percussion device or changes in the plane or velocity of the rotational forces). A number of experimental studies performed recently have revealed a close relationship between injury severity and the posttraumatic responses and rate of recovery following experimental TBI. As a result, a classification for the severity of experimental TBI has been developed and established, and it is similar to the clinical categories of mild, moderate, and severe.

4.2.2 Types of Injury

Human TBI is not a single pathophysiologic entity, and the majority of patients suffering from TBI display more than one lesion upon careful diagnostic evaluation. Because the clinical situation is seldom as controlled as the experimental setting, different injury models have been developed and characterized to elucidate the main characteristics of TBI that include focal and diffuse damage. Focal abnormalities involve contusions and lacerations not always accompanied by skull fracture or hematoma formation. This type of damage occurs in the direct vicinity of the site of the mechanical impact to the head and typically involves the underlying cortical and, in the case of injury of higher severity, subcortical structures. Several experimental models have been established that mimic these aspects of focal TBI over a wide range of injury severity (weight drop closed head injury, fluid percussion brain injury, and rigid indentation injury, vide infra). However, most models are associated with concussive events, and if the injury severity exceeds a certain threshold, substantial displacement of the brain occurs that adds a new, more remote component of axonal injury to the predominantly focal damage. Diffuse injuries may include concussions and diffuse axonal pathology. This type of injury is sometimes more difficult to detect in the clinical setting, but appears to occur more commonly than previously believed and is presumably present in the whole range from mild to severe head injury. Diffuse brain injuries are thought to occur primarily from the tissue distortion, or shear, caused by inertial forces that are present at the moment of injury. Experimental models that predominantly mimic this type of damage (e.g., models of inertial acceleration, lateral fluid percussion, and to a lesser extent, impact acceleration, vide infra) lead to substantial diffuse injury in the absence of profound focal damage. These changes are usually observed peripheral to the vicinity of the impact, but also remote to the injury site.

Evaluation of patients suffering from moderate-severe TBI has made it unequivocally clear that behavioral impairment, particularly with respect to neurologic motor function and cognitive deficits, comprises the most persistent deficits after TBI, lasting for months and even years. Additionally, a number of radiologic and histologic studies have revealed that morphologic damage to the brain tissue represents another consistent feature of human TBI that remains unresolved and persists over time. In contrast, a broad variety of different posttraumatic events occur in the immediate and acute phase after TBI (e.g., changes in electrophysiology, blood-brain barrier dysfunction, edema formation, changes in cerebral perfusion and intracranial pressure, activation of ion channels and ion-shift, genomic changes, production of free radicals, inflammation, etc.). Some of these changes are transient, but the duration of others has yet to be determined. To follow and describe injury-induced changes in the experimental setting, laboratory techniques have been developed and refined to determine many of the reversible and persistent posttraumatic sequelae after experimental TBI. Although the great majority of experimental studies have been conducted with posttraumatic survival times of hours or days, a smaller number of studies have successfully identified persistent neurobehavioral impairment and histologic changes in the chronic phase after experimental TBI up to 1 year or more postinjury.

4.3 IN VITRO MODELS OF TBI

To complement in vivo studies, in vitro models of TBI have been developed and utilized over the years (Tecoma et al., 1989; Shepard et al., 1991; Ellis et al., 1995; LaPlaca et al., 2005; Cater et al., 2006; Effgen et al., 2012; Dollé et al., 2013) (Figure 4.1). In vitro TBI models allow for a reduction in animal use, isolation of specific tissue components, and, if designed correctly, better control over injury biomechanics. However, a common limitation of many is that they are not able to achieve this level of precise biomechanical control (Morrison et al., 1998, 2011). A question often asked is "how representative are in vitro TBI models of in vivo TBI models?" Because we cannot know all aspects of the posttraumatic sequelae, it is difficult to

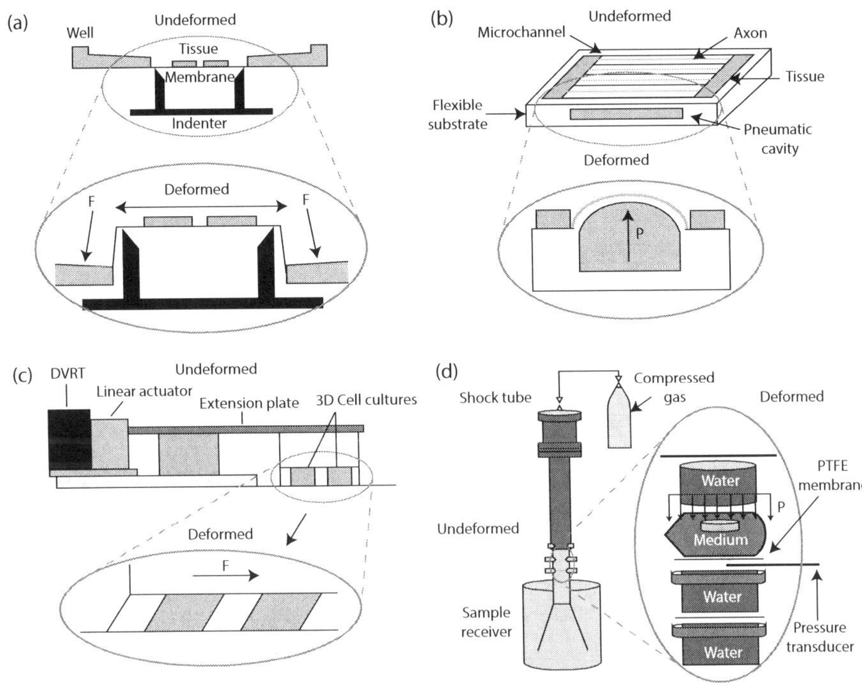

FIGURE 4.1 Examples of in vitro models of traumatic brain injury (TBI). (a) Tissue cultures are stretched equibiaxially through deformation of the culture substrate by pulling the clamped membrane over a hollow cylindrical indenter. (b) Uniaxial strain is delivered to either individual axons or bundles of axons that extend from an organotypic hippocampal slice and connect to an adjacent organotypic slice within microchannels. Strain is applied by pressurizing a cavity beneath the microchannels. (c) Simple shear is applied to 3D cell constructs with the magnitude and rate of the applied strain under closed-loop feedback control, using a differential variable reluctance transducer (DVRT), a linear actuator, and an extension plate. (d) Blast exposure is delivered by a compressed gas-driven shock tube to cell/tissue cultures sterilely housed within a fluid-filled receiver. The receiver immobilizes the culture surrogate enabling primary blast exposure, simulates the skull/fluid surrounding the brain tissue, and prevents secondary pressure loading from reflections of the pressure wave. (Adapted from [a] Cater, H.L., L.E. Sundstrom, B. Morrison III, *J Biomech*, 2006. 39(15), p. 2810–2818; Cater, H.L. et al., *J Neurochem*, 2007. 101(2), p. 434–447; Elkin, B.S., B. Morrison III, *Stapp Car Crash J*, 2007. 51, p. 127–138; Choo, A.M. et al., *Brain*, 2013. 136, p. 65–80; [b] Dollé, J. et al., *Lab Chip*, 2013. 13, p. 432–442; [c] LaPlaca, M.C. et al., *J Biomech*, 2005. 38(5), p. 1093–1105; Cullen, D.K. et al., *J Neurotrauma*, 2011. 28, p. 2219–2233; [d] Effgen, G.B. et al., *Front Neurol*, 2012. 3(23), p. 1–10; Hue, C.D. et al., *J Neurotrauma*, 2013. 30, p. 1652–1663.).

evaluate in vitro models with a priori criteria. Instead, we propose a surrogate measure to determine the correspondence between the two classes of models of whether pharmacologic interventions after in vitro TBI produce the same outcomes after in vivo TBI. The purpose of this section will be to review how well recent findings using in vitro models of TBI reproduce the pathobiology of experimental in vivo TBI models.

One advantage of in vitro injury models is that they provide the opportunity to precisely control and verify the initiating injury biomechanics. If the biomechanical input of the in vitro injury is not consistent with the in vivo injury (e.g., tissue strain and strain rate), the likelihood that the ensuing pathobiology in vitro diverges from that in vivo is greater. Another advantage of in vitro models is that they avoid the pharmacokinetic or pharmacodynamic issues that arise in vivo which may stifle the early discovery process. In vitro models also allow for increased access to the tissue so that the postinjury response can be monitored immediately after the mechanical stimulus and over time to unravel the complex interplay between multiple cascades. Below is a brief description of the in vitro TBI models currently utilized.

4.3.1 Substrate Strain

A class of in vitro TBI models imparts strain to the cultured tissue or cells by the controlled deformation of the underlying substrate. One method for deforming the substrate uses compressed gas (Ellis et al., 1995; Morrison et al., 1998; Smith et al., 1999a,b; Shahlaie et al., 2010; Weber et al., 2012). A particular model in this class is commercially available and can be implemented with relative ease (Ellis et al., 1995, 2007). Limitations to this model are that it delivers a heterogeneous strain over the surface of the culture substrate, and because the membrane deforms out of plane, it is difficult to verify the substrate strains, especially at higher deformation rates. Another limitation with this stretch model is that the cell or tissue cultures must be well attached to the substrate to ensure transfer of the mechanical stretch (Morrison et al., 1998, 2006). If the substrate is stretched upward (into the culture well) by a positive pressure, then cell adhesion to the substrate is less critical (Dollé et al., 2013). A second variant of the substrate stretch model utilizes a circular hollow indenter that is moved under closed-loop computer control to deform the substrate and adherent cells (Cater et al., 2006, 2007; Elkin and Morrison 2007; Choo et al., 2013). This model is advantageous because the user has precise control over the injury parameters, high-speed video allows for verification of tissue strain, and the injury is equibiaxial and homogenous across the substrate. These properties were critical for the development of tolerance criteria for living hippocampal and cortical tissue to identify safe limits of tissue strain (Cater et al., 2006; Elkin and Morrison 2007). Although this model is scientifically appealing, it is technically challenging, not commercially available, and deformation is dependent on culture adhesion to the substrate. A third variant of this class utilizes uniaxial strain of microchannels to specific cellular compartments (Tang-Schomer et al., 2010, 2012; Dollé et al., 2013). These models recreate in vivo–like axonal injury, allow for isolation of stretch to specific cellular areas, and eliminate the confounding effects of cell body deformation (Smith et al., 1999a,b). One limitation of this microchannel model is that deformation of the cultures requires their adhesion to the substrate, and verification of this adhesion, as well as applied deformation, is difficult. Although there are several variations to the substrate strain injury models, the most biomechanically realistic and precisely controllable model is the equibiaxial, homogeneous stretch model (Cater et al., 2006, 2007; Elkin and Morrison 2007; Choo et al., 2013).

4.3.2 Shear Strain

During an injurious impact event, brain tissue experiences shear strain, a complex pattern of deformation without volume change (Holbourn, 1945; Gennarelli et al., 1982; Zhang et al., 2001; Kleiven et al., 2002; Hardy et al., 2007). Two recently described in vitro models of TBI produce deformation principally through the application of shear strain. In one model, shear strain was produced through inertial forces due to linear acceleration induced by impact (Bottlang et al., 2007). Culture deformation was monitored with high-speed video, which allowed for validation of the deformation. However, the shear rates are unrealistically large, ranging from about 2075 to 13,625 s^{-1}. Previous studies have reported clinically relevant shear strain rates are in the range of 10–50 s^{-1}(Margulies et al., 1990; Meaney and Thibault 1990). Another model directly applied shear to three-dimensional cell cultures via a linear actuator and extension plates under feedback motion control (LaPlaca et al., 2005; Cullen et al., 2011). This model can impart shear strain greater than 50% at strain rates up to 30 s^{-1} (LaPlaca et al., 2005; Cullen et al., 2011). However, a disadvantage of this model is that samples could not be monitored in real time, preventing exact characterization of tissue deformation and acute cellular response.

4.3.3 Compression

Several in vitro models that use an impactor have been designed to mimic the most popular in vivo models of controlled cortical impact (CCI) or weight drop (Shi and Blight 1996; Harris et al., 2013). Primary damage to the cultured tissue or cells is generated in the region directly below the focal compression, with a secondary injury penumbra extending into the surrounding area (Church et al., 2005). Altering impactor speed, depth, shape, and the duration of contact have been proposed to control injury severity, albeit measuring the applied force or resultant tissue deformation is difficult. Injury severity is most likely dependent on the quantity of cells undergoing frank disruption at the time of impact, similar to the transection models (vide infra). Although compressive injuries do not mimic the biomechanics behind diffuse axonal or inertial injuries, they have been reported to serve as a valuable representation of contusions in spinal cord tissue cultures (Shi and Blight 1996).

4.3.4 Fluid Pressure

Fluid percussion injury (FPI) is a widely used animal model of TBI [vide infra] that has been adapted to in vitro models through the application of either transient or static fluid pressures to cultures (Shepard et al., 1991; Murphy et al., 1993; Wallis and Panizzon, 1995). Due to the incompressible nature of brain tissue and cells, static models require large pressure levels (nearly 1 MPa) to cause injury. Hydrostatic pressure application is technically challenging due to the large pressures involved (Morrison et al., 2011). These pressures greatly exceed those that occur during TBI of approximately 160 kPa (Kleiven et al., 2002; Hardy et al., 2007). The clinical correlate of the static pressure model is unclear, due to the necessity to apply high pressures for extended durations to generate trauma (Morrison et al., 2011). In the dynamic model, transient pressure waves are generated by utilizing a FPI model to apply fluid pressure to a chamber (Shepard et al., 1991). These pressure waves more closely resemble the pressure histories of closed-head TBI (Kleiven et al., 2002; Hardy et al., 2007). Another model generating transient pressure waves does so by dropping a weight directly onto a sealed chamber; however, pressure levels were not reported, making it a challenge to verify the biomechanics (Wallis and Panizzon, 1995). As the pressure wave travels through the medium and cells, the resulting pressure gradients induce deformation, which in turn leads to injury. Although precise control over the applied pressure is theoretically possible, quantifying the subsequent tissue deformations remains challenging with this weight drop technique.

4.3.5 Transection

Laceration of brain tissue via penetrating objects causes focal axotomy, subsequent cell loss, and inflammation. These injuries can be replicated at multiple scales: single cell, multiple cell, or tissue levels, using a plastic stylet (Tecoma et al., 1989), rotating (Mukhin et al., 1997) or static scribes (Mukhin et al., 1998), or a scalpel blade (Ng et al., 2012), respectively. Following primary axotomy or cell disruption, the injury site is subjected to excitotoxicity followed by secondary injury responses such as glial activation, inflammation, and cell repair (Ng et al., 2012; Blizzard et al., 2013). Transection models can be scaled to perform high-throughput drug discovery and testing (Faden et al., 2005). Although this injury is easy to implement, it remains difficult to control the biomechanics; severity is graded by the quantity of cells or processes undergoing primary transection, as opposed to mechanical inputs (e.g., strain, strain rate). Clinically, the incidence of penetrating brain injury is increasing, particularly in the United States, due to the use of firearms in violence-related injuries (Maas et al., 2008). However, penetrating injuries account for less than 10% of brain injuries overall (Blank-Reid and Reid, 2000; Javouhey et al., 2006). The low incidence of penetrating TBI and uncertainty as to whether its pathobiology translates to closed-head injuries temper enthusiasm for these models.

4.3.6 Blast Induced

Blast-induced TBI is a complex, multimodal form of injury [vide infra]. Most in vitro studies utilize an air/gas-driven shock tube (Effgen et al., 2012, 2014; Hue et al., 2013). In vivo studies have used similar shock tubes (Rafaels et al., 2011; Goldstein et al., 2012) or live explosives placed near subjects (Rubovitch et al., 2011). Recent efforts have focused on whether primary blast contributes to the overall injury or if secondary and tertiary mechanisms predominate (Effgen et al., 2012, 2014; Hue et al., 2013). Finite element models have predicted increased pressure distributions within hundreds of microseconds following shock wave exposure (Panzer et al., 2012). However, the effects of the shock wave on the ensuing injury cascades and neuronal function remain elusive.

A benefit of in vitro blast models is that precise control of the input injury parameters is possible, e.g., peak over pressure, duration, and impulse. One limitation is that most blast studies utilize the simplest pressure history, a Friedlander curve, which rarely occurs in real-world blasts. Another criticism of some in vitro blast models is the unrealistic exposure of cell or tissue cultures to the shock wave directly (Arun et al., 2011; Ravin et al., 2012). In reality, the shock wave must traverse the skull, tissue, and fluid barriers before reaching brain areas of interest (e.g., the hippocampus). For an in vitro model, this transition through the tissue has been accomplished with a fluid-filled receiver, which serves as a physiologically relevant barrier between the applied shock wave and the cell or tissue cultures (Effgen et al., 2012). Although it is simple to measure the pressure loading from the shock wave, it is challenging to measure the induced tissue and cell deformations to effectively characterize the biomechanics. In vitro blast TBI models are relatively new, and their utility is still being evaluated.

4.3.7 Recapitulating In Vivo Pathobiology with In Vitro Models

A strategy to assess the utility of in vitro models is to compare the pathologic changes in vitro with those in vivo. However, this comparison is difficult because the full pathobiology is unknown. As a surrogate measure, it may be possible to quantify their congruence on whether pharmacologic interventions produce the same outcome in both model types. Correlation of these responses would be interpreted as a high likelihood that the same pathways are operant in both classes of models. One example of this strategy is provided below.

4.3.7.1 Glutamate Receptors

Following TBI, glutamate excitotoxicity is hypothesized to be a common secondary injury mechanism that, in turn, initiates additional secondary injury cascades. Broad-spectrum antagonists of *N*-methyl-D-aspartate receptors (NMDARs), like MK-801 and APV, have previously been shown to afford protection in vivo and in vitro by reducing postinjury intracellular calcium concentration, depression of mitochondria membrane potential, and cell death (Morrison et al., 2011). Recent advances in understanding the role of NMDARs in the TBI cascade have been focused on receptor subtypes and subunits.

4.3.7.1.1 N-Methyl-D-Aspartate Receptors Containing NR2B Subunits

The NR2B subunit of the NMDAR has drawn much research interest for its potential as a therapeutic target with fewer side effects than indiscriminant NMDAR antagonism. A higher percentage of NR2B subunit–containing NMDARs are located on the extrasynaptic membrane (Singh et al., 2012). Stimulation of extrasynaptic NMDARs led to cell death, while stimulation of synaptic NMDARs induced prosurvival pathways (Hardingham et al., 2002), suggesting antagonism of the NR2B subunit as an attractive therapeutic strategy. Human embryonic kidney 293 (HEK) cells transfected with the NR1-subunit and different combinations of either the NR2A or NR2B-subunit, or both, varied in their mechanosensitivity in response to moderate stretch injury (Singh et al., 2012). Eliminating the majority of the C-terminal tail on the NR2B subunit also reduced mechanosensitivity but did not affect glutamate signaling of the receptor (Singh et al., 2012). Ro25-6981, an NR2B-subunit antagonist, abolished stretch-induced calcium influx in HEK cells (Ferrario et al., 2013). The same antagonist also prevented the postinjury increase in whole cell currents in organotypic hippocampal slice cultures after stretch injury (Ferrario et al., 2013). The importance of NR2B-containing NMDAR has been corroborated in vivo. Pretreatment with Ro25-6981 reduced cerebral edema in rats 24 hours following CCI injury (Laird et al., 2014). Reducing NR2B subunit expression with glutathione administration 6 hours after CCI in the rat significantly reduced caspase-3 expression, an indicator of apoptotic cell death (Arifin et al., 2011). Other compounds targeting NR2B-containing NMDA subunits, like CP101606 and Eliprodil, have reduced lesion volume (Toulmond et al., 1993), memory deficits, edema (Okiyama et al., 1997), and behavioral deficits (Hogg et al., 1998) after FPI in vivo. The important role of nonsynaptic NMDARs after TBI appears to be preserved in vitro.

4.3.7.2 Conclusions

In summary, there are many different models for in vitro TBI that apply mechanical stimuli by different means, as described above. For modeling TBI due to falls, sports injuries, or motor vehicle accidents, the applied strains should range from 0.1 to 0.5 at strain rates between 10 and 50 s^{-1} (Margulies et al., 1990; Meaney and Thibault 1990; Kleiven et al., 2002; Hardy et al., 2007). An ideal model should also provide a means to verify the injury biomechanics; interpretation of experimental results is simplified if that mechanical stimulus is consistent across the culture well. Of the models described above, the equibiaxial strain model from Morrison et al. meets these criteria, although it is not commercially available (Morrison et al., 2006).

Although there are many different in vitro injury models in use, as a whole, the models do faithfully reproduce the pathobiology initiated by in vivo models of TBI. We specifically reviewed NR2B NMDAR as a potential therapeutic target, but other targets (voltage-sensitive calcium channels [VSCC], extracellular signal-regulated kinases [ERK], c-Jun N-terminal kinases [JNK], p38 kinase, microtubule-associated protein 2 [MAP-2], tau, neurofilament, calpain, and blood-brain barrier [BBB]) have previously been reviewed (Morrison et al., 2011). In total, 6 out of 10 targets produced similar pathobiologic changes from both in vitro and in vivo models. For 6 out of 10 of these targets, pharmacologic interventions produced the same outcome in both in vitro and in vivo models.

However, if the four biologic targets that have not been investigated in detail either in vitro or in vivo (JNK, MAP-2, neurofilament, and tau) are removed, then the effects of interventions against five out of six (83%) targets were successfully replicated between in vivo and in vitro models, with modulation of p38 kinase producing mixed results in both classes of models.

Given the variety of in vitro injury models in use and the fact that these faithfully reproduce the in vivo pathology with similar pharmacologic findings, there appears to be little need for the development of new injury models. Instead, efforts could be better applied to utilizing the current in vitro models to understand the pathobiology in greater detail, identify novel therapeutic targets, develop new therapies directed at those targets, and begin to translate findings into a much-needed pharmacologic therapy for treating TBI victims.

4.4 CONTROLLED CORTICAL IMPACT AND IMPACT-ACCELERATION MODELS OF TBI IN RODENTS

4.4.1 Controlled Cortical Impact Brain Injury

The CCI models of TBI can be classified as rigid indentation models that produce direct brain deformation. These models utilize a rigid impactor to generate mechanical energy to induce TBI, generally delivering impacts to the intact dura while the head of the animal is restrained in a head holder/stereotaxic frame. Key advantages of the CCI models include the use of a known impact interface with ability to control the severity of injury via precise control over impact velocity (meter/second) and the depth and duration (dwell time) of tissue compression. The central CCI model initially developed for use in the ferret (Lighthall, 1988; Lighthall et al., 1989, 1990) and adapted for use in the rat (Lighthall et al., 1989; Dixon et al., 1991) employed a dual-action, stroke-constrained pneumatic cylinder that was mounted vertically on a rigid crossbar. A rounded or spherically shaped impactor tip on the lower end of the piston of the pneumatic cylinder was used to induce CCI injury through craniotomies located on the midline and centrally located between bregma and lambda. Soon thereafter, this type of device was modified to induce lateral CCI in the rat using various-diameter (5–9 mm) rounded impact tips with either vertical or angled impacts (Palmer et al., 1993; Cherian et al., 1994; Goodman et al., 1994; Prasad et al., 1994; Smith et al., 1994; Kochanek et al., 1995). Other investigators made adaptations to secure a pneumatic cylinder with a 5-mm diameter impactor tip with a flattened surface and rounded edges to a stereotaxic micromanipulator to induce lateral CCI to the sensorimotor cortex (Sutton et al., 1993) or to induce bilateral frontal CCI after creation of a midline craniotomy anterior to bregma (Roof et al., 1993; Hoffman et al., 1994) in the rat. To increase the extent of tissue movement and axonal damage, investigators have created separate craniotomies with or without dural opening over each hemisphere and then induced lateral CCI through one craniotomy site in the rat (Meaney et al., 1994; Dixon et al., 1996). Variations of these CCI models were also developed for use in the mouse (Smith et al., 1995; Fox et al., 1998a; Hannay et al., 1999; Natale et al., 2003). More recently, electromagnetically controlled CCI devices have been developed (Bilgen, 2005; Brody et al., 2007; Onyszchuk et al., 2007).

Readers should be cognizant of multiple factors that potentially influence the primary injury characteristics and outcome measures evaluated in the numerous variants of the rodent CCI models. Such factors include not only the generally well-known effects of altering impact velocity and impact depth, but also the dwell time, site of impact, size and geometric shape of the impactor tip, size and shape of the craniotomy or craniectomy (e.g., circular versus rectangular, distance between edges of craniotomy/craniectomy and the impactor tip), direction of the impact relative to surface of dura (vertical versus angled), use of secondary (e.g., bilateral) craniotomies, whether attempts to replace the bone flap or seal the cranial defect are made (craniotomy) or not (craniectomy), anesthetics used (including the absence or presence of ventilator support), and the use of postoperative analgesics or antibiotics.

In general, the rodent models of CCI models have been shown to produce impact velocity- and depth-dependent hemorrhagic cortical contusions that progress to form glial-lined cortical cavities (mild-to-moderate) or total loss of contused cortex and underlying white matter with more severe injuries (Dixon et al., 1991; Sutton et al., 1993; Goodman et al., 1994; Hoffman et al., 1994; Saatman et al., 2006; Taylor et al., 2008; Washington et al., 2012), and bilateral damage or contrecoup injury is increased using lateral CCI with an angled impact or bilateral craniotomies (Meaney et al., 1994; Dixon et al., 1998). Hemorrhagic lesions within the parenchyma and at the gray-white matter interface, as well as subdural hematoma, subarachnoid hemorrhage, or subcortical or intraventricular hemorrhages are likely to occur with increased injury severity (Dixon et al., 1991; Sutton et al., 1993; Meaney et al., 1994). Importantly, the shape of the impactor tip alters injury biomechanics and the extent and rate of cortical and subcortical neurodegeneration (Mao et al., 2010, 2011; Pleasant et al., 2011).

Neuronal cell loss (both necrotic and apoptotic) has also been shown to be impact dependent in the CCI models. Neuronal loss gradually extends into peri-contusional cortical regions after CCI (Sutton et al., 1993; Dunn-Meynell and Levin, 1997; Fox et al., 1998a; Newcomb et al., 1999; Saatman et al., 2006; Pleasant et al., 2011), and damage to hippocampal neurons (hilar region of the dentate gyrus, CA3 and CA1 regions) has been well characterized in the CCI models (Goodman et al., 1994; Smith et al., 1995; Colicos et al., 1996; Baldwin et al., 1997; Scheff et al., 1997; Dixon et al., 1998; Fox et al., 1998a; Saatman et al., 2006; Pleasant et al., 2011). Controlled cortical impact also produces acute damage and delayed retrograde degeneration of neurons in thalamic nuclei (Hoffman et al., 1994; Smith et al., 1995; Dunn-Meynell and Levin, 1997; Chen et al., 2003; Hall et al., 2005, 2008; Igarashi et al., 2007), with neuronal injury or loss also being reported within the amygdala, hypothalamus, and cerebellum (Colicos et al., 1996; Igarashi et al., 2007; Taylor et al., 2013). Progressive hemispheric tissue loss from 3 weeks to 12 months occurs with lateral CCI (Dixon et al., 1999). A limited extent of dendritic and axonal injury is most often observed with lateral or central CCI (Dixon et al., 1991; Dunn-Meynell and Levin, 1997; Chen et al., 2003) and after lateral CCI with bilateral craniotomy (Meaney et al., 1994; Posmantur et al., 1996, 2000). Silver staining methods have shown that the extent of neurodegeneration and traumatic axonal injury after lateral CCI is much more widespread than is often appreciated, extending not only into bilateral cortical and hippocampal regions but also into striatum, thalamus, substantia nigra, red nucleus, optic chiasm and optic tracts, and the cerebellum (Soblosky et al., 1996a; Matthews et al., 1998; Hall et al., 2005, 2008; Chauhan et al., 2010). However, diffuse axonal injury of the type and extent often seen in human head injury is not well duplicated in the rodent CCI models. Seizure-like activity and posttraumatic epilepsy do occur after CCI (Hunt et al., 2009; Yang et al., 2010; Bolkvadze and Pitkanen, 2012; Guo et al., 2013), and these clinically described phenomenon may impact on the extent of neuronal survival.

Disruption of the blood-brain barrier (BBB) (increased permeability to intravascular dyes or proteins) occurs within minutes of CCI, with bimodal peaks reported within 10 hours and at 3 days postinjury (Dhillon et al., 1994; Smith et al., 1994; Baskaya et al., 1997; Whalen et al., 1999). Blood-brain barrier disruption was found to be most severe in the ipsilateral cortex followed by the hippocampus, striatum, and/or dorsal thalamus over 7 days after lateral CCI (Smith et al., 1995; Saatman et al., 2006) and is associated with formation of cerebral edema (Baskaya et al.,

1997; Duvdevani et al., 1995). Cerebral edema generally peaks 1 to 2 days after CCI (Kochanek et al., 1995; Roof et al., 1996; Kiening et al., 2002; Schuhmann et al., 2003a), proportional to injury severity (Elliot et al., 2008). The increased intracranial pressure (ICP) with decreased cerebral perfusion (CPP) and cerebral blood flow (CBF) contralateral to injury are also dependent on injury severity (Cherian et al., 1994), and marked reductions in CBF within the contused cortex, with moderate-to-severe reductions in striatum, hippocampus, and thalamus, occur over the first 2 to 6 hours after CCI (Sutton et al., 1994; Bryan et al., 1995; Kochanek et al., 1995). Reduced cortical blood flow in the first few hours after CCI is associated with vasoconstriction, microcirculatory stasis, and reduced number of perfused capillaries (Thomale et al., 2002; Lundblad et al., 2004), and cerebral blood volume is reduced (Immonen et al., 2010). Improved CBF returns to the hippocampus within 24 hours and to the injured cortex in 7–10 days (Sutton et al., 1994; Kochanek et al., 1995), although moderate reductions in cortical and hippocampal regions with considerable tissue loss are detected 1 year after CCI (Kochanek et al., 2002).

The CCI models have proven useful for elucidating dysfunctions in cerebral metabolism and energy production that occur with TBI. Acute increases in glucose utilization and glycolysis are evident after CCI (Prasad et al., 1994; Krishnappa et al., 1999; Statler et al., 2003; Thomale et al., 2007; Fukushima et al., 2009; Clausen et al., 2011), but widespread reductions in cerebral metabolic rates of glucose (CMRGlc) occur within hours and persist for weeks (Sutton et al., 1994; Lee et al., 1999; Prins and Hovda, 2009; Moro et al., 2011) or months postinjury (Dunn-Meynell and Levin, 1995). Uncoupling of CBF and CMRGlc after CCI (Kelly et al., 2000) has been shown to impact on the extent of axonal injury (Harris NG et al., 2012). Studies using the CCI models have increased our understanding of the roles of calcium and reactive oxygen/nitrogen species in mitochondrial dysfunctions and guided use and development of mitochondrial targeted neuroprotective therapies (Xiong et al., 1998; Sullivan et al., 2000; Pandya et al., 2007; Robertson et al., 2007; Mirzayan et al., 2008; Gilmer et al., 2010; Mustafa et al., 2010; Sauerbeck et al., 2011; Singh et al., 2013). Consistent with the preceding studies, 1H- and 13C-nuclear magnetic resonance (NMR) spectroscopy work in the CCI models has shown alterations in glycolytic and oxidative metabolic pathways and cellular energy metabolism (Schuhmann et al., 2003b; Deng-Bryant et al., 2011; Xu et al., 2011; Harris JL et al., 2012; Bartnik-Olson et al., 2013; Robertson et al., 2013).

The CCI models have also been extensively used for the study of complex and interrelated molecular and cell signaling pathways that contribute to histopathology after TBI. These TBI models have been used for studies describing increased release of excitatory amino acids (Palmer et al., 1993; Rose et al., 2002) and increased reactive oxygen/nitrogen species that modify functions of multiple proteins and reduce antioxidant reserves (Smith et al., 1994; Tyurin et al., 2000; Deng et al., 2007; Opii et al., 2007; Ansari et al., 2008). The CCI-induced release of cytochrome c from mitochondria, activation of protein phosphatases, calpains, and caspases (Newcomb et al., 1997; Pike et al., 1998; Beer et al., 2001; Sullivan et al., 2002; Hall et al., 2005; Liu et al., 2006; Thompson et al., 2006), and alterations in Bcl-2 family genes affecting cell death/survival (Raghupathi et al., 1998; Strauss et al., 2004; Tehranian et al., 2006) have been studied extensively. The CCI models are known to rapidly activate multiple proinflammatory mediators/modulators (Strauss et al., 2000; Mori et al., 2002; Schuhmann et al., 2003a; He et al., 2004; Harting et al., 2008; Israelsson et al., 2008; Dalgard et al., 2012; Lagraoui et al., 2012; Redell et al., 2013), and tissue infiltrations of neutrophils (4 to 48 h after CCI) and leukocytes are associated with upregulation of cell adhesion molecules and chemokines (Clark et al., 1994; Whalen et al., 1997, 1999; Clausen et al., 2007; Szmydynger-Chodobska et al., 2009; Jin et al., 2012). Robust increases in markers of astrogliosis (GFAP, S100beta, vimentin) are reported in cortex, corpus callosum, hippocampus, striatum, and/or thalamus within hours of injury, with astrogliosis peaking in 4–7 days and persisting as long as 1 month after CCI (Hoffman et al., 1994; Meaney et al., 1994; Smith et al., 1995; Baldwin and Scheff., 1996; Dunn-Meynell and Levin, 1997; Hinkle et al., 1997; Chen et al., 2003; Saatman et al., 2006). A role for proliferating astrocytes in neuronal protection and suppression of infiltrating inflammatory cells has been reported (Myer et al., 2006). Macrophage infiltration and acute and chronic activation of microglia within damaged regions has also been well documented in CCI models of TBI (Hoffman et al., 1994; Chen et al., 2003; Schuhmann et al., 2003a; Igarashi et al., 2007; Sauerbeck et al., 2011; Jacobowitz et al., 2012; Jin X et al., 2012; Bedi et al., 2013; Kumar et al., 2013; Wang et al., 2013). Chronic activation of microglia has been associated with progressive deterioration and suppressed neurogenesis after CCI (Acosta et al., 2013), and novel treatments to attenuate proinflammatory actions of reactive microglia have been developed using a lateral CCI model (Byrnes et al., 2012).

The CCI models have been used extensively for studies revealing multiple alterations to classical neurotransmitter systems that interact with many of the TBI-induced pathophysiologic and cellular pathologies described above, and influence neurobehavioral outcomes (described below). In addition to studies already cited, some key CCI studies demonstrating alterations of neurotransmitter systems include for excitatory amino acids and adenosine (Bell et al., 1998; Rao et al., 1998; Kumar et al., 2002; Stover et al., 2004; Kharlamov et al., 2011; Verrier et al., 2012); for the cholinergic system (Donat et al., 2008; Hoffmeister et al., 2011, 2013); for the monoamines, including dopamine (see Bales et al., 2009 for review), norepinephrine (Prasad et al., 1994; Levin et al., 1995; Kroppenstedt et al., 2003; Kobori et al., 2006), and serotonin (Kline et al., 2001; Wang et al., 2011; Olsen et al., 2012); and for the GABAergic system (Kobori and Dash 2006; Mtchedlishvili et al., 2010; Lee et al., 2011).

Behavioral deficits have been extensively characterized in the rodent CCI models. Transient suppression of acute somatomotor reflexes occurs after CCI, and the duration of suppressed reflexes may be useful for evaluating injury severity (Dixon et al., 1991). A variety of tasks have been used to evaluate sensorimotor impairments, revealing deficits that are often shown to be dependent on either injury site or injury severity. Deficits have been reported for a variety of neurologic reflexes or composite neurologic severity scores (Hoffman et al., 1994; Fox et al., 1998a; Markgraf et al., 2001; Whishaw et al., 2004; Saatman et al., 2006; Shelton et al., 2008; Yu et al., 2009; Washington et al., 2012), beam balance and beam walking abilities (Dixon et al., 1991; Dunn-Meynell and Levin 1995; Fox et al., 1998a; Markgraf et al., 2001; Chauhan et al., 2010; Ajao et al., 2012; Zhao et al., 2012), overall locomotor activity (Chauhan et al., 2010; Ajao et al., 2012; Zhao et al., 2012), foot-fault or forepaw/hindpaw placement accuracy (Soblosky et al., 1996a; Baskin et al., 2003; Shelton et al., 2008; Goffus et al., 2010; Ajao et al., 2012), rotarod performance (Lindner et al., 1998; Fox et al., 1998a; Markgraf et al., 2001; Yu et al., 2009; Ajao et al., 2012), the cylinder task test of forelimb use asymmetry (Soblosky et al., 1996a; Baskin et al., 2003; Whishaw et al., 2004; Taylor et al., 2008; Goffus et al., 2010), forepaw deficits revealed during food pellet retrieval tasks (Hoane et al., 2003; Whishaw et al., 2004; Harris et al., 2010), and measures of sensory neglect assessed by latency to remove adhesive tape from limbs or vibrissae (Dunn-Meynell and Levin 1995; Hoane et al., 2003; Goffus et al., 2010). Kinematic analyses have revealed gait impairments in mice after CCI, including disturbances of stride and step length, gait velocity, and interlimb coordination (Neumann et al., 2009; Ueno and Yamashita, 2011; Wang et al., 2011).

The CCI-induced spatial learning and memory deficits (both reference and working memory tasks having been utilized) have been most often evaluated using the Morris water maze (MWM), with deficits often reported to be dependent on injury severity (Hamm et al., 1992; Dixon et al., 1994, 1996; Hoffman et al., 1994; Scheff et al., 1997; Fox et al., 1998a; Markgraf et al., 2001; Saatman et al., 2006; Hoskison et al., 2009; Yu et al., 2009; Washington et al., 2012; Zhao et al., 2012). Working memory and reference memory deficits after CCI have also been examined using the eight-arm radial arm maze (RAM) (Soblosky et al., 1996b), T-maze and plus maze (Taylor et al., 2008; Hoskison et al., 2009; Chauhan et al., 2010; Moro et al., 2011), the Barnes maze (Fox et al., 1998b), and the novel object recognition task (Scafidi et al., 2010; Wakade et al., 2010; Han et al., 2011; Zhao et al., 2012). Deficits in MWM, RAM, and operant conditioning tasks may endure for several months, dependent on whether lateral or bilateral frontal CCI is induced (Lindner et al., 1998; Dixon et al., 1999). Deficits in passive avoidance (an amygdala-dependent nonspatial learning task) have also been reported after CCI (Zhao et al., 2012). The CCI-induced effects on emotional behaviors such as depression or anxiety have been revealed using the open field test and free choice novelty tasks (Wagner et al., 2007), the elevated plus maze or zero maze (Chauhan et al., 2010; Ajao et al., 2012; Washington et al., 2012), the tail suspension task (Zhao et al., 2012), and in the forced swim test as well as for prepulse inhibition of the acoustic startle response (Washington et al., 2012).

The addition of secondary insults after CCI has been demonstrated to exacerbate injury and worsen outcomes. Secondary ischemia after CCI increases reductions in brain tissue PO_2 and CBF, increases extracellular lactate while decreasing glucose, and increases neuropathology (Cherian et al., 1996; Krishnappa et al., 1999; Giri et al., 2000). Secondary hypoxia alone can exacerbate neuropathologic and functional deficits (Clark et al., 1997). The combination of secondary hypoxia and hypotension after CCI worsens BBB disruption and cerebral edema, disrupts ionic homeostasis, and blunts CCI-induced upregulation of aquaporin-4 (Portella et al., 2000; Taya et al., 2010). Controlled cortical impact combined with hemorrhagic shock increases reductions in CBF, worsens functional outcomes, and exacerbates histopathology (Dennis et al., 2009; Hemerka et al., 2012; Foley et al., 2013).

4.4.2 Impact Acceleration Models of TBI in Rodents

Impact acceleration models of closed head injury in rodents have been developed to better produce the diffuse axonal injury that occurs with high frequency in human cases of TBI. The impact acceleration model developed by Marmarou and colleagues employs dropping of brass weights of varying mass (e.g., 450 or 500 gm) from various heights (1 or 2 m) down a Plexiglas guide tube to fall on a stainless steel disc (10 mm diameter, 2 mm thick; to reduce skull fracture) that is cemented on the vertex of the intact skull. The body and head of the rat are maintained in a prone position on a foam cushion of known spring constant, so that upon impact the head is allowed to accelerate/decelerate in a linear fashion (Foda and Marmarou, 1994; Marmarou et al., 1994). At high severity levels, this TBI model can produce coma in rats, and to reduce mortality from respiratory failure, mechanical ventilation is necessary (Marmarou et al., 1994; Adelson et al., 1996). Although investigators have varied weights and heights, this model has been shown to produce graded TBI with the mass of the weight and the height from which it is dropped being directly related to the resultant injury severity in adult (Foda and Marmarou, 1994; Marmarou et al., 1994; Povlishock et al, 1997; De Mulder et al., 2000, Kallakuri et al., 2003) and immature rats (Adelson et al., 1996). Variants of this model have also been used in mice (Nawashiro et al., 1995; Ohta et al., 2013). Alternative models providing linear impact acceleration include those of Engelborghs et al. (1998), where a weight with an attached silicone disc is dropped from various heights to impact the skull vertex and acceleration/deceleration is enabled using springs of variable elasticity, and of Cernak et al (2004), where a pneumatic impactor (with varied velocity and travel distance) is used to strike a metal disc affixed to the skull vertex while the head is supported on a gel-filled base with known compressibility.

Histologic studies from the impact acceleration model reveal a general absence of focal brain lesions, with petechial hemorrhages in brainstem only after severe injury, but widespread bilateral damage to neuronal structure, axons, dendrites, and the microvasculature (with edema, vascular congestion) that is dependent on injury severity (Foda and Marmarou, 1994; Folkerts et al., 1998; Buki et al., 1999; Okonkwo and Povlishock, 1999; Stone et al., 2001, 2004; Kallakuri et al., 2003, 2012). Impact acceleration of mild severity can result in mild subarachnoid hemorrhage (Folkerts et al., 1998), with more extensive subarachnoid and petechial hemorrhage occurring with more severe impact (Foda and Marmarou, 1994; Heath and Vink, 1995; Engelborghs et al., 1998). It is commonly found that axonal injury is diffuse after impact acceleration injury, with damage apparent within the white matter tracts of the corpus callosum, internal and external capsule, optic tracts, cerebral and cerebellar peduncles, rubrospinal and corticospinal tracts (Foda and Marmarou, 1994; Marmarou et al., 1994; Rafols et al., 2007; Kallakuri et al., 2003, 2012; Li et al., 2011; Li et al., 2013). Neuronal death may be entirely absent, but if present it is primarily reported within the neocortex directly underlying the impact site (Foda and Marmarou, 1994; Yamamoto et al., 1999; Rafols et al., 2007), with some investigators reporting neuronal death or atrophy in the hippocampus and/or thalamus (Yang and Cui, 1998; Cernak et al., 2002; Thornton et al., 2006; Hellewell et al., 2010; Yan et al., 2011; Alwis et al., 2012).

The impact acceleration model has also been shown to produce widespread increases in CBF in conjunction with impaired cerebral autoregulation acutely after impact (Nawashiro et al., 1995; Ito et al., 1996), with subsequent reductions in CBF (Sharma et al., 1998) and reductions in CPP (Engelborghs et al., 1998; Goren et al.,

2001). Impaired vasoreactivity of the microvasculature after injury (Suehiro et al., 2003; Baranova et al., 2008) is likely related to injury-induced changes in endothelin and nitric oxide levels (Sharma et al., 1998; Tuzgen et al., 2003), neurogenic damage in perivascular regions (Ueda et al., 2006), or oxidative stress (Baranova et al., 2008). Disruption of the BBB is reported to occur acutely after impact acceleration injury (Barzo et al., 1996; Cernak et al., 2004) and some authors have reported increased ICP (Ito et al., 1996; Engelborghs et al., 1998; De Mulder et al., 2000; Goren et al., 2001). Cerebral edema acutely postinjury has been shown to be vasogenic in nature, with cytotoxic edema and predominant cellular swelling developing over time after impact (Barzo et al., 1997; Cernak et al., 2004). This increased brain edema may be related to increases in the concentrations of brain cell osmolytes, taurine and myo-inositol (Pascual et al., 2007), and has been shown to be related to injury-induced increases in HIF-1α and aquaporin proteins (Shenaq et al., 2012).

The impact acceleration model of TBI has been shown to increase oxidative stress and to reduce endogenous antioxidants (Vagnozzi et al., 1999; Santos et al., 2005; Tavazzi et al., 2005, 2007; Prieto et al., 2011; Ohta et al., 2013). This TBI model is also reported to activate multiple proinflammatory mediators/modulators (Hans et al., 1999; Cernak I et al., 2001; Rancan et al., 2001; Rhodes et al., 2002; Lu et al., 2005a,b; Rooker et al., 2006; Yan et al., 2011; Bitto et al., 2012), with minimal tissue infiltration of neutrophils (Rancan et al., 2001) but a robust activation of astrogliosis and macrophage infiltration in damaged regions (Csuka et al., 2000; Rooker et al., 2006; Hellewell et al., 2010; Bye et al., 2011). Molecular biology and pharmacologic studies have implicated calcium-mediated pathways in neuronal and axonal injury, including activation of calcineurin (Singleton et al., 2001), calpains, and caspases (Buki et al., 1999, 2000; Cernak et al., 2002, 2004; Thornton et al., 2006; Bitto et al., 2012), and impact acceleration injury results in mitochondrial release of cytochrome c (Buki et al., 2000) and proapoptotic Bcl-2 family members (Cernak et al., 2002; Bitto et al., 2012).

The effects of impact acceleration TBI on neurotransmitter systems has not been as well studied as has been the case for CCI or FPI models. Decreased brain choline uptake 1 week after impact acceleration injury in young adult rats has been reported (Schmidt et al., 2000). Other authors found no such changes 1–5 weeks following injury for young adult rats, although forebrain choline uptake was decreased in rats 20–23 months of age at the time of injury (Maughan et al., 2000). Noradrenergic neurons and axons have been reported to be swollen acutely following injury, subsequently showing atrophy 1–7 weeks after impact acceleration injury, during which time NE turnover is reduced (Fujinaka et al., 2003). Extracellular levels of glutamate were reported to be unchanged in the first 2–5 hours following injury in cerebral microdialysis studies (Geeraerts et al., 2006; Blanie et al., 2012), although ^{1}H-MRS analyses of ex vivo brain homogenates detected progressive increases in glutamate, GABA, alanine, taurine, threonine, and glycine over 24 hours, with the higher levels of glutamate, GABA, and alanine being maintained to 48 hours after impact acceleration injury (Pascual et al., 2007).

Impact acceleration TBI has also been shown to alter several aspects of cerebral metabolism and energy production. Morphologic studies have revealed swollen/abnormal mitochondria associated with damaged axons, and the ability of cyclosporin A to reduce mitochondrial swelling and axonal injury has indicated permeability transition pore opening after injury (Okonkwo and Povlishock, 1999; Buki et al., 2000). Reductions in brain free magnesium concentrations and the cytosolic phosphorylation ratio, with increased rates of mitochondrial oxidative phosphorylation, have been reported after impact acceleration injury (Heath and Vink, 1995; Cernak et al., 2004). Extracellular levels of glucose, lactate, or pyruvate and the lactate/glucose ratio (reflective of glycolysis) were not affected by impact acceleration injury over a 2–5 hour monitoring period (Geeraerts et al., 2006; Blanie et al., 2012), although the lactate/pyruvate ratio (reflective of TCA activity and brain redox state) was increased with a more severe level of TBI (Geeraerts et al., 2006). The neuronal glucose transporter (Glut3) is upregulated from 4 to 48 hours following injury (Hamlin et al., 2001), suggestive of reduced glucose supply or increased metabolic demands within neurons. Cerebral energy metabolism studies using the impact acceleration model have reported severity-dependent reductions in ATP, GTP, nicotinic enzymes (NAD, NADP) involved in maintaining redox states, as well as reductions in N-acetylaspartate (Vagnozzi et al., 1999; Signoretti et al., 2001; Tavazzi et al., 2005; Prieto et al., 2011).

In young adult rats, impact acceleration injury has been shown to produce sensorimotor impairments on tests of neurologic reflexes (Folkerts et al., 1998; Rancan et al., 2001; Ucar et al., 2006; Alwis et al., 2012; Zhang et al., 2013), inclined plane (Beaumont et al., 2000; Berman et al., 2000), adhesive tape removal (Alwis et al., 2012; Zhang et al., 2013), the whisker nuisance task (Alwis et al., 2012), beam balance and beam walking (Beaumont et al., 2000; Berman et al., 2000; Kovesdi et al., 2010; Alwis et al., 2012), a foot-fault task (Zhang et al., 2013), and rotarod performance (Heath and Vink, 1995, 1999; Cernak et al., 2001; O'Connor et al., 2003, 2006; Vink et al., 2003; Thornton et al., 2006; Alwis et al., 2012). Cognitive deficits in the first month after injury have been shown for the MWM (Beaumont et al., 2000; Berman et al., 2000; Schmidt et al., 2000; Zhang et al., 2013), the RAM (Berman et al., 2000; Shenaq et al., 2012), and the Barnes maze (Cernak et al., 2001; O'Connor et al., 2003, 2006; Vink et al., 2003). Impact acceleration-induced effects on depression or anxiety-like behaviors are suggested by impairments revealed in open field tests (O'Connor et al., 2003; Vink et al., 2003; Fromm et al., 2004; Kovesdi et al., 2010) and the elevated plus maze (Kovesdi et al., 2010). Novel object recognition deficits have been reported for mice with impact acceleration TBI at 9 days following injury (Ohta et al., 2013), but were not apparent at 7 weeks following injury in the rat (Alwis et al., 2012). In the young (postnatal day 17) rat, impact acceleration injury produces severity-dependent deficits in beam balance, inclined plane, and MWM performance (Adelson et al., 1997), and the cognitive deficits revealed by the MWM task endure to 90 days following injury (Adelson et al., 2000). Older rats (20–23 months of age) have been reported to exhibit impaired MWM performance compared to young adult rats (2–3 months old) after impact acceleration TBI (Maughan et al., 2000).

The addition of secondary insults after impact acceleration injury has demonstrated the worsening of several outcome measures. Secondary hypoxia alone has been reported to increase cerebral edema and inflammation (Yan et al., 2011); exacerbate sensorimotor outcomes in multiple tests including inclined plane, beam balance and/or beam walking, adhesive tape removal, and rotarod (Beaumont et al.,

1999; Hallam et al., 2004; Yan et al., 2011); decrease open field exploration (Yan et al., 2011); and worsen MWM performance in one study (Beaumont et al., 1999), but not in another (Hallam et al., 2004). Secondary ischemia impairs ion homeostasis and is associated with elevated ICP after impact acceleration injury (Stiefel et al., 2005). Addition of an autologous subdural hematoma after impact acceleration injury increases the acute reductions in MABP and CBF, raises the ICP, and increases cerebral edema (Sawauchi et al., 2003), and addition of hypoxia to this combined injury progressively decreases CBF and increases cerebral edema and ICP (Sawauchi et al., 2003, 2004). Use of progressive hemorrhage or induced hypotension following injury has revealed impaired cerebral autoregulation in the impact acceleration model (Engelborghs et al., 2000; Fujita et al., 2012). Addition of hypoxia and hypotension has been shown to worsen neuronal and dendritic damage (Nawashiro et al., 1995; Yamamoto et al., 1999); alter hemodynamic responses to pressor drugs (Holtzer et al., 2001); increase BBB disruption, CPP, and edema while increasing ICP (Barzo et al., 1996; Ito et al., 1996; Beaumont et al., 2001); and increase oxidative stress and the alterations or reductions in cerebral energy metabolism and N-Acetylaspartate (NAA) (Al-Samsam et al., 2000; Signoretti et al., 2001; Tavazzi et al., 2005; Geeraerts et al., 2006, 2008; Blanie et al., 2012).

4.4.3 Conclusions

In summary, the rodent models of CCI have been shown to produce a wide range of structural, physiologic, and functional outcomes consistent with those reported with human TBI. Strengths of the CCI models are their high reproducibility and control of mechanical factors with low mortality and lack of rebound injury. Disadvantages are the use of skull removal prior to injury, the lack of prolonged coma, and diffuse axonal injury seen clinically. Although the focus of this review has been on rodent models of CCI, it should be pointed out that large animal models of CCI have now been developed in adult swine (Alessandri et al, 2003; Manley et al., 2006; Jin G et al., 2012) and monkey (King et al., 2010), as well as in juvenile and neonatal swine piglets (Durham et al., 2000; Duhaime et al., 2003; Mytar et al., 2012). These large animal models, although costly, should facilitate monitoring and collection of data similar to those of intensive care units and translation of promising therapeutics into clinical trials.

Specific advantages of the rodent impact acceleration model include relatively low cost and ease of operation and that it produces widespread bilateral axonal damage. The model has been particularly useful for studies of cerebral energy metabolism and the complex processes underlying damage to the axolemma, neurofilament compaction, and axonal disconnection. Disadvantages of the model include variability of injuries, risk of rebound ("second hit") inherent in weight drop models, and relatively high mortality at higher impact levels. The model also primarily employs linear acceleration, and models combining linear (with impact on vertex of the skull), and angular accelerations (Wang et al., 2010) or using an impact directed frontally to produce anterior-posterior plus sagittal rotational acceleration (Kilbourne et al., 2009) have been recently developed. Additional work will be needed to fully evaluate these models.

4.5 FLUID PERCUSSION MODELS OF TBI IN RODENTS

Fluid percussion injury (FPI) is among the most well-characterized and commonly used TBI models, having originated in the 1970s and been adapted to rodents in the 1980s (Thompson et al., 2005). FPI is scalable to induce highly reproducible TBI that models the clinical conditions of concussion through to complex TBI. Using this model, brain injury is induced by a 20 ms fluid pulse delivered onto the intact dura through a craniectomy (Lifshitz et al., 2008). The clinical and pathologic outcomes from FPI vary based on the placement of the craniectomy. If the craniectomy lies along the midline suture, the result is a diffuse brain injury, whereas a lateral craniectomy results in a focal injury with a diffuse component (Thompson et al., 2005). Additionally, shifting between parasagittal and lateral craniectomy location can induce more subtle but detectable differences, specifically in lesion development, cognitive performance, hippocampal cell loss, and reactive astrocytosis (Vink et al., 2001; Floyd et al., 2002). Although FPI necessitates breaching the cranial vault, the skull is sealed to the injury device, recreating a closed system, which approximates a closed head injury.

Certain concessions and assumptions predominate in studies involving FPI. As with most experimental brain injury, FPI is conducted primarily in male rodents to reflect the 3:1 demographics of TBI in males over females (Faul et al., 2010). Focusing on male rats eliminates confounding effects of hormone cycling females but overlooks any gender-related effects on pathophysiology or therapeutic efficacy (Stein, 2007). As indicated below, rodents with FPI regain neurologic function rapidly, with little intervention. Therefore, it is generally acknowledged that FPI most closely models mild–moderate focal and diffuse TBI, with a Glasgow coma score (GCS) of 9–13, in which patients are generally responsive, but likely disoriented. Severe TBI (GCS <9), on the other hand, involves brainstem damage. A rodent model of severe TBI would necessitate intensive clinical care, which is neither practical nor warranted for rodent studies.

A posttraumatic neurologic deficit is a transient consequence of injury-related pathology that impairs brain function, including cognitive, motor and sensory domains, which naturally recover with time after injury. Animals subjected to FPI experience neurologic deficits immediately after injury that typically resolve within a few weeks after injury.

4.5.1 Immediate Transient Deficits

On impact from FPI, the reflexes of the injured animal are suppressed; whether this is a loss of consciousness remains a philosophical argument. Brain injury immediately suppresses neurologic reflexes, including corneal, pinnae, and righting, of which the righting reflex involves a noninvasive evaluation and its duration correlates with injury severity (Dixon et al., 1988; Hosseini and Lifshitz, 2009; McIntosh et al., 1989; Chauhan et al., 2010). Reflex suppression may be accompanied by apnea and seizure, where the duration and prevalence also correlate with injury severity. These acute reflex suppression and physiologic responses to the injury forces more faithfully reflect injury severity than the physical parameters of the device itself. In this way, the acute neurologic responses to injury reflect the human responses captured by the GCS.

Physiologic responses to FPI injury are immediately evident in cerebrovascular changes. After injury, there is a severity-dependent pattern of hypertension and elevated intracranial pressure (ICP) (Lafrenaye et al., 2012). The increased ICP is accompanied by an increase in mean arterial pressure (MAP), irregular breathing, and reduced heart rate, known as the Cushing reflex. Regional cerebral blood flow shows an initial global suppression within 1 hour following injury, and at the trauma site a more persistent focal reduction, which resolves within 4–24 hours following injury (Hillered et al., 2005; Thompson et al., 2005). Concomitant with physiologic alterations, blood-brain barrier permeability reflects the location of the craniotomy, with a mix of local and diffuse effects, particularly in the cervicomedullary junction (Schmidt and Grady, 1993). Similar dynamic responses in ICP and local cerebral blood flow (lCBF) have been documented in humans and FPI animal models (Hillered et al., 2005; Thompson et al., 2005), albeit on different time scales. Cerebrovascular changes evident early after injury typically resolve within hours to days following injury.

After FPI, potassium efflux releases excitatory amino acids (EAAs), which activate glutamate receptors and secondary calcium signaling pathways. In reestablishing homeostasis, an initial period of glucose hypermetabolism is followed by a period of glucose hypometabolism (Yoshino et al., 1991). Although the time course of glucose utilization may vary between human and animal models, both follow similar temporal courses (Bergsneider et al., 2000). Sustained posttraumatic glucose hypometabolism correlates with injury severity and may pretend vulnerability to secondary insult (Ip et al., 2003; Barkhoudarian et al., 2011).

As an extension of physiologic deficits, prevailing hypotheses suggest that sleep may aid in cellular repair and be beneficial in recovery following injury. After FPI, sleep significantly increases for up to 6 hours regardless of injury severity or time of day (Rowe et al., 2014). The temporal profile of secondary injury cascades, particularly inflammatory cascades (Raghupathi et al., 1995), may be driving the significant increase in posttraumatic sleep, which could contribute to or confound the natural course of recovery. Further studies are needed to fully understand the cellular benefit or detriment, if any, of acute posttraumatic sleep on recovery following TBI, as well as other neurologic conditions.

4.5.2 Behavioral Deficits

Motor deficits have been reported in rats using the beam balance, beam walk, rotarod, and incline plane tests (Hamm, 2001). Cognitive deficits have been shown in rats with the radial arm maze and the Morris water maze. These deficits are evident over the first few days following injury and may persist for months to a year, depending on the severity of injury. Retrograde and anterograde cognitive deficits are evident with FPI, depending on training either prior to or following the injury (Hamm et al., 2001). In many cases, cognitive deficits are most prominent when the tasks are sufficiently difficult, as achieved with underexposure to training trails. Regardless, most retrograde and anterograde hippocampal-dependent cognitive deficits typically resolve by 1 month following injury.

4.5.3 Chronic Morbidities

Posttraumatic morbidity is the long-term neurologic consequence of injury-related pathologic processes that impair brain circuit function and activation, including problems with cognition, sensory processing, communication, and behavior/mental health (McAllister, 1992; Chen and D'Esposito, 2010). Morbidities evolve over time in clinical conditions, and effective models of TBI would demonstrate injury-specific morbidities.

4.5.4 Enduring Physiologic Morbidities

The endocrine system is particularly vulnerable to TBI. Endocrine dysfunction arises from structural damage, hypopituitarism, or adrenal insufficiency, and may contribute to morbidity and occur at any time after TBI (Behan et al., 2008; Dusick et al., 2012). Endocrine dysfunction can contribute to the array of neurologic, somatic, and emotional morbidities after TBI, but systematic investigations in FPI models are sparse. Acute injury-induced increases in the stress response, measured by serum hormone levels, are evident (Grundy et al., 2001; Griesbach et al., 2012; Griesbach et al., 2014), but few studies have investigated enduring endocrine dysfunction. The mild–moderate nature of FPI may warrant specific challenges (e.g., dexamethasone, whisker stimulation) to unmask endocrine dysfunction (McNamara et al., 2010).

Traumatic brain injury can result in impaired vasodilatory responses, which contribute to secondary insults in the injured brain (Povlishock and Katz, 2005; Fujita et al., 2012). Fluid percussion injury disturbs the dilation of cerebral arterioles in the face of decreasing blood pressure in the first 4 hours after TBI (Fujita et al., 2012), without investigation into the more chronic time periods. Although vasoreactivity endures in the clinic, vasodilatory responses in FPI appear to resolve within 2 weeks of injury (Sviri et al., 2009; Fujita et al., 2012).

Epilepsy is a common comorbidity of TBI, with nearly 50% of TBI cases also presenting with epilepsy (Yeh et al., 2013). Evidence for spontaneous electrographic seizures exists after FPI, although the incidence ranges from 50% to 92% and requires up to 8 weeks for phenotypic penetrance (D'Ambrosio et al., 2004; Pitkanen et al., 2006). Further, FPI increases susceptibility to seizure and decreases seizure threshold (Pitkanen et al., 2012). In fact, a single lateral FPI is capable of causing posttraumatic epilepsy (PTE) in the rat, with electrophysiologic and structural sequelae paralleling changes seen in human PTE (D'Ambrosio et al., 2004). However, it is likely that genetic or chemical predisposition may be necessary for a reliable model of PTE.

4.5.5 Behavioral Morbidities

Months to years after clinical TBI, neuromotor deficits can develop into chronic morbidities, including difficulties with coordination, posture, and steadiness of movement (Thompson et al., 2005). Once neuromotor symptoms develop into morbidities, function may never be completely restored. After FPI, rodents demonstrate neuromotor morbidities, including poor performance in rotarod, beam walk, catwalk, and the rope hang, indicating dysfunction in coordination, balance, gait, and grip strength, respectively (Hamm et al., 2001). Based on the gross motor tasks available, few motor morbidities are evident; however, more sensitive behavioral tests or analyses may be necessary to uncover the morbidities.

Somatic signs of postconcussion symptoms include sensory sensitivity, particularly to light and sound, which can persist for months after injury (McAllister et al., 1992). Similarly, midline FPI produces a sensory sensitivity to whisker stimulation in rats. In the whisker nuisance task, brain-injured rodents show an aggravated response to whisker stimulation that develops by 28 days following injury, while sham animals are ambivalent to the same stimulation (McNamara et al., 2010). The sticky paper and limb placement tests also detect sensorimotor dysfunction after FPI (Riess et al., 2001), but the clinical relevance of these tests is unclear.

Patients also suffer from enduring cognitive morbidities following TBI, which are among the lasting symptoms following a range of types and severities of injury (Thompson et al., 2005). In experimental models, evaluating rodent cognition typically relies on performance in hippocampal-dependent spatial mazes (Thompson et al., 2005), such as the Lashley, Morris water, and 8-arm radial mazes. Long-term deficits, some indicating partial natural recovery, can be detected, given specific testing conditions and parameters (Pierce et al., 1998). However, spatial navigation approximates but does not replicate the verbal, mathematic, and executive functions evaluated in a clinical setting. More recently, the multivariate concentric square test (MCSF) has been applied to account for more complex cognitive behavioral analyses and tease apart enduring behavioral deficits after FPI (Ekmark-Lewen et al., 2013).

Psychiatric disorders are yet another consequence of TBI that can emerge acutely and endure chronically. Up to 28.8% of TBI patients experienced posttraumatic stress disorder (PTSD) at 1 month after trauma, with a 17.7% occurrence at 12 months after injury (Santiago et al., 2013). Rodent behavioral tests can evaluate psychiatric symptoms, where elevated stress and anxiety are found at 1, 3, and 6 months after FPI using the open field, elevated plus maze, the forced swim, and the sucrose preference tests (Jones et al., 2008; Reger et al., 2012). Avoidance tasks are complex tests of cognition using aversive stimuli, which necessarily induce stress; after FPI, rats showed less passive avoidance and more active avoidance than sham animals, indicating cognitive abilities that are differentially affected by stress (Hogg et al., 1998).

4.5.6 Anatomic and Pathologic Correlates

Diffuse axonal injury is the primary persisting histologic feature of mild–moderate TBI (Adams et al., 2011). However, the low mortality with mild–moderate TBI precludes postmortem histologic evaluation. Brains of boxers and football players have been studied for pathology associated with repetitive brain injury, where atrophy and degeneration, as well as abundant Tau staining have been described

(Mez et al., 2013). The most prominent imaging findings in diffuse TBI, where available, include hematoma. In FPI, subdural hematoma occurs with impact and generally resolves by 1 week after injury. Subarachnoid and intraventricular hemorrhage, sequelae of severe TBI, do not occur in FPI, which further defines FPI as a mild–moderate injury.

Because anatomic features of mild–moderate TBI are limited, research focus has shifted to molecules with biomarker potential in the clinic and laboratory. Leading biomarker candidates are αII-spectrin and its breakdown products (SBDPs), glial fibrillary acidic protein (GFAP), ubiquitin c-terminal hydrolase-L1 (UCHL1), and S100B (Berger et al., 2012). Elevation of calpain- and caspase-specific SBDPs in CSF following TBI has been established in human subject studies, and correlates with the severity of injury, imaging findings, and clinical outcomes (Yan et al., 2012). Likewise, SBDPs increase in the acute phase following FPI (Reeves et al., 2010). Other reports have demonstrated that serum GFAP is a specific marker of brain damage after brain injury (Mondello et al., 2011), without investigation in FPI. The UCHL1 is upregulated after TBI both in the clinic and the laboratory, and is correlated with Glasgow outcome score (GOS) in humans (McGinn et al., 2009; Berger et al., 2012). S100B serum levels predict TBI better than psychologic symptoms, and correlate with both GCS scores and neuroradiologic findings at hospital admission (Ingebrigtsen et al., 2000; Mondello et al., 2011). Serum S100B is also increased after FPI (Kleindienst et al., 2005); however, the direct relationship to neuropathology remains to be investigated in animal models.

4.5.7 Conclusions

Animal research holds significant validity toward the human condition but cannot faithfully replicate every feature (Thompson et al., 2005). For example, TBI results in chronic sleep disorders in humans, but FPI and other rodent models, to date, have failed to report chronic sleep issues (Rowe et al., 2014). Additionally, acute and chronic vestibular deficits and morbidities remain challenging to investigate in four-legged animals with prehensile tails; vestibular deficits may have to be far greater in the animal model to elicit phenotypic penetration. Along these lines, compensatory actions and movements may permit injured rodents to accomplish the contrived behavioral tasks in the laboratory, while significant underlying changes in circuitry actually exist. Attention to detail in testing animal performance is critical to elucidate the consequences of brain injury in the rodent.

After 30 years of characterization and development, FPI is recognized as a clinically relevant model of TBI, particularly diffuse brain injury, such as concussion. This model is valuable for translational research and can be analyzed at various levels. Repetitive brain injury, common in sports-related injury, has yet to be routinely modeled in FPI (Shultz et al., 2012). Additionally, TBI, whether it occurs in a motor vehicle accident, sports injury, or assault, often occurs concomitant with other injuries. As a result, there are endless opportunities to study combined injury, such as TBI with broken bone, TBI with hypoxia, or TBI with hypoglycemia. Additionally, age-at-injury has been largely overlooked in the FPI model (Prins et al., 2003; Hawkins et al., 2013), where continued investigation is warranted. It is essential that females gain representation in brain-injury research to address anatomic and physiologic differences between genders.

4.6 EXPERIMENTAL MODELS OF PEDIATRIC TBI

Pediatric traumatic brain injury (TBI) remains one of the leading causes of acquired disability and death, with the highest combined rates of TBI-related emergency room visits, hospitalizations, and deaths occurring in infants and young children less than 5 years of age (Coronado et al., 2011). Survivors in this age group have worse chronic cognitive and disability outcome compared to older children (Levin et al., 1992; Anderson et al., 2005).The first version of evidence-based guidelines for the medical management of severe TBI in this population highlighted the need for improved supportive neurocritical care (Adelson et al., 2003). However, a recent update suggested that no definitive treatments exist due to our lack of understanding of the underlying mechanisms of the injured pediatric brain (Kochanek et al., 2012). Brain trauma in the pediatric population results in responses to TBI compared to adults. Although the chronic behavioral problems are similar between brain-injured adults and children, the pathophysiology appears to be different; diffuse brain swelling leading to brain atrophy is more common in the pediatric patients (Bruce et al., 1981; Aldrich et al. 1992; Duhaime and Durham, 2007). Consequently, it becomes imperative that age-appropriate animal models be developed and used to evaluate the pathogenesis of pediatric TBI with a view to developing age- and injury-dependent treatment strategies. A summary of these models supported by key references is presented in Table 4.1.

4.6.1 Behavior

The majority of studies in models of pediatric TBI have focused on cognitive function, as this has been identified as a significant behavioral deficit in man (Levin et al., 1992). Either noncontusive or contusive trauma in rodents between the ages of 11 and 28 days results in spatial learning (acquisition) deficits within the first 2 weeks following injury and retention (memory) deficits at 1 month (Prins and Hovda, 1998; Adelson et al., 1997, 2000; Gurkoff Ozdemir et al., 2005; Gukoff et al., 2006; Pullela et al., 2006; Huh and Raghupathi, 2007; Raghupathi and Huh, 2007; Huh et al., 2008, 2011; Cernak et al., 2010). The complexity of the modeling pediatric TBI is highlighted by the fact that in certain models (e.g., impact-acceleration), learning deficits were observed only at the most severe level (Adelson et al., 1997, 2000), while a relatively mild impact to the intact skull with a rigid indentor resulted in learning deficits up to the third week following injury (Huh et al., 2008). Similarly, in the 21-day-old mouse, only spatial learning deficits (not retention) were observed after contusive trauma (Pullela et al., 2006). More recently, cognitive deficits have been described in a large animal model of rotational injury, which underscores the translational value of this model (Raghupathi and Margulies, 2002; Friess et al., 2007). In addition to cognition, recent studies have revealed that immature rodents between 17 and 21 days old subjected to CCI contusive brain injury revealed emerging behavioral problems with social

TABLE 4.1
Target/Drug, In Vitro, In Vivo

Target/Drug	In Vitro	In Vivo
NR2B-containing NMDARs		Agreement
Ro25-6981	Protective	Protective
Glutathione	N/A	Protective
CP101606	N/A	Protective
Eliprodil	N/A	Protective
VSCC		Agreement
SNX-185	Protective	Protective
CTX MVIIA	Mixed	Protective
CTX MVIIC	Protective	N/A
Cd^{2+}	Protective	N/A
Nifedipine	Mixed	N/A
ERK		Agreement
PD98059	Protective	Protective
U0126	Protective	Protective
Cyclophilin A	Protective	N/A
Cyclosporin A	N/A	Protective
JNK		More in vivo studies necessary
SP600125	Protective	N/A
TAT-JNK Inhibitor	Protective	N/A
p38 Kinase		Disagreement: More studies necessary
SB203580	Mixed	No effect
p38 α deletion	N/A	Beneficial
MAP-2		More in vivo studies necessary
TTX	Protective	N/A
MK801	Protective	N/A
Oxyresveratrol	Protective	N/A
Tau		Pharmacological studies necessary
Neurofilament		Pharmacological studies necessary
Calpain		Agreement
MDL28170	Protective	Protective
Calpain Inhibitor 1	Protective	N/A
AK295	N/A	Protective
Calpain Inhibitor 2	N/A	Protective
Calpistatin	N/A	Protective
Calpain-1 deletion	N/A	Protective
BBB		Agreement: More in vitro studies
PEG-SOD	Protective	N/A
EpoE	N/A	Protective
Ghrelin	N/A	Protective
Poloaxmer-188	N/A	Protective
Cannabanoid Receptor Agonist 2	N/A	Protective

interactions such as aggression and anxiety at adulthood (Ajao et al., 2012; Semple et al., 2012).

4.6.2 Histopathologic Alterations

Diffuse brain injury, marked by hemorrhage, edema, and axonal damage, is a hallmark of the neuropathology of pediatric TBI. These alterations have, for the most part, been identified in both large and small animal models (Adelson et al., 2001; Raghupathi and Margulies, 2002; Gurkoff et al., 2006; Huh et al., 2006; Huh and Raghupathi, 2007; Raghupathi and Huh, 2007; Clause et al., 2011). Importantly, the acute hemorrhagic tissue tears developed into a profound reduction in the area/volume of the cortical mantle and the underlying white matter (Huh and Raghupathi, 2007; Raghupathi and Huh, 2007; Clause et al., 2011), a feature reminiscent of the holohemispheric atrophy reported in infants (Duhaime and Durham, 2007). In contrast, contusive trauma in piglets resulted in a greater cortical lesion volume in the oldest (4-month-old) animals compared to the 4–5-day-old piglet (Duhaime et al., 2000). In part, this may reflect that the lesion in the youngest animal peaked early and resolved more quickly compared to the older piglets (Duhaime et al., 2003).

It is as yet unclear if traumatic axonal injury (TAI), the accumulation of amyloid precursor protein (APP) within axons, is influenced by the age-at-injury (Adelson et al., 2001; Huh et al., 2006, 2008, 2011; Raghupathi and Huh, 2007; Dikranian et al., 2008). In contrast, rotational-acceleration head injury in 3–5-day-old piglets resulted in more injured axons per unit area in the cerebrum compared to the brain-injured 1-month-old or adult pig (Smith et al., 2000; Raghupathi and Margulies, 2002; Ibrahim et al., 2010). In addition to accumulation of APP, neurofilament compaction (NFC) has been reported in the injured neonate brain, albeit with minimal evidence of co-localization (DiLeonardi et al., 2009), as observed in the adult rat brain (Marmarou et al., 2005). However, the amplitude of the compound action potential of axons decreased acutely in both the adult (Reeves et al., 2005) and 17-day-old rat (DiLeonardi et al., 2012), but recovered in the adult animal.

In models of adult TBI, both diffuse and contusive brain trauma was associated with neuron loss in both the acute and chronic posttraumatic period (Raghupathi, 2004). In contrast, diffuse brain injury in either the 11-day-old or the 17-day-old rat did not result in overt neuronal loss (Adelson et al., 2001; Gurkoff et al., 2006; Raghupathi and Huh, 2007; Huh et al., 2008). Axonal injury leading to secondary axotomy was associated with neuronal apoptosis delayed cell death in the 7-day-old mouse (Dikranian et al., 2008) but not in the 17-day-old rat (Huh et al., 2008). More recently, we reported evidence of caspase activation in neurons (within the gray matter) and oligodendrocytes (in white matter tracts) following contusive brain trauma in the 17-day-old rat (Huh et al., 2011); interestingly, contusive TBI resulted in more extensive apoptotic cell death in 3- and 7-day-old rats compared to their older (14- and 30-day-old) counterparts (Bittigau et al., 1999).

4.6.3 Plasticity

Convention has long dictated that the immature brain has a high degree of plasticity and therefore is relatively invulnerable to environmental (nongenetic) damage. Even though this may partially be true, clinical and experimental evidence from studies of pediatric TBI suggest otherwise. Rearing brain-injured 17–20-day-old rats in an enriched environment failed to prevent an increase in cortical thickness and dendritic arborization or an improvement in cognitive performance (Fineman et al., 2000; Ip et al., 2002). In part, this may be explained by the observation that the NR2A subunit of the *N*-methyl-D-aspartate receptor, which increases with normal maturation (Monyer et al., 1994; Tovarand Westbrook, 1999) was reduced following TBI (Osteen et al., 2004; Giza et al., 2006). Needless to say, the paucity of data with respect to mechanisms underlying recovery of function predicates the need for additional studies.

4.6.4 Limitations of Experimental Pediatric TBI Models

Modeling TBI in the immature animal poses many challenges. The biggest question facing researchers is the "appropriate" age of the animal that accurately reflects the maturational stages of the developing brain in the human. This has to take into account the diverse aspects of brain maturation such as neurotransmission, proliferation, migration, synaptogenesis, and myelination, which may develop at different maturational stages in different parts of the developing brain and at different rates in different species. When trying to determine if the younger brain is more or less vulnerable to the traumatic insult, one has to produce comparable injuries across ages and therefore faces issues of "scaling"—where the force of the injury needs to be standardized to the size, mass, or stiffness of the brain. A third issue surrounding the development of a clinically relevant model is that the typical pediatric patient presents with more than one intracranial pathology (TAI, contusion, and intracranial hemorrhage) and comorbidities, such as hypoxia, hypotension, and hypercarbia, which may promote secondary brain damage. Finally, data comparison across laboratories becomes problematic due to the variations in the actual device(s) and the parameters used to induce the injury, and the dose and type of anesthetic used, and different definitions of injury "severity," with some investigators using length of apnea, righting reflex, or mortality, while others use tissue pathology at a specific survival time.

Despite the limitations of animal models of pediatric TBI, there is increasing evidence to suggest that the injured developing brain responds differently at individual maturational stages; a fact that mirrors the human condition. Continuing to use and refine current injury models, by varying injury severities in multiple age groups supported by multiple outcome measures, with the addition of other comorbidities such as hypoxia, hypotension, and hypercarbia will undoubtedly enhance our understanding of the unique injury responses of the developing brain at different maturational stages, so that we can discover specific treatment strategies in the hopes that the injured pediatric patient will have the best chances for maximal recovery and to allow ongoing neuromaturational development (Table 4.2).

TABLE 4.2
Summary of Animal Models of Pediatric TBI

Injury Model	Craniotomy	Species	Age	Outcome Measures	References
Cortical impact	No	Rat Rat Mice	11 and 17d 7d 7d	Spatial learning and memory, histology Histology Histology	Raghupathi and Huh (2007) Bayly et al. (2006) Dikranian et al. (2008)
Cortical impact	Yes	Rat Rat Mice	7 and 17d 16–17d 21d	Spatial learning, metabolism Metabolism Histology Motor and cognition, histology	Hickey et al. (2007) Robertson et al. (2007) Tong et al. (2002) Pullela et al. (2006)
Cortical impact	Yes	Pig	5d, 1 and 4 mo	Histology	Duhaime et al. (2000)
Fluid percussion	Yes	Rat	17 and 28d	Spatial learning and memory, histology Metabolism	Prins and Hovda (1998) Osteen et al. (2001)
Fluid percussion	Yes	Pig	1–5d, 3–4 wk	Cerebral physiology	Armstead and Kurth (1994)
Weight-drop	No	Rat Rat	17d 3–30d	Spatial learning and memory, histology Histology	Adelson et al. (2000, 2001) Bittigau et al. (1999)
Weight-drop	Yes	Rat	3.5–4.5 wk	Cerebral physiology	Grundl et al. (1994)
Nonimpact, rotational	No	Pig	3–5d	Histology Behavior, histology	Raghupathi and Margulies (2002) Friess et al. (2007)

4.7 EXPERIMENTAL MODELS OF EXPLOSIVE/BLAST TBI

4.7.1 Overview

Explosive blast TBI is a leading combat casualty. As improvised explosive devices (IEDs) are the preferred weapon of our adversaries, more of our service members are injured by them than by any other weapon, including bullets. Since the wars in Iraq and Afghanistan began, there have been close to 290,000 service members injured by explosive blasts of which 68% suffered TBI (Masel et al., 2012; Center et al., 2013). Of significant concern are the acute and chronic long-term impacts of blast-related TBI. This has engendered a significant scientific and clinical effort to better characterize the clinical spectrum, understand the mechanisms of injury, and elucidate the pathobiology of blast-related TBI. The objective of which is to improve its prevention, detection, and treatment. This effort is particularly relevant to combat-related mild TBI (mTBI), where blasts account for 72% of cases in the war theater (Wilk et al., 2010).

There are four categories of explosive blast injury. Primary blast injury refers to damage that results directly from blast-created overpressure waves impacting the head and brain. Secondary blast injury is from being struck by weapon case fragments or debris cast by the detonation. Tertiary injury is a result of the victim striking an object or the ground after being bodily thrown through the air by the blast. Quaternary injury refers to any other mechanism attributable to the weapon such as the explosive fireball causing burns, toxic fumes, etc. (Phillips, 1986; Ling et al., 2009; Ling and Ecklund, 2011).

It is hypothesized that, in primary blast injury, the brain is injured by explosive blast–created pressure waves. These waves are initiated by the chemical detonation leading to a rapid displacement of air moving through the atmosphere. As the explosive-created pressure waves travel faster than the characteristic wave speed in the host medium (air), they compress to become shock waves. When the shock waves strike the victim's head, the wave both engulfs and passes through the skull. Waves that transit into brain parenchyma cause tissue acceleration and deformation resulting in shearing of white matter tracks and neuron and glia cell injury and death. The extent of tissue damage depends on the blast shock wave's characteristics, peak overpressure, and pulse duration, as well as the brain's natural resonant frequencies (Cullis, 2001; Desmoulin and Dionne, 2009; Magnuson et al., 2012).

The classic description of an explosive blast wave is the Friedlander curve, which is a pressure-time relationship. In brief, there is an initial rapid rise in pressure (shock wave) followed by slow decay. At the end of the decay phase, there may be a negative pressure phase. Thereafter, the pressure rises back to baseline. This all occurs in a matter of 200–300 milliseconds (Baker et al., 1973).

Another possible mechanism is transmission of shock waves from the torso to the brain. In particular, it has been proposed that the blast shock waves impact the abdomen and thorax. The blast shock waves compress the abdomen and thorax, which creates oscillating waves inside the fluid column contained in large vascular vessels. The oscillating waves transit via the vessels into the brain resulting in both morphologic and functional damage (Camm and Garratt, 1991; Cernak et al., 1999, 2001, 2011; Courtney and Courtney, 2009).

The scientific evidence supporting these hypotheses is incomplete. Some preclinical models using explosive-driven gas tubes provide findings that explosive blast waves, when of sufficient severity, lead to brain pathology whereas others do not. Those that do reveal neuropathologic findings show multifocal axonal and neuronal injuries, astroglial alterations, inflammation with elevated cytokine and reactive oxygen species activity, blood-brain barrier anomalies, and intracranial hemorrhages. These pathologic findings correlate with behavior alterations such as impaired spatial and cognitive performance and coordination (Desnoyers et al., 1984; de Lanerolle et al., 2011; Lu et al., 2012).

One important clinical issue is identifying patients with mild TBI as they often do not know that they have been injured. The most commonly used screening tool is that devised by the Defense and Veterans Brain Injury Center (DVBIC) called the military acute concussion evaluation (MACE) (DVBIC, 2008, 2011, 2013). When using this tool, in addition to a neurologic examination, cognitive testing of attention, orientation, immediate memory, and memory recall using the Standardized Assessment of Concussion are used to help identify patients. This is not specific for explosive blast TBI as it can be used for TBI from any etiology. Formal diagnosis is made by an advanced medical care provider who must integrate all available clinical data including history and general physical examination, as TBI remains a clinical diagnosis.

Clinically, explosive TBI can result in mild self-limited symptoms such as headache, dizziness, or confusion. At higher levels of exposures, the clinical condition may become more severe, such as altered level of sensorium with focal neurologic deficits to coma and death (Ling et al., 2009; Ling and Ecklund, 2011). A possibly unique clinical consequence of explosive-blast TBI is prolonged symptomatic cerebral vasospasm risk. This develops soon after blast exposure (with a day or 2) and may persist up to 4 weeks (Armonda et al., 2006). This differs from vasospasm encountered following aneurysmal subarachnoid hemorrhage which lasts only 2 weeks.

Mild TBI cases typically respond well to reassurance, rest, and symptoms management. The U.S. military experience has been that early intervention, i.e., within 24 hours of the event, has the highest likelihood of complete recovery. Thus, the Veterans Administration (VA) and Department of Defense (DoD) and the American Academy of Neurology created clinical practice guidelines for management of mild TBI or concussion (VA/DOD, 2009; Giza et al., 2013). Moderate-severe TBI patients require advance neurotrauma care, often in an intensive care setting, and possibly neurosurgical intervention as well. The clinical practice guidelines for managing moderate to severe TBI developed by the Congress of Neurologic Surgeons and the American Association of Neurological Surgery help guide clinical care of these more seriously injured patients (The Brain Foundation, 2007).

4.7.2 Models of Blast TBI

Blast emerged as the predominant cause of casualties in Operation Iraqi Freedom (OIF) and Operation Enduring Freedom (OEF), with the majority of these injuries resulting from blast propagated by improvised explosive devices (IEDs) (Defense Manpower Data Center; Military Health System; Tanielian and Jaycox, 2008; Gulf War and Health Institute of Medicine, 2009). It is likely that the high incidence of blast-induced traumatic brain injury (bTBI) is in part due to the widespread use of body armor as well as unprecedented rapid medevac and advanced treatment at forward positions, which have allowed war-fighters to survive blasts that would otherwise be fatal. Blast is also now a significant civilian medical concern due to widespread use of IEDs in terrorist and insurgent activities (Cooper et al., 1983; Leibovici et al., 1996; Lucci 2006; Turégano-Fuentes et al., 2008; Bochicchio et al., 2008). Collectively, these casualty statistics highlight the need to advance medical care targeting bTBI, in part through the critical consideration of preclinical modeling requirements.

A chemical explosion is the release of energy resulting from the nearly instantaneous conversion of a reactive solid or liquid into gases at extremely high pressure and temperature by the process of detonation (Wightman and Gladish, 2001;

Leung et al., 2008; Champion et al., 2009; Ling et al., 2009; Benzinger et al., 2009; Cernak et al., 2010). The gaseous detonation products expand rapidly and compress the surrounding medium (such as air, water, or soil), generating a blast wave. Thermodynamic "work" is done by the expanding gas on the surrounding medium, and transient compressive and translational forces are imparted to objects exposed to the blast wave. In air, the most distinctive feature of the blast wave is the supersonic shock front, which is the leading element of the pressure disturbance, through which there is a nearly instantaneous "step" increase in all gas-dynamic conditions of the air (pressure, density, flow velocity, and temperature). With time and distance, the peak pressure and velocity of the blast wave weaken. Near the source of the explosion, the overpressure decreases approximately with the inverse cube of the distance from the origin, but at greater distances it decays inversely with distance as an acoustic wave (Leung et al., 2008; Bass et al., 2012). For an idealized explosion in the free field, the time record of static overpressure can be approximated by the "modified Friedlander equation" (U.S. Department of Defense, 2008), which has the form of an initial instantaneous rise in pressure followed by an exponential-like decay that typically continues to fall below ambient pressure and produce a negative overpressure phase, thereafter rising back to baseline (Wightman and Gladish, 2001; Leung et al., 2008; Benzinger et al., 2009; Ling et al., 2009; Cernak et al., 2010). The shock wave is often characterized by its peak amplitude, duration, and impulse, which is the integration of overpressure with respect to time (i.e., area under the pressure-time curve). Rise time and decay rate are also important parameters affecting viscoelastic biologic materials. Real-world effects due to irregular charge shape, buried charges, nonideal detonations, and shock wave reflections due to surrounding structures yield considerably more complex waveforms. Although it is rare that victims of blast are exposed to the idealized "Friedlander" shock waveform, it represents an important reference benchmark blast exposure condition for research studies (Ritzel et al., 2011; Bass et al., 2012).

Despite its prominence in biomedical blast literature, it is important to recognize that the static overpressure condition described by the Friedlander waveform is but one parameter of a traveling shock wave in which all the gas-dynamic conditions vary (Ritzel et al., 2011) and affect damage and injury. The dynamic pressure (blast wind) is a measure of the kinetic energy imparted to the air as it is traversed by the shock wave (Wightman and Gladish, 2001; Leung et al., 2008; Benzinger et al., 2009; Ling et al., 2009; Cernak et al., 2010; Long et al., 2010; Bass et al., 2012). Both the static and dynamic pressure conditions determine the blast loading imparted to an object (Long et al., 2010). Thus, an individual exposed to a blast wave will be subjected to a step increase in static pressure as well as an abrupt wind of increased air density (Wightman and Gladish, 2001; Leung et al., 2008; Benzinger et al., 2009; Ling et al., 2009; Cernak et al., 2010). The loading imparted to an object from a particular incident shock wave is highly nonuniform as the shock wave reflects and diffracts around its shape. Albeit compressible flow, the blast-wave encounter with a target is somewhat analogous to that of a breaking water wave striking an obstruction such as a pier piling, whereby high reflected pressures develop on frontal areas and the wave "wraps around" the shape. In simplest terms, the head is abruptly enveloped by high pressure within a millisecond without necessarily imparting significant acceleration or inflicting overt external wounding as with a localized impact of comparable severity. The distinction regarding the incident blast flow conditions (i.e., static and dynamic pressures) and target loading have important implications with regard to the mechanisms for blast injury, imparted loading, and cellular stresses as well as the proper simulation of blast in the laboratory (Benzinger et al., 2009).

At present, there are many uncertainties and competing hypotheses concerning virtually all aspects of bTBI, including the means by which blast waves impart stresses through the brain (e.g., direct transcranial propagation [Clemedson 1956; Clemedson and Pettersson, 1953; Clemedson et al., 1956; Saljo et al., 2008], skull deformation [Bolander et al., 2011; Mediavilla Varas et al., 2011], vascular transmission [Cernak et al., 2001, 2010]), injury etiology (e.g., shock or compression waves, cavitation, head acceleration [Nakagawa et al., 2011; Goldstein et al., 2012; Ganpule et al., 2013]), blast wave effects on brain function (e.g., interplay with and differentiation from PTSD and other psychiatric disorders [Hoge et al., 2008, 2009; Stein and McAllister, 2009; Rosenfeld and Ford, 2010; Wilk et al., 2010]), injury etiology (e.g., shock or compression waves, cavitation, head acceleration [Nakagawa et al., 2011; Goeller et al., 2012; Goldstein et al., 2012; Ganpule et al., 2013]), blast wave effects on brain function (e.g., interplay with and differentiation from PTSD and other psychiatric disorders [Hoge et al., 2008, 2009; Stein et al., 2009; Rosenfeld et al., 2010; Wilk et al., 2010]), and cellular changes resulting from single or repeated blast exposures yielding long-term sequelae such as chronic traumatic encephalopathy (Goldstein et al., 2012; Lakis et al., 2013; Petrie et al., 2013; Weiner et al., 2013). Additionally, although it is well established that the amplitude and duration of the blast wave (i.e., the time that an object in the path of the shock wave is subjected to the pressure effects) depend on the nature of the explosive and the distance from the point of detonation (Clemedson 1956), the particular shock-wave parameters (i.e., rise time, peak positive amplitude, duration, impulse, or negative phase) that most greatly influence the extent of brain injury remain unknown, and along with the other considerations identified above, require careful evaluation through appropriate and validated preclinical bTBI modeling.

Animal models of other forms of closed and penetrating TBI have been developed and validated based on comparison with substantial human data to demonstrate that they reproduce features relevant to human injuries (Hicks et al., 2010). These models were not intended to directly replicate the circumstances in which human TBI occurs with the substitution of animal subjects; rather, they generate loading conditions by other means to *simulate* the mechanisms thought to underlie human head injury (Hicks et al., 2010). Much less is currently known about the neuropathologic outcomes and biomechanical underpinnings of blast TBI in humans to provide similar guidelines for model development. Human blast exposure conditions can be complex, and despite recent advances in the development of blast sensors and acceleration gauges in helmets (Rigby et al., 2011; Chu et al., 2012), to date they generally remain poorly described, and validated models that replicate the biomechanical and neurobiologic complexity of bTBI encountered on the battlefield remain a work in progress. Moreover, because of a lack of understanding of blast physics and the absence of well-defined biomechanical and injury criteria, experimental approaches have varied widely, making comparison and interpretation of the experimental results extremely challenging (Cernak et al., 2010; Nakagawa et al., 2011). In many cases, although brain injuries and disrupted outcomes have been described, they may have little if any relevance to bTBI. Even when BOP has been incorporated into the bTBI evaluation, exposure conditions have often been incompletely or inappropriately described; when maximal peak pressures are compared as a basis to establish and judge injury severities across models and laboratories, they have ranged over several orders of magnitude (Nakagawa et al., 2011). This lack of common description with wide-ranging experimental approaches has impeded meaningful comparisons and progress, yielding a somewhat nascent state of current research, to date. Because of the unique physics of the insult and the currently limited understanding of the critical parameters underlying human blast TBI, credible preclinical models of blast TBI have generally placed animal subjects in recreations of blast exposure conditions, using open field or closed structural exposures to explosive detonations, shock tube testing, in which compressed gases are used to generate a shock wave, and blast tube testing, in which an explosive charge is used to generate a shock wave.

4.7.3 Free-Field Explosives Testing

In free-field testing, blast waves are generated using explosives in an open field (Cernak et al., 1991; Richmond 1991; Axelsson et al., 2000; Saljo and Hamberger, 2004; Bauman et al., 2009; Rubovitch et al., 2011; Lu et al., 2012). Structures or vehicle mock-ups can be employed to better simulate actual war-fighting scenarios entailing exposures to complex blast waveforms (Bauman et al., 2009). Placement of the charge can be in the ground, yielding ejecta, or above ground, causing the "Mach reflection" phenomena in which the peak overpressure and impulse of the blast can far exceed twice that of the incident wave. For idealized free-field blast, the peak overpressure and positive-phase duration of the exposure are determined by the size of the charge and standoff distance (U.S. Department of Defense, 2008).

Free-field explosive blast testing can closely replicate real-world blast conditions, and large-scale experiments of this nature are historically important, having provided much fundamental data to study lung injuries and to determine injury and mortality thresholds (Bowen et al., 1965, 1968). Field experiments are also more amenable to tests with larger animal species that may better approximate humans. However, blast field experiments require specialized test sites with qualified personnel, and they are expensive, difficult to schedule for systematic studies, subject to uncontrolled environmental conditions, and challenging for application of advanced diagnostics.

4.7.4 Shock and Blast Tube Testing

In its simplest form, a shock tube is cylindrical pipe that is separated into a "driver" section charged with high-pressure gas and a low-pressure test or "driven" section which can either be capped or open to ambient air conditions (Celander et al., 1955; Ritzel et al., 2011; Bass et al., 2012). The two sections are separated by a frangible diaphragm such as polyester plastic or thin metal foil. Upon rupture of the diaphragm either by perforation or simple overpressurization, the high-pressure gas expands and drives a shock wave that propagates through the test section. Anesthetized experimental subjects are secured in the test section in holders that are typically designed to minimize movement in response to blast and, thus, reduce or control tertiary effects of the shock or blast wave (Cernak et al., 2010).

Although this concept appears straightforward, the wave dynamics developed in a conventional shock tube are complex, and the particular waveforms produced at a test position in the tube are dependent on the measurement location along with details of the configuration such as the type of driver gas, its pressure and temperature, the relative lengths of the driver and driven sections, and the end conditions such as being open or closed (Celander et al., 1955; Ritzel et al., 2011; Bass et al., 2012). Through careful adjustment of driver length, driver gas, driver pressure, and measurement location along the axis of the tube, it is possible to produce shock-wave conditions approximating those of the positive phase of the free-field explosive blast. However, it is important to note that the shock wave developed in a conventional shock tube is a composite of several interacting wave systems unlike explosive blast. A shock tube without a reflection eliminator at its end will expose test specimens to multiple wave reflections uncharacteristic of explosive blast. It is also important to minimize obstruction presented by the experimental subject or test apparatus within the tube; a test article should not block more than 20% of the cross-sectional area of the test section.

Nearby or immediately outside the mouth of an open shock tube, flow conditions are unstable and complex involving development of an "end-jet" with high flow gradients such that slight variations in position impart large changes in static and dynamic pressure conditions (Long et al., 2010; Ritzel et al., 2011). As a consequence, despite the appearance of injuries and pathophysiologic responses, experimental subjects placed outside the mouth of the shock are exposed to very different loading and injury phenomena than those resulting from explosive blast, and might also yield disparate neuropathologic changes. It is therefore essential that positioning and exposure conditions be carefully monitored and considered to validate the fidelity of the experimental model. Interestingly, a new approach to laboratory blast simulation, based on replicating the actual wave-dynamics of explosive blast by means of specially shaped shock tube geometry, offers a means to improve and standardize blast-test methodology across laboratories (Ritzel et al., 2011).

Combustion-based drivers or blast tubes offer advantages in that high driver pressures and volumes can be efficiently generated, no diaphragm system is required, and some manner of "blast-like" pressure profile can be directly transmitted into the test section (Engin 1969; Ritzel et al., 2011; Bass et al., 2012). Combustion drivers initially entailed use of distributed high-explosive or propellants being set within a heavy-walled small-diameter driver section expanding into a test section of wider diameter to accommodate test articles. However, this approach yields combustion products and residue that are dispersed into the test section; also, due to the charge and driver configuration strong transverse waves can be generated within the driver or upon expansion into the wider test section. Furthermore, explosive handling and storage impose appreciable costs, security, and safety requirements.

4.7.5 Species Selection and Scaling Concerns

Much of the biomedically related modeling of bTBI has been conducted with laboratory rats and swine (Nakagawa et al., 2011). In addition to the typical neuroanatomical considerations that influence species selection for preclinical TBI modeling (e.g., brain geometry and mass, white matter/gray matter ratio, gyrencephalic versus lissencephalic cortical surfaces), because of the need to employ BOP exposures with bTBI models, several additional species and scaling concerns must be addressed that consider both biologic and blast-physics aspects of the modeling. Unlike any other form of head trauma, extreme stresses can be induced by shock-wave exposure without significant global acceleration. Indeed, the process of the loading by shock-wave reflection and diffraction around the skull, and consequently how, where, and what stresses are imparted within the brain remain the foremost phenomena to resolve. The skull acts as a "transfer function" for the external loading to a particular stress-wave condition within the brain. The rat skull evolved to be pliable for access through tight spaces, whereas the swine skull evolved as exceedingly resilient for combat and as a foundation for tusks; clearly neither of these would at first consideration be expected to replicate the stress transfer to the human brain for a given loading, whether blast or otherwise. As noted by Nakagawa et al. (2011), unlike the CCI and fluid percussion models, the nature of the skull in bTBI studies influences blast-induced damage. The impact of the shock front on the skull, the transfer of the compression wave across the skull, and the movement of this wave within the calvarium are all affected by the size, geometry, and thickness of the skull, and also probably by the physical characteristics of bone, such as the ratio of compact to trabecular bone, and the degree of mineralization (Bauman et al., 2009). The extremely small, thin, ellipsoid-shaped rat skull, and the more angular, wedge-shaped, thicker, and more mineralized swine skull, present significant challenges for modeling the events that ultimately result in blast injury to the human.

As a "first-order" simplification, the human head might be modeled as a fluid-filled elastic spherical shell so as to resolve its mechanical response dynamics through development of a mathematical model of its natural frequencies (Engin, 1969). The dynamic structural response of the coupled skull/brain system (and hence imparted stress to the brain) is affected by the shock-wave loading that is a complex process in itself; the shock-wave loading may be specially coupled to the structural dynamics of a particular skull type. For example, a shock wave of certain strength will pass over the human head in a prescribed time, which may be an important factor in the response dynamics of the skull/brain system, possibly related to its natural period of vibration. Therefore, to replicate response dynamics with respect to natural frequency, the material and thickness of the elastic shell representing the skull will have to be altered from the actual system. For phenomena governed by physical processes, such as the dynamic mechanical stress imparted in a structure from loading, there are "rules" for the development of a physical model, and criteria relevant to one phenomenon may have to be chosen over another as priority to simulate.

As illustration, Figure 4.2 shows a simple cubic block of material of uniform density under the force of gravity. A 1/4-scaled version of the same block will have

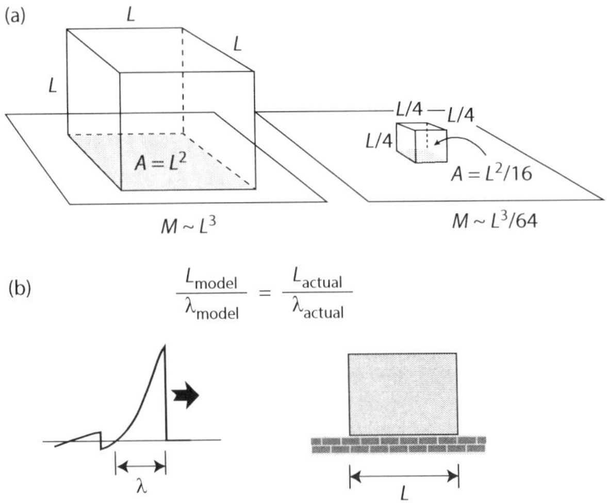

FIGURE 4.2 (a,b) Scaling considerations for animal models of bTBI and wavelength scaling.

a volume, hence weight, 1/64th that of the original. However, the surface area of the model in contact with the ground has decreased to only 1/16th. Therefore, the bearing weight per unit area or compressive stress from self-weight in the scaled system is one-quarter that of the original system. Although the objects have been scaled as physical replicas, they do not replicate the same bearing stresses. This aspect of scaling of structures explains why the leg-bones of a modern gecko are not "scaled" versions of those for a massive dinosaur having the same shape. Similarly, drag forces on a sphere relate to its cross-sectional area ($\sim r^2$) whereas its mass relates to volume ($\sim r^3$); hence, a grain of sand is blown by wind, whereas a boulder of the same material and shape subjected to the same wind will not. Bodily acceleration from a blast is a function of the applied force per unit mass for which the total force is a function of presented cross-sectional area. Therefore, a replica scale model of the same material density which is $\mathbf{1/x}$ the size of the actual case will have $\boldsymbol{x}$-fold the acceleration of the actual case for the same pressure loading.

Because blast loading over a target shape is a complex process, some blast experiments employ nonresponding replica scale models exclusively to resolve the spatially and time-variant pressure loading over the surfaces without attempting to resolve the structure's response dynamics such as imparted stresses or bodily acceleration. Although "replica" scale modeling (i.e., linear scaling of all geometric features) will give the appearance of simulating the actual case, potentially critical response behavior will likely not match or scale accordingly. Consider the natural frequency of a simple circular plate $\boldsymbol{f} \sim \sqrt{(\boldsymbol{t}^3/\boldsymbol{d}^4)}$, where $\boldsymbol{t}$ is the plate thickness and $\boldsymbol{d}$ its diameter. For all other factors such as plate material and boundary conditions being the same, maintaining the ratio $\boldsymbol{t/d}$ as in replica scaling will result in differences in response amplitude and frequency of the model system to dynamic loads. The relevance here is that even should an animal head be assumed to be a scaled replica of the human head (which it certainly is not), from a biomechanical perspective it would not be expected to impart the same stress condition to the brain for a particular loading. However, even if not directly replicating the real-world system as a direct analog, scale models to investigate the basic phenomenology of an event can nevertheless be extremely useful to validate computational model simulations that can be applied to the real system with greater confidence.

4.7.6 Conclusion

Many salient features of the biophysics of BOP can be recreated in the laboratory for biomedical research, and some manner of validated biologic response is ultimately required to establish an injury model. Human brain injury from a blast is likely governed by the particular scale, anatomy, and physiology of the human; one must therefore approach preclinical models judiciously to investigate bTBI phenomena. Rules developed for modeling responses of structural/mechanical systems merit consideration in the development of a biomechanical model for blast TBI. Due to scaling restrictions, an approach that combines animal, biofidelic physical replicas, postmortem human subjects, and computational finite-element models will likely be required. If the blast-induced brain stress conditions for a test species such as the rat can be resolved and associated with cognitive or functional deficits in a bTBI model, using those results to look for similar stress conditions in human cadaveric or validated biofidelic headforms is possible, even if the comparable local stress conditions may not result from the same blast exposure. Conversely, when human injury biomechanics and imparted brain stress have been resolved by use of biofidelic physical replicas of the head supplemented with computational simulations or cadaveric studies, animal experiments can be devised to appropriately replicate damaging stress conditions in rats and other species through tailored shock-wave profiles.

REFERENCES

Section 4.3: In Vitro Models

Arifin, M.Z., A. Faried, M.N. Shahib, K. Wiriadisastra, T. Bisri, Inhibition of activated NR2B gene- and caspase-3 protein-expression by glutathione following traumatic brain injury in a rat model. *Asian J Neurosurg*, 2011. 6(2), p. 72–77.

Arun, P., J. Spadaro, J. John, R.B. Gharavi, T.B. Bentley, M.P. Nambiar, Studies on blast traumatic brain injury using in-vitro model with shock tube. *NeuroReport*, 2011. 22, p. 379–384.

Blank-Reid, C., P.C. Reid, Penetrating trauma to the head. *Crit Care Nurs Clin North Am*, 2000. 12(4), p. 477–487.

Blizzard, C.A., A.E. King, J. Vickers, T. Dickson, Cortical murine neurons lacking the neurofilament light chain protein have an attenuated response to injury in vitro. *J Neurotrauma*, 2013. 30, p. 1908–1918.

Bottlang, M., M.B. Sommers, T.A. Lusardi, J.J. Miesch, R.P. Simon, Z.G. Xiong, Modeling neural injury in organotypic cultures by application of inertia-driven shear strain. *J Neurotrauma*, 2007. 24(6), p. 1068–1077.

Cater, H.L., D.P. Gitterman, S.M. Davis, C.D. Benham, B. Morrison III, L.E. Sundstrom, Stretch-induced injury in organotypic hippocampal slice cultures reproduces in vivo post-traumatic neurodegeneration: Role of glutamate receptors and voltage-dependent calcium channels. *J Neurochem*, 2007. 101(2), p. 434–447.

Cater, H.L., L.E. Sundstrom, B. Morrison III, Temporal development of hippocampal cell death is dependent on tissue strain but not strain rate. *J Biomech*, 2006. 39(15), p. 2810–2818.

Choo, A.M. et al., Antagonism of purinergic signalling improves recovery from traumatic brain injury. *Brain*, 2013. 136, p. 65–80.

Church, A.J., R.D. Andrew, Spreading depression expands traumatic injury in neocortical brain slices. *J Neurotrauma*, 2005. 22(2), p. 277–290.

Cullen, D.K., V.N. Vernekar, M.C. LaPlaca, Trauma-induced plasmalemma disruptions in three-dimensional neural cultures are dependent on strain modality and rate. *J Neurotrauma*, 2011. 28, p. 2219–2233.

Dollé, J., B. Morrison III, R.S. Schloss, M.L. Yarmush, An organotypic uniaxial strain model using microfluidics. *Lab Chip*, 2013. 13, p. 432–442.

Effgen, G.B., C.D. Hue, E. Vogel III, M.B. Panzer, C.R. Bass, D.F. Meaney, B. Morrison III, A multiscale approach to blast neurotrauma modeling: Part II: Methodology for inducing blast injury to in vitro models. *Front Neurol*, 2012. 3(23), p. 1–10.

Effgen, G.B., E.W. Vogel III, K.A. Lynch, A. Lobel, C.D. Hue, D.F. Meaney, C.R. Bass, B. Morrison III, Isolated primary blast alters neuronal function with minimal cell death in organotypic hippocampal slice cultures. *J Neurotrauma*. 2014. 31, p. 1202–1210.

Elkin, B.S., B. Morrison III, Region-specific tolerance criteria for the living brain. *Stapp Car Crash J*, 2007. 51, p. 127–138.

Ellis, E.F., J.S. McKinney, K.A. Willoughby, S. Liang, J.T. Povlishock, A new model for rapid stretch-induced injury of cells in culture: Characterization of the model using astrocytes. *J Neurotrauma*, 1995. 12(3), p. 325–339.

Ellis, E.F., K.A. Willoughby, S.A. Sparks, T. Chen, S100B protein is released from rat neonatal neurons, astrocytes, and microglia by in vitro trauma and anti-S100 increases trauma-induced delayed neuronal injury and negates the protective effect of exogenous S100B on neurons. *J Neurochem*, 2007. 101(6), p. 1463–1470.

Faden, A.I., V.A. Movsesyan, S.M. Knoblach, F. Ahmed, I. Cernak, Neuroprotective effects of novel small peptides in vitro and after brain injury. *Neuropharmacology*, 2005. 49, p. 410–424.

Ferrario, C.R., B.O. Ndukew, J. Ren, L.S. Satin, P.B. Goforth, Stretch injury selectively enhances extrasynaptic GluN2B-containing NMDA receptor function in cortical neurons. *J Neurophysiol*, 2013. 110(1), p. 131–140.

Folkerts, M.M., R.F. Berman, J.P. Muizelaar, J.A. Rafols, Disruption of MAP-2 immunostaining in rat hippocampus after traumatic brain injury. *J Neurotrauma*, 1998. 15(5), p. 349–363.

Gennarelli, T.A., L.E. Thibault, J.H. Adams, D.I. Graham, C.J. Thompson, R.P. Marcincin, Diffuse axonal injury and traumatic coma in the primate. *Ann Neurol*, 1982. 12, p. 564–574.

Goldstein, L.E. et al., Chronic traumatic encephalopathy in blast-exposed military veterans and a blast neurotrauma mouse model. *Sci Transl Med*, 2012. 4, p. 134–160.

Hardingham, G.E., Y. Fukunaga, H. Bading, Extrasynaptic NMDARs oppose synaptic NMDARs by triggering CREB shut-off and cell death pathways. *Nat Neurosci*, 2002. 5(5), p. 405–414.

Hardy, W.N. et al., A study of the response of the human cadaver head to impact. *Stapp Car Crash J*, 2007. 51, p. 17–80.

Harris, K., S.P. Armstrong, R. Pires-Campos, N.P. Franks, R. Dickinson, Neuroprotection against traumatic brain injury by xenon, but not argon, is mediated by inhibition at the *N*-Methyl-D-Aspartate receptor glycine site. *Anesthesiology*, 2013. 119(5), p. 1137–1148.

Hogg, S., C. Perron, P. Barneoud, D.J. Sanger, P.C. Moser, Neuroprotective effect of eliprodil: Attenuation of a conditioned freezing deficit induced by traumatic injury of the right parietal cortex in the rat. *J Neurotrauma*, 1998. 15(7), p. 545–553.

Holbourn, A.H.S., The mechanics of brain injuries. *Br Med Bull*, 1945. 3, p. 147–149.

Hue, C.D., S. Cao, S.F. Haider, K.V. Vo, G.B. Effgen, E. Vogel III, M.B. Panzer, C.R.D. Bass, D.F. Meaney, B. Morrison III, Blood-brain barrier dysfunction after primary blast injury in vitro. *J Neurotrauma*, 2013. 30, p. 1652–1663.

Javouhey, E., A. Guerin, M. Chiron, Incidence and risk factors of severe traumatic brain injury resulting from road accidents: A population-based study. *Accid Anal Prev*, 2006. 38, p. 225–233.

Kleiven, S.K., W.N. Hardy, Correlation of an FE model of the human head with local brain motion—Consequences for injury prediction. *Stapp Car Crash J*, 2002. 46, p. 123–144.

Laird, M.D., J.S. Shields, S. Sukumari-Ramesh, D.E. Kimbler, R.D. Fessler, B. Shakir, P. Youssef, N. Yanasak, J.R. Vender, K.M. Dhandapani, High mobility group box protein-1 promotes cerebral edema after traumatic brain injury via activation of toll-like receptor 4. *Glia*, 2014. 62, p. 26–38.

LaPlaca, M.C., D.K. Cullen, J.J. McLoughlin, R.S. Cargill, High rate shear strain of three-dimensional neural cell cultures: A new in vitro traumatic brain injury model. *J Biomech*, 2005. 38(5), p. 1093–1105.

Maas, A.I., N. Stocchetti, R. Bullock, Moderate and severe traumatic brain injury in adults. *Lancet Neurol*, 2008. 7(8), p. 728–741.

Margulies, S.S., L.E. Thibault, T.A. Gennarelli, Physical model simulations of brain injury in the primate. *J Biomech*, 1990. 23(8), p. 823–836.

Meaney, D.F., L.E. Thibault, Physical model studies of cortical brain deformation in response to high strain rate inertial loading. *Proc Int Conf Biomech Impacts*, 1990, p. 215–224.

Morrison III, B., B.S. Elkin, J. Dolle, M.L Yarmush, In vitro models of traumatic brain injury. *Ann Rev Biomed Eng*, 2011. 13, p. 91–126.

Morrison III, B., D.F. Meaney, T.K. McIntosh, Mechanical characterization of an in vitro device to quantitatively injure living brain tissue. *Ann Biomed Eng*, 1998. 26(3), p. 381–390.

Morrison III, B., H.L. Cater, C.D. Benham, L.E. Sundstrom, An in vitro model of traumatic brain injury utilising two-dimensional stretch of organotypic hippocampal slice cultures. *J Neurosci Meth*, 2006. 150(2), p. 192–201.

Morrison III, B., K.E. Saatman, D.F. Meaney, T.K. McIntosh, In vitro central nervous system models of mechanically induced trauma: A review. *J Neurotrauma*, 1998. 15(11), p. 911–928.

Mukhin, A.G., S.A. Ivanova, J.W. Allen, A.I. Faden, Mechanical injury to neuronal/glial cultures in microplates: Role of NMDA receptors and pH in secondary neuronal cell death. *J Neurosci Res*, 1998. 51(6), p. 748–758.

Mukhin, A.G., S.A. Ivanova, S.M. Knoblach, A.I. Faden, New in vitro model of traumatic neuronal injury: Evaluation of secondary injury and glutamate receptor-mediated neurotoxicity. *J Neurotrauma*, 1997. 14(9), p. 651–663.

Murphy, E.J., L.A. Horrocks, A model for compression trauma: Pressure induced injury in cell cultures. *J Neurotrauma*, 1993. 10, p. 431–444.

Ng, J.M.J. et al., Transcriptional insights on the regenerative mechanics of axotomized neurons in vitro. *J Cell Mol Med*, 2012. 4, p. 789–811.

Okiyama, K., D.H. Smith, W.F. White, K. Richter, T.K. McIntosh, Effects of the novel NMDA antagonists CP-98,113, CP-101,581 and CP-101,606 on cognitive function and regional cerebral edema following experimental brain injury in the rat. *J Neurotrauma*, 1997. 14(4), p. 211–222.

Panzer, M.B., B.S. Myers, B.P. Capehart, C.R. Bass, Development of a finite element model for blast brain injury and the effects of CSF cavitation. *Ann Biomed Eng*, 2012. 40(7), p. 1530–1544.

Posmantur, R., A. Kampfl, R. Siman, J. Liu, X. Zhao, G.L. Clifton, R.L. Hayes, A calpain inhibitor attenuates cortical cytoskeletal protein loss after experimental traumatic brain injury in the rat. *Neuroscience*, 1997. 77(3), p. 875–888.

Rafaels, K., C.R. Bass, R.S. Salazar, M.B. Panzer, W. Woods, S. Feldman, T. Cummings, B. Capehart, Survival risk assessment for primary blast exposures to the head. *J Neurotrauma*, 2011. 28, p. 2319–2328.

Ravin, R., P.S. Blank, A. Steinkamp, S.M. Rappaport, N. Ravin, L. Bezrukov, H. Guerrero-Cazares, A. Quinones-Hinojosa, S.M. Bezrukov, J. Zimmerberg, Shear forces during blast, not abrupt changes in pressure alone, generate calcium activity in human brain cells. *PLoS One*, 2012. 7(6), p. e39421.

Rubovitch, V., M. Ten-Bosch, O. Zohar, C.R. Harrison, C. Tempel-Brami, E. Stein, B.J. Hoffer, C.D. Balaban, S. Schreiber, W. Chiu, C.G. Pick, A mouse model of blast-induced mild traumatic brain injury. *Exp Neurol*, 2011. 232, p. 280–289.

Saatman, K.E., H. Murai, R.T. Bartus, D.H. Smith, N.J. Hayward, B.R. Perri, T.K. McIntosh, Calpain inhibitor AK295 attenuates motor and cognitive deficits following experimental brain injury in the rat. *Proc Natl Acad Sci USA*, 1996. 93(8), p. 3428–3433.

Shahlaie, K., B.G. Lyeth, G.G. Gurkoff, J.P. Muizelaar, R.F. Berman, Neuroprotective effects of selective N-type VGCC blockade on stretch-injury-induced calcium dynamics in cortical neurons. *J Neurotrauma*, 2010. 27(1), p. 175–187.

Shepard, S.R., J.B.G. Ghajar, R. Giannuzzi, S. Kupferman, R.J. Hariri, Fluid percussion barotrauma chamber: A new in vitro model for traumatic brain injury. *J Surg Res*, 1991. 51(5), p. 417–424.

Shi, R., A.R. Blight, Compression injury of mammalian spinal cord in vitro and the dynamics of action potential conduction failure. *J Neurophysiol*, 1996. 76(3), p. 1572–1580.

Singh, P., S. Doshi, J.M. Spaethling, A.J. Hockenberry, T.P. Patel, D.M. Geddes-Klein, D.R. Lynch, D.F. Meaney, N-methyl-D-aspartate receptor mechanosensitivity is governed by C terminus of NR2B subunit. *J Biol Chem*, 2012. 287, p. 4348–4359.

Smith, D.H., J.A. Wolf, T.A. Lusardi, V.M. Lee, D.F. Meaney, High tolerance and delayed elastic response of cultured axons to dynamic stretch injury. *J Neurosci*, 1999a. 19(11), p. 4263–4269.

Smith, D.H., X.H. Chen, M. Nonaka, J.Q. Trojanowski, V.M. Lee, K.E. Saatman, M.J. Leoni, B.N. Xu, J.A. Wolf, D.F. Meaney, Accumulation of amyloid beta and tau and the formation of neurofilament inclusions following diffuse brain injury in the pig. *J Neuropathol Exp Neurol*, 1999b. 58(9), p. 982–992.

Tang-Schomer, M.D., A.R. Patel, P.W. Baas, D.H. Smith, Mechanical breaking of microtubules in axons during dynamic stretch injury underlies delayed elasticity, microtubule disassembly and axon degeneration. *FASEB J*, 2010. 24, p. 1401–1410.

Tang-Schomer, M.D., V.E. Johnson, P.W. Baas, W. Stewart, D.H. Smith, Partial interruption of axonal transport due to microtubule breakage accounts for the formation of periodic varicosities after traumatic axonal injury. *Exp Neurol*, 2012. 233, p. 364–372.

Tecoma, E.S., H. Monyer, M.P. Goldberg, D.W. Choi, Traumatic neuronal injury in vitro is attenuated by NMDA antagonists. *Neuron*, 1989. 2(6), p. 1541–1545.

Toulmond, S., A. Serrano, J. Benavides, B. Scatton, Prevention by eliprodil (SL 82.0715) of traumatic brain damage in the rat. Existence of a large (18 h) therapeutic window. *Brain Res*, 1993. 620(1), p. 32–41.

Wallis, R.A., K.L. Panizzon, Felbamate neuroprotection against CA1 traumatic neuronal injury. *Eur J Pharmacol*, 1995. 294(2–3), p. 475–482.

Weber, J.T., M. Lamont, L. Chibrikova, D. Fekkes, A.S. Vlug, P. Lorenz, P. Kreutzmann, J.E. Slemmer, Potential neuroprotective effects of oxyresveratrol against traumatic injury. *Eur J Pharmacol*, 2012. 680, p. 55–62.

Zhang, L., K.H. Yang, A.I. King, Comparison of brain responses between frontal and lateral impacts by finite element modeling. *J Neurotrauma*, 2001. 18(1), p. 21–30.

Section 4.4: Controlled Cortical Impact

Acosta, S.A., N. Tajiri, K. Shinozuka, H. Ishikawa, B. Grimmig, D. Diamond, P.R. Sanberg, P.C. Bickford, Y. Kaneko, C.V. Borlongan, Long-term upregulation of inflammation and suppression of cell proliferation in the brain of adult rats exposed to traumatic brain injury using the controlled cortical impact model. *PLoS One*, 2013. 8, p. e53376.

Adelson, P.D., P. Robichaud, R.L. Hamilton, P.M. Kochanek, A model of diffuse traumatic brain injury in the immature rat. *J Neurosurg.*, 1996. 85, p. 877–884.

Adelson, P.D., C.E. Dixon, P. Robichaud, P.M. Kochanck, Motor and cognitive functional deficits following diffuse traumatic brain injury in the immature rat. *J Neurotrauma*, 1997. 14, p. 99–108.

Adelson, P.D., C.E. Dixon, P.M. Kochanek, Long-term dysfunction following diffuse traumatic brain injury in the immature rat. *J Neurotrauma*, 2000. 17, p. 273–282.

Ajao, D.O. et al., Traumatic brain injury in young rats leads to progressive behavioral deficits coincident with altered tissue properties in adulthood. *J Neurotrauma*, 2012. 29, p. 2060–2074.

Al-Samsam, R.H., B. Alessandri, R. Bullock, Extracellular N-acetyl-aspartate as a biochemical marker of the severity of neuronal damage following experimental acute traumatic brain injury. *J Neurotrauma*, 2000. 17, p. 31–39.

Alessandri, B., A. Heimann, R. Filippi, L. Kopacz, O. Kempski, Moderate controlled cortical contusion in pigs: Effects on multi-parametric neuromonitoring and clinical relevance. *J Neurotrauma*, 2003. 20, p. 1293–1305.

Alwis, D.S., E.B. Yan, M.C. Morganti-Kossmann, R. Rajan, Sensory cortex underpinnings of traumatic brain injury deficits. *PLoS One*, 2012. 7, p. e52169.

Ansari, M.A., K.N. Roberts, S.W. Scheff, Oxidative stress and modification of synaptic proteins in hippocampus after traumatic brain injury. *Free Radic Biol Med*, 2008. 45, p. 443–452.

Baldwin, S.A., S.W. Scheff, Intermediate filament change in astrocytes following mild cortical contusion. *Glia*, 1996. 16, p. 266–275.

Baldwin, S.A., T. Gibson, C.T. Callihan, P.G. Sullivan, E. Palmer, S.W. Scheff, Neuronal cell loss in the CA3 subfield of the hippocampus following cortical contusion utilizing the optical disector method for cell counting. *J Neurotrauma*, 1997. 14, p. 385–398.

Bales, J.W., A.K. Wagner, A.E. Kline, C.E. Dixon, Persistent cognitive dysfunction after traumatic brain injury: A dopamine hypothesis. *Neurosci Biobehav Rev*, 2009. 33, p. 981–1003.

Baranova, A.I., E.P. Wei, Y. Ueda, M.M. Sholley, H.A. Kontos, J.T. Povlishock, Cerebral vascular responsiveness after experimental traumatic brain injury: The beneficial effects of delayed hypothermia combined with superoxide dismutase administration. *J Neurosurg*, 2008. 109, p. 502–509.

Bartnik-Olson, B.L., N.G. Harris, K. Shijo, R.L. Sutton, Insights into the metabolic response to traumatic brain injury as revealed by C NMR spectroscopy. *Front Neuroenergetics*, 2013. 5, 8.

Barzo, P., A. Marmarou, P. Fatouros, F. Corwin, J. Dunbar, Magnetic resonance imaging-monitored acute blood-brain barrier changes in experimental traumatic brain injury. *J Neurosurg*, 1996. 85, p. 1113–1121.

Barzo, P., A. Marmarou, P. Fatouros, K. Hayasaki, F. Corwin, Contribution of vasogenic and cellular edema to traumatic brain swelling measured by diffusion-weighted imaging. *J Neurosurg*, 1997. 87, p. 900–907.

Baskaya, M.K., A.M. Rao, A. Dogan, D. Donaldson, R.J. Dempsey, The biphasic opening of the blood-brain barrier in the cortex and hippocampus after traumatic brain injury in rats. *Neurosci Lett*, 1997. 226, p. 33–36.

Baskin, Y.K., W.D. Dietrich, E.J. Green, Two effective behavioral tasks for evaluating sensorimotor dysfunction following traumatic brain injury in mice. *J Neurosci Meth*, 2003. 129, p. 87–93.

Beaumont, A., A. Marmarou, A. Czigner, M. Yamamoto, K. Demetriadou, T. Shirotani, C.R. Marmarou, J.G. Dunbar, The impact-acceleration model of head injury: Injury severity predicts motor and cognitive performance after trauma. *Neurol Res*, 1999. 21, p. 742–754.

Beaumont, A., C. Marmarou, A. Marmarou, The effects of human corticotrophin releasing factor on motor and cognitive deficits after impact acceleration injury. *Neurol Res*, 2000. 22, p. 665–673.

Beaumont, A., K. Hayasaki, A. Marmarou, P. Barzo, P. Fatouros, F. Corwin, Contrasting effects of dopamine therapy in experimental brain injury. *J Neurotrauma*, 2001. 18, p. 1359–1372.

Bedi, S.S., P. Smith, R.A. Hetz, H. Xue, C.S. Cox, Immunomagnetic enrichment and flow cytometric characterization of mouse microglia. *J Neurosci Meth*, 2013. 219, p. 176–182.

Beer, R. et al., Temporal and spatial profile of caspase 8 expression and proteolysis after experimental traumatic brain injury. *J Neurochem*, 2001. 78, p. 862–873.

Bell, M.J., P.M. Kochanek, J.A. Carcillo, Z. Mi, J.K. Schiding, S.R. Wisniewski, R.S. Clark, C.E. Dixon, D.W. Marion, E. Jackson, Interstitial adenosine, inosine, and hypoxanthine are increased after experimental traumatic brain injury in the rat. *J Neurotrauma*, 1998. 15, p. 163–170.

Berman, R.F., B.H. Verweij, J.P. Muizelaar, Neurobehavioral protection by the neuronal calcium channel blocker ziconotide in a model of traumatic diffuse brain injury in rats. *J Neurosurg*, 2000. 93, p. 821–828.

Bilgen, M., A new device for experimental modeling of central nervous system injuries. *Neurorehabil Neural Repair*, 2005. 19, p. 219–226.

Bitto, A. et al., Protective effects of melanocortins on short-term changes in a rat model of traumatic brain injury. *Crit Care Med*, 2012. 40, p. 945–951.

Blanie, A., B. Vigue, D. Benhamou, J. Duranteau, T. Geeraerts, The frontal lobe and thalamus have different sensitivities to hypoxia-hypotension after traumatic brain injury: A microdialysis study in rats. *J Neurotrauma*, 2012. 29, p. 2782–2790.

Bolkvadze, T., A. Pitkanen, Development of post-traumatic epilepsy after controlled cortical impact and lateral fluid-percussion-induced brain injury in the mouse. *J Neurotrauma*, 2012. 29, p. 789–812.

Brody, D.L., D.C. Mac, C.C. Kessens, C. Yuede, M. Parsadanian, M. Spinner, E. Kim, K.E. Schwetye, D.M. Holtzman, P.V. Bayly, Electromagnetic controlled cortical impact device for precise, graded experimental traumatic brain injury. *J Neurotrauma*, 2007. 24, p. 657–673.

Bryan, R.M., Jr., L. Cherian, C. Robertson, Regional cerebral blood flow after controlled cortical impact injury in rats. *Anesth Analg*, 1995. 80, p. 687–695.

Buki, A., D.O. Okonkwo, J.T. Povlishock, Postinjury cyclosporin A administration limits axonal damage and disconnection in traumatic brain injury. *J Neurotrauma*, 1999. 16, p. 511–521.

Buki, A., D.O. Okonkwo, K.K. Wang, J.T. Povlishock, Cytochrome c release and caspase activation in traumatic axonal injury. *J Neurosci*, 2000. 20, p. 2825–2834.

Bye, N., S. Carron, X. Han, D. Agyapomaa, S.Y. Ng, E. Yan, J.V. Rosenfeld, M.C. Morganti-Kossmann, Neurogenesis and glial proliferation are stimulated following diffuse traumatic brain injury in adult rats. *J Neurosci Res*, 2011. 89, p. 986–1000.

Byrnes, K.R., D.J. Loane, B.A. Stoica, J. Zhang, A.I. Faden, Delayed mGluR5 activation limits neuroinflammation and neurodegeneration after traumatic brain injury. *J Neuroinflammation*, 2012. 9, 43.

Cernak, I., C. O'Connor, R. Vink, Activation of cyclo-oxygenase-2 contributes to motor and cognitive dysfunction following diffuse traumatic brain injury in rats. *Clin Exp Pharmacol Physiol*, 2001. 28, p. 922–925.

Cernak, I., S.M. Chapman, G.P. Hamlin, R. Vink, Temporal characterisation of pro- and anti-apoptotic mechanisms following diffuse traumatic brain injury in rats. *J Clin Neurosci*, 2002. 9, p. 565–572.

Cernak, I., R. Vink, D.N. Zapple, M.I. Cruz, F. Ahmed, T. Chang, S.T. Fricke, A.I. Faden, The pathobiology of moderate diffuse traumatic brain injury as identified using a new experimental model of injury in rats. *Neurobiol Dis*, 2004. 17, p. 29–43.

Chauhan, N.B., R. Gatto, M.B. Chauhan, Neuroanatomical correlation of behavioral deficits in the CCI model of TBI. *J Neurosci Methods*, 2010. 190, p. 1–9.

Chen, S., J.D. Pickard, N.G. Harris, Time course of cellular pathology after controlled cortical impact injury. *Exp Neurol*, 2003. 182, p. 87–102.

Cherian, L., C.S. Robertson, C.F. Contant, R.M. Bryan, Lateral cortical impact injury in rats: Cerebrovascular effects of varying depth of cortical deformation and impact velocity. *J Neurotrauma*, 1994. 11, p. 573–585.

Cherian, L., C.S. Robertson, J.C. Goodman, Secondary insults increase injury after controlled cortical impact in rats. *J. Neurotrauma*, 1996. 13, p. 371–383.

Clark, R.S., P.M. Kochanek, C.E. Dixon, M. Chen, D.W. Marion, S. Heineman, S.T. Dekosky, S.H. Graham, Early neuropathologic effects of mild or moderate hypoxemia after controlled cortical impact injury in rats. *J Neurotrauma*, 1997. 14, p. 179–189.

Clark, R.S. B., J.K. Schiding, S.L. Kaczorowski, D.W. Marion, P.M. Kochanek, Neutrophil accumulation after traumatic brain injury in rats: Comparison of weight drop and controlled cortical impact models. *J Neurotrauma*, 1994. 11, p. 499–506.

Clausen, F., T. Lorant, A. Lewen, L. Hillered, T lymphocyte trafficking: A novel target for neuroprotection in traumatic brain injury. *J Neurotrauma*, 2007. 24, p. 1295–1307.

Clausen, F., L. Hillered, J. Gustafsson, Cerebral glucose metabolism after traumatic brain injury in the rat studied by 13C-glucose and microdialysis. *Acta Neurochir (Wien)*, 2011. 153, p. 653–658.

Colicos, M.A., C.E. Dixon, P.K. Dash, Delayed, selective neuronal death following experimental cortical impact injury in rats: Possible role in memory deficits. *Brain Res*, 1996. 739, p. 111–119.

Csuka, E., V.H. Hans, E. Ammann, O. Trentz, T. Kossmann, M.C. Morganti-Kossmann, Cell activation and inflammatory response following traumatic axonal injury in the rat. *Neuro Report*, 2000. 11, p. 2587–2590.

Dalgard, C.L., J.T. Cole, W.S. Kean, J.J. Lucky, G. Sukumar, D.C. McMullen, H.B. Pollard, W.D. Watson, The cytokine temporal profile in rat cortex after controlled cortical impact. *Front Mol Neurosci*, 2012. 5, 6.

De Mulder, G., K. Van Rossem, J. Van Reempts, M. Borgers, J. Verlooy, Validation of a closed head injury model for use in long-term studies. *Acta Neurochir*, 2000. Suppl 76, p. 409–413.

Deng-Bryant, Y., M.L. Prins, D.A. Hovda, N.G. Harris, Ketogenic diet prevents alterations in brain metabolism in young but not adult rats after traumatic brain injury. *J Neurotrauma*, 2011. 28, p. 1813–1825.

Deng, Y., B.M. Thompson, X. Gao, E.D. Hall, Temporal relationship of peroxynitrite-induced oxidative damage, calpain-mediated cytoskeletal degradation and neurodegeneration after traumatic brain injury. *Exp Neurol*, 2007. 205, p. 154–165.

Dennis, A.M., M.L. Haselkorn, V.A. Vagni, R.H. Garman, K. Janesko-Feldman, H. Bayir, R.S. Clark, L.W. Jenkins, C.E. Dixon, P.M. Kochanek, Hemorrhagic shock after experimental traumatic brain injury in mice: Effect on neuronal death. *J Neurotrauma*, 2009. 26, p. 889–899.

Dhillon, H.S., D. Donaldson, R.J. Dempsey, M.R. Prasad, Regional levels of free fatty acids and Evans blue extravasation after experimental brain injury. *J Neurotrauma*, 1994. 11, p. 405–415.

Dixon, C.E., G.L. Clifton, J.W. Lighthall, A.A. Yaghami, R.L. Hayes, A controlled cortical impact model of traumatic brain injury in the rat. *J Neurosci Methods*, 1991. 39, p. 253–262.

Dixon, C.E., R.J. Hamm, W.C. Taft, R.L. Hayes, Increased anticholinergic sensitivity following closed skull impact and controlled cortical impact traumatic brain injury in the rat. *J Neurotrauma*, 1994. 11, p. 275–287.

Dixon, C.E., J. Bao, D.A. Long, R.L. Hayes, Reduced evoked release of acetylcholine in the rodent hippocampus following traumatic brain injury. *Pharmacol Biochem Behav*, 1996. 53, p. 679–686.

Dixon, C.E., C.G. Markgraf, F. Angileri, B.R. Pike, B. Wolfson, J.K. Newcomb, M.M. Bismar, A.J. Blanco, G.L. Clifton, R.L. Hayes, Protective effects of moderate hypothermia on behavioral deficits but not necrotic cavitation following cortical impact injury in the rat. *J Neurotrauma*, 1998. 15, p. 95–103.

Dixon, C.E., P.M. Kochanek, H.Q. Yan, J.K. Schiding, R.G. Griffith, E. Baum, D.W. Marion, S.T. Dekosky, One-year study of spatial memory performance, brain morphology, and cholinergic markers after moderate controlled cortical impact in rats. *J Neurotrauma*, 1999. 16, p. 109–122.

Donat, C.K., M.U. Schuhmann, C. Voigt, K. Nieber, W. Deuther-Conrad, P. Brust, Time-dependent alterations of cholinergic markers after experimental traumatic brain injury. *Brain Res*, 2008. 1246, p. 167–177.

Duhaime, A.C., J.V. Hunter, L.L. Grate, A. Kim, J. Golden, E. Demidenko, C. Harris, Magnetic resonance imaging studies of age-dependent responses to scaled focal brain injury in the piglet. *J Neurosurg*, 2003. 99, p. 542–548.

Dunn-Meynell, A.A., B.E. Levin, Lateralized effect of unilateral somatosensory cortex contusion on behavior and cortical reorganization. *Brain Res*, 1995. 675, p. 143–156.

Dunn-Meynell, A.A., B.E. Levin, Histological markers of neuronal, axonal and astrocytic changes after lateral rigid impact traumatic brain injury. *Brain Res*, 1997. 761, p. 25–41.

Durham, S.R., R. Raghupathi, M.A. Helfaer, S. Marwaha, A.C. Duhaime, Age-related differences in acute physiologic response to focal traumatic brain injury in piglets. *Pediatr Neurosurg*, 2000. 33, p. 76–82.

Duvdevani, R., R.L. Roof, Z. Fulop, S.W. Hoffman, D.G. Stein, Blood-brain barrier breakdown and edema formation following frontal cortical contusion: Does hormonal status play a role? *J Neurotrauma*, 1995. 12, p. 65–75.

Elliott, M.B., J.J. Jallo, R.F. Tuma, An investigation of cerebral edema and injury volume assessments for controlled cortical impact injury. *J Neurosci Methods*, 2008. 168, p. 320–324.

Engelborghs, K., J. Verlooy, J. Van Reempts, B. Van Deuren, V. de Van, M. Borgers, Temporal changes in intracranial pressure in a modified experimental model of closed head injury. *J Neurosurg*, 1998. 89, p. 796–806.

Engelborghs, K., M. Haseldonckx, J. Van Reempts, K. van Rossem, L. Wouters, M. Borgers, J. Verlooy, Impaired autoregulation of cerebral blood flow in an experimental model of traumatic brain injury. *J Neurotrauma*, 2000. 17, p. 667–677.

Foda, M.A., A. Marmarou, A new model of diffuse brain injury in rats: Part II: Morphological characterization. *J Neurosurg*, 1994. 80, p. 301–313.

Foley, L.M., A.M. Iqbal O'Meara, S.R. Wisniewski, H.T. Kevin, J.A. Melick, C. Ho, L.W. Jenkins, P.M. Kochanek, MRI assessment of cerebral blood flow after experimental traumatic brain injury combined with hemorrhagic shock in mice. *J Cereb Blood Flow Metab*, 2013. 33, p. 129–136.

Folkerts, M.M., R.F. Berman, J.P. Muizelaar, J.A. Rafols, Disruption of MAP-2 immunostaining in rat hippocampus after traumatic brain injury. *J Neurotrauma*, 1998. 15, p. 349–363.

Fox, G.B., L. Fan, R.A. LeVasseur, A.I. Faden, Sustained sensory motor and cognitive deficits with neuronal apoptosis following controlled cortical impact brain injury in the mouse. *J Neurotrauma*, 1998a. 15, p. 599–614.

Fox, G.B., L. Fan, R.A. LeVasseur, A.I. Faden, Effect of traumatic brain injury on mouse spatial and nonspatial learning in the Barnes circular maze. *J Neurotrauma*, 1998b. 15, p. 1037–1046.

Fromm, L., D.L. Heath, R. Vink, A.J. Nimmo, Magnesium attenuates post-traumatic depression/anxiety following diffuse traumatic brain injury in rats. *J Am Coll Nutr*, 2004. 23, p. 529S–533S.

Fujinaka, T., E. Kohmura, T. Yuguchi, T. Yoshimine, The morphological and neurochemical effects of diffuse brain injury on rat central noradrenergic system. *Neurol Res*, 2003. 25, p. 35–41.

Fujita, M., E.P. Wei, J.T. Povlishock, Effects of hypothermia on cerebral autoregulatory vascular responses in two rodent models of traumatic brain injury. *J Neurotrauma*, 2012. 29, p. 1491–1498.

Fukushima, M., S.M. Lee, N. Moro, D.A. Hovda, R.L. Sutton, Metabolic and histologic effects of sodium pyruvate treatment in the rat after cortical contusion injury. *J Neurotrauma*, 2009. 26, p. 1095–1110.

Geeraerts, T., C. Ract, M. Tardieu, O. Fourcade, J.X. Mazoit, D. Benhamou, J. Duranteau, B. Vigue, Changes in cerebral energy metabolites induced by impact-acceleration brain trauma and hypoxic-hypotensive injury in rats. *J Neurotrauma*, 2006. 23, p. 1059–1071.

Geeraerts, T., A. Friggeri, J.X. Mazoit, D. Benhamou, J. Duranteau, B. Vigue, Posttraumatic brain vulnerability to hypoxia-hypotension: The importance of the delay between brain trauma and secondary insult. *Intensive Care Med*, 2008. 34, p. 551–560.

Gilmer, L.K., M.A. Ansari, K.N. Roberts, S.W. Scheff, Age-related mitochondrial changes after traumatic brain injury. *J Neurotrauma*, 2010. 27, p. 939–950.

Giri, B.K., I.K. Krishnappa, R.M. Bryan, Jr., C. Robertson, J. Watson, Regional cerebral blood flow after cortical impact injury complicated by a secondary insult in rats. *Stroke*, 2000. 31, p. 961–967.

Goffus, A.M., G.D. Anderson, M. Hoane, Sustained delivery of nicotinamide limits cortical injury and improves functional recovery following traumatic brain injury. *Oxid Med Cell Longev*, 2010. 3, p. 145–152.

Goodman, J.C., L. Cherian, R.M. Bryan, C.S. Robertson, Lateral cortical impact injury in rats: Pathologic effects of varying cortical compression and impact velocity. *J Neurotrauma*, 1994. 11, p. 587–597.

Goren, S., N. Kahveci, T. Alkan, B. Goren, E. Korfali, The effects of sevoflurane and isoflurane on intracranial pressure and cerebral perfusion pressure after diffuse brain injury in rats. *J Neurosurg Anesthesiol*, 2001. 13, p. 113–119.

Guo, D., L. Zeng, D.L. Brody, M. Wong, Rapamycin attenuates the development of posttraumatic epilepsy in a mouse model of traumatic brain injury. *PLoS One*, 2013. 8, p. e64078.

Hall, E.D., P.G. Sullivan, T.R. Gibson, K.M. Pavel, B.M. Thompson, S.W. Scheff, Spatial and temporal characteristics of neurodegeneration after controlled cortical impact in mice: More than a focal brain injury. *J Neurotrauma*, 2005. 22, p. 252–265.

Hall, E.D., Y.D. Bryant, W. Cho, P.G. Sullivan, Evolution of post-traumatic neurodegeneration after controlled cortical impact traumatic brain injury in mice and rats as assessed by the de olmos silver and fluorojade staining methods. *J Neurotrauma*, 2008. 25, p. 235–247.

Hallam, T.M., C.L. Floyd, M.M. Folkerts, L.L. Lee, Q.Z. Gong, B.G. Lyeth, J.P. Muizelaar, R.F. Berman, Comparison of behavioral deficits and acute neuronal degeneration in rat lateral fluid percussion and weight-drop brain injury models. *J Neurotrauma*, 2004. 21, p. 521–539.

Hamlin, G.P., I. Cernak, J.A. Wixey, R. Vink, Increased expression of neuronal glucose transporter 3 but not glial glucose transporter 1 following severe diffuse traumatic brain injury in rats. *J Neurotrauma*, 2001. 18, p. 1011–1018.

Hamm, R.J., C.E. Dixon, D.M. Gbadebo, A.K. Singha, L.W. Jenkins, B.G. Lyeth, R.L. Hayes, Cognitive deficits following traumatic brain injury produced by controlled cortical impact. *J Neurotrauma*, 1992. 9, p. 11–20.

Han, X., J. Tong, J. Zhang, A. Farahvar, E. Wang, J. Yang, U. Samadani, D.H. Smith, J.H. Huang, Imipramine treatment improves cognitive outcome associated with enhanced hippocampal neurogenesis after traumatic brain injury in mice. *J Neurotrauma*, 2011. 28, p. 995–1007.

Hannay, H.J., Z. Feldman, P. Phan, A. Keyani, N. Panwar, J.C. Goodman, C.S. Robertson, Validation of a controlled cortical impact model of head injury in mice. *J Neurotrauma*, 1999. 16, p. 1103–1114.

Hans, V.H., T. Kossmann, P.M. Lenzlinger, R. Probstmeier, H.G. Imhof, O. Trentz, M.C. Morganti-Kossmann, Experimental axonal injury triggers interleukin-6 mRNA, protein synthesis and release into cerebrospinal fluid. *J Cereb Blood Flow Metab*, 1999. 19, p. 184–194.

Harris, J.L., H.W. Yeh, I.Y. Choi, P. Lee, N.E. Berman, R.H. Swerdlow, S.C. Craciunas, W.M. Brooks, Altered neurochemical profile after traumatic brain injury: (1)H-MRS biomarkers of pathological mechanisms. *J Cereb Blood Flow Metab*, 2012. 32, p. 2122–2134.

Harris, N.G., Y.A. Mironova, D.A. Hovda, R.L. Sutton, Chondroitinase ABC enhances pericontusion axonal sprouting but does not confer robust improvements in behavioral recovery. *J Neurotrauma*, 2010. 27, p. 1971–1982.

Harris, N.G., Y.A. Mironova, S.F. Chen, H.K. Richards, J.D. Pickard, Preventing flow-metabolism uncoupling acutely reduces axonal injury after traumatic brain injury. *J Neurotrauma*, 2012. 29, p. 1469–1482.

Harting, M.T., F. Jimenez, S.D. Adams, D.W. Mercer, C.S., Cox, Jr. Acute, regional inflammatory response after traumatic brain injury: Implications for cellular therapy. *Surgery*, 2008. 144, p. 803–813.

He, J., C.O. Evans, S.W. Hoffman, N.M. Oyesiku, D.G. Stein, Progesterone and allopregnanolone reduce inflammatory cytokines after traumatic brain injury. *Exp Neurol*, 2004. 189, p. 404–412.

Heath, D.L., R. Vink, Impact acceleration-induced severe diffuse axonal injury in rats: Characterization of phosphate metabolism and neurologic outcome. *J. Neurotrauma*, 1995. 12, p. 1027–1034.

Heath, D.L., R. Vink, Improved motor outcome in response to magnesium therapy received up to 24 hours after traumatic diffuse axonal brain injury in rats. *J Neurosurg*, 1999. 90, p. 504–509.

Hellewell, S.C., E.B. Yan, D.A. Agyapomaa, N. Bye, M.C. Morganti-Kossmann, Post-traumatic hypoxia exacerbates brain tissue damage: Analysis of axonal injury and glial responses. *J Neurotrauma*, 2010. 27, p. 1997–2010.

Hemerka, J.N. et al., Severe brief pressure-controlled hemorrhagic shock after traumatic brain injury exacerbates functional deficits and long-term neuropathological damage in mice. *J Neurotrauma*, 2012. 29, p. 2192–2208.

Hinkle, D.A., S.A. Baldwin, S.W. Scheff, P.M. Wise, GFAP and S100beta expression in the cortex and hippocampus in response to mild cortical contusion. *J Neurotrauma*, 1997. 14, p. 729–738.

Hoane, M.R., S.L. Akstulewicz, J. Toppen, Treatment with vitamin B3 improves functional recovery and reduces GFAP expression following traumatic brain injury in rats. *J Neurotrauma*, 2003. 20, p. 1189–1199.

Hoffman, S.W., Z. Fulop, D.G. Stein, Bilateral frontal cortical contusion in rats: Behavioral and anatomic consequences. *J Neurotrauma*, 1994. 11, p. 417–431.

Hoffmeister, P.G., C.K. Donat, M.U. Schuhmann, C. Voigt, B. Walter, K. Nieber, J. Meixensberger, R. Bauer, P. Brust, Traumatic brain injury elicits similar alterations in alpha7 nicotinic receptor density in two different experimental models. *Neuromolecular Med*, 2011. 13, p. 44–53.

Holschneider, D.P., Y. Guo, Z. Wang, M. Roch, O.U. Scremin, Remote brain networks changes after unilateral cortical impact injury and their modulation by acetylcholinesterase inhibition. *J Neurotrauma*, 2013. 30, p. 907–919.

Holtzer, S., B. Vigue, C. Ract, K. Samii, P. Escourrou, Hypoxia-hypotension decreases pressor responsiveness to exogenous catecholamines after severe traumatic brain injury in rats. *Crit Care Med*, 2001. 29, p. 1609–1614.

Hoskison, M.M., A.N. Moore, B. Hu, S. Orsi, N. Kobori, P.K. Dash, Persistent working memory dysfunction following traumatic brain injury: Evidence for a time-dependent mechanism. *Neuroscience*, 2009. 159, p. 483–491.

Hunt, R.F., S.W. Scheff, B.N. Smith, Posttraumatic epilepsy after controlled cortical impact injury in mice. *Exp Neurol*, 2009. 215, p. 243–252.

Igarashi, T., M.B. Potts, L.J. Noble-Haeusslein, Injury severity determines Purkinje cell loss and microglial activation in the cerebellum after cortical contusion injury. *Exp Neurol*, 2007. 203, p. 258–268.

Immonen, R. et al., Cerebral blood volume alterations in the perilesional areas in the rat brain after traumatic brain injury—Comparison with behavioral outcome. *J Cereb Blood Flow Metab*, 2010. 30, p. 1318–1328.

Israelsson, C., H. Bengtsson, A. Kylberg, K. Kullander, A. Lewen, L. Hillered, T. Ebendal, Distinct cellular patterns of upregulated chemokine expression supporting a prominent inflammatory role in traumatic brain injury. *J Neurotrauma*, 2008. 25, p. 959–974.

Ito, J., A. Marmarou, P. Barzo, P. Fatouros, F. Corwin, Characterization of edema by diffusion-weighted imaging in experimental traumatic brain injury. *J Neurosurg*, 1996. 84, p. 97–103.

Jacobowitz, D.M., J.T. Cole, D.P. McDaniel, H.B. Pollard, W.D. Watson, Microglia activation along the corticospinal tract following traumatic brain injury in the rat: A neuroanatomical study. *Brain Res*, 2012. 1465, p. 80–89.

Jin, G. et al., Traumatic brain injury and hemorrhagic shock: Evaluation of different resuscitation strategies in a large animal model of combined insults. *Shock*, 2012. 38, p. 49–56.

Jin, X., H. Ishii, Z. Bai, T. Itokazu, T. Yamashita, Temporal changes in cell marker expression and cellular infiltration in a controlled cortical impact model in adult male C57BL/6 mice. *PLoS One*, 2012. 7, p. e41892.

Kallakuri, S., J.M. Cavanaugh, A.C. Ozaktay, T. Takebayashi, The effect of varying impact energy on diffuse axonal injury in the rat brain: A preliminary study. *Exp. Brain Res*, 2003. 148, 419–424.

Kallakuri, S., Y. Li, R. Zhou, S. Bandaru, N. Zakaria, L. Zhang, J.M. Cavanaugh, Impaired axoplasmic transport is the dominant injury induced by an impact acceleration injury device: An analysis of traumatic axonal injury in pyramidal tract and corpus callosum of rats. *Brain Res*, 2012. 1452, p. 29–38.

Kelly, D.F., D.A. Kozlowski, E. Haddad, A. Echiverri, D.A. Hovda, S.M. Lee, Ethanol reduces metabolic uncoupling following experimental head injury. *J Neurotrauma*, 2000. 17, p. 261–272.

Kharlamov, E.A., E. Lepsveridze, M. Meparishvili, R.O. Solomonia, B. Lu, E.R. Miller, K.M. Kelly, Z. Mtchedlishvili, Alterations of GABA(A) and glutamate receptor subunits and heat shock protein in rat hippocampus following traumatic brain injury and in post-traumatic epilepsy. *Epilepsy Res*, 2011. 95, p. 20–34.

Kiening, K.L., F.K. van Landeghem, S. Schreiber, U.W. Thomale, D.A. von Deimling, A.W. Unterberg, J.F. Stover, Decreased hemispheric Aquaporin-4 is linked to evolving brain edema following controlled cortical impact injury in rats. *Neurosci Lett*, 2002. 324, p. 105–108.

Kilbourne, M., R. Kuehn, C. Tosun, J. Caridi, K. Keledjian, G. Bochicchio, T. Scalea, V. Gerzanich, J.M. Simard, Novel model of frontal impact closed head injury in the rat. *J Neurotrauma*, 2009. 26, p. 2233–2243.

King, C., T. Robinson, C.E. Dixon, G.R. Rao, D. Larnard, C.E. Nemoto, Brain temperature profiles during epidural cooling with the ChillerPad in a monkey model of traumatic brain injury. *J Neurotrauma*, 2010. 27, p. 1895–1903.

Kline, A.E., J. Yu, E. Horvath, D.W. Marion, C.E. Dixon, The selective 5-HT1A receptor agonist Repinotan HCL attenuates histopathology and spatial learning deficits following traumatic brain injury in rats. *Neuroscience*, 2001. 106, p. 547–555.

Kobori, N., G.L. Clifton, P.K. Dash, Enhanced catecholamine synthesis in the prefrontal cortex after traumatic brain injury: Implications for prefrontal dysfunction. *J Neurotrauma*, 2006. 23, p. 1094–1102.

Kobori, N., P.K. Dash, Reversal of brain injury-induced prefrontal glutamic acid decarboxylase expression and working memory deficits by D1 receptor antagonism. *J Neurosci*, 2006. 26, p. 4236–4246.

Kochanek, P.M. et al., Severe controlled cortical impact in rats: Assessment of cerebral edema, blood blow, and contusion volume. *J Neurotrauma*, 1995. 12, p. 1015–1025.

Kochanek, P.M., K. Hendrich, C.E. Dixon, J.K. Schiding, D.S. Williams, C. Ho, Cerebral blood flow at one year after controlled cortical impact in rats: Assessment by magnetic resonance imaging. *J Neurotrauma*, 2002. 19, p. 1029–1037.

Kovesdi, E., P. Bukovics, V. Besson, J. Nyiradi, J. Luckl, J. Pal, B. Sumegi, T. Doczi, I. Hernadi, A. Buki, A novel PARP inhibitor L-2286 in a rat model of impact acceleration head injury: An immunohistochemical and behavioral study. *Int J Mol Sci*, 2010. 11, p. 1253–1268.

Krishnappa, I.K., C.F. Contant, C.S. Robertson, Regional changes in cerebral extracellular glucose and lactate concentrations following severe cortical impact injury and secondary ischemia in rats. *J Neurotrauma*, 1999. 16, p. 213–224.

Kroppenstedt, S.N., U.W. Thomale, M. Griebenow, O.W. Sakowitz, K.D. Schaser, P.S. Mayr, A.W. Unterberg, J.F. Stover, Effects of early and late intravenous norepinephrine infusion on cerebral perfusion, microcirculation, brain-tissue oxygenation, and edema formation in brain-injured rats. *Crit Care Med*, 2003. 31, p. 2211–2221.

Kumar, A., L. Zou, X. Yuan, Y. Long, K. Yang, N-methyl-D-aspartate receptors: Transient loss of NR1/NR2A/NR2B subunits after traumatic brain injury in a rodent model. *J Neurosci Res*, 2002. 67, p. 781–786.

Kumar, A., B.A. Stoica, B. Sabirzhanov, M.P. Burns, A.I. Faden, D.J. Loane, Traumatic brain injury in aged animals increases lesion size and chronically alters microglial/macrophage classical and alternative activation states. *Neurobiol Aging*, 2013. 34, p. 1397–1411.

Lagraoui, M., J.R. Latoche, N.G. Cartwright, G. Sukumar, C.L. Dalgard, B.C. Schaefer, Controlled cortical impact and craniotomy induce strikingly similar profiles of inflammatory gene expression, but with distinct kinetics. *Front Neurol*, 2012. 3, 155.

Lee, S., M. Ueno, T. Yamashita, Axonal remodeling for motor recovery after traumatic brain injury requires downregulation of gamma-aminobutyric acid signaling. *Cell Death*, 2011. Dis. p. 2, e133.

Lee, S.M., D.A. Wong, A. Samii, D.A. Hovda, Evidence for energy failure following irreversible traumatic brain injury. *Ann NY Acad Sci*, 1999. 893, p. 337–340.

Levin, B.E., K.L. Brown, G. Pawar, A. Dunn-Meynell, Widespread and lateralized effects of acute traumatic brain injury on norepinephrine turnover in the rat brain. *Brain Res*, 1995. 674, p. 307–313.

Li, S., Y. Sun, D. Shan, B. Feng, J. Xing, Y. Duan, J. Dai, H. Lei, Y. Zhou, Temporal profiles of axonal injury following impact acceleration traumatic brain injury in rats—A comparative study with diffusion tensor imaging and morphological analysis. *Int J Legal Med*, 2013. 127, p. 159–167.

Li, Y., L. Zhang, S. Kallakuri, R. Zhou, J.M. Cavanaugh, Quantitative relationship between axonal injury and mechanical response in a rodent head impact acceleration model. *J Neurotrauma*, 2011. 28, p. 1767–1782.

Lighthall, J.W. Controlled cortical impact: A new experimental brain injury model. *J Neurotrauma*, 1988. 5, p. 1–15.

Lighthall, J.W., C.E. Dixon, T.E. Anderson, Experimental models of brain injury. *J Neurotrauma*, 1989. 6, p. 83–97.

Lighthall, J.W., H.G. Goshgarian, C.R. Pinderski, Characterization of axonal injury produced by controlled cortical impact. *J Neurotrauma*, 1990. 7, p. 65–76.

Lindner, M.D., M.A. Plone, C.K. Cain, B. Frydel, J.M. Francis, D.F. Emerich, R.L. Sutton, Dissociable long-term cognitive deficits after frontal versus sensorimotor cortical contusions. *J Neurotrauma*, 1998. 15, p. 199–216.

Liu, M.C., V. Akle, W. Zheng, J.R. Dave, F.C. Tortella, R.L. Hayes, K.K. Wang, Comparing calpain- and caspase-3-mediated degradation patterns in traumatic brain injury by differential proteome analysis. *Biochem J*, 2006. 394, p. 715–725.

Lu, K.T., Y.W. Wang, J.T. Yang, Y.L. Yang, H.I. Chen, Effect of interleukin-1 on traumatic brain injury-induced damage to hippocampal neurons. *J Neurotrauma*, 2005a. 22, p. 885–895.

Lu, K.T., Y.W. Wang, Y.Y. Wo, Y.L. Yang, Extracellular signal-regulated kinase-mediated IL-1-induced cortical neuron damage during traumatic brain injury. *Neurosci Lett*, 2005b. 386, p. 40–45.

Lundblad, C., P.O. Grande, P. Bentzer, A mouse model for evaluation of capillary perfusion, microvascular permeability, cortical blood flow, and cortical edema in the traumatized brain. *J Neurotrauma*, 2004. 21, p. 741–753.

Manley, G.T., G. Rosenthal, M. Lam, D. Morabito, D. Yan, N. Derugin, A. Bollen, M.M. Knudson, S.S. Panter, Controlled cortical impact in swine: Pathophysiology and biomechanics. *J Neurotrauma*, 2006. 23, p. 128–139.

Mao, H., K.H. Yang, A.I. King, K. Yang, Computational neurotrauma-design, simulation, and analysis of controlled cortical impact model. *Biomech Model Mechanobiol*, 2010. 9, p. 763–772.

Mao, H., F. Guan, X. Han, K.H. Yang, Strain-based regional traumatic brain injury intensity in controlled cortical impact: A systematic numerical analysis. *J Neurotrauma*, 2011. 28, p. 2263–2276.

Markgraf, C.G., G.L. Clifton, M. Aguirre, S.F. Chaney, C. Knox-Du Bois, K. Kennon, N. Verma, Injury severity and sensitivity to treatment after controlled cortical impact in rats. *J Neurotrauma*, 2001. 18, p. 175–186.

Marmarou, A., M.A. Foda, W. Brink, J. Campbell, H. Kita, K. Demetriadou, A new model of diffuse brain injury in rats: Part I: Pathophysiology and biomechanics. *J Neurosurg*, 1994. 80, p. 291–300.

Matthews, M.A., M.E. Carey, J.S. Soblosky, J.F. Davidson, S.L. Tabor, Focal brain injury and its effects on cerebral mantle, neurons, and fiber tracks. *Brain Res*, 1998. 794, p. 1–18.

Maughan, P.H., K.J. Scholten, R.H. Schmidt, Recovery of water maze performance in aged versus young rats after brain injury with the impact acceleration model. *J Neurotrauma*, 2000. 17, p. 1141–1153.

Meaney, D.F., D.T. Ross, B.A. Winkelstein, J. Brasko, D. Goldstein, L.B. Bilston, L.E. Thibault, T.A. Gennarelli, Modification of the cortical impact model to produce axonal injury in the rat cerebral cortex. *J Neurotrauma*, 1994. 11, p. 599–612.

Mirzayan, M.J., P.M. Klinge, S. Ude, A. Hotop, M. Samii, T. Brinker, Z. Korkmaz, G.J. Meyer, W.H. Knapp, A. Samii, Modified calcium accumulation after controlled cortical impact under cyclosporin A treatment: A 45Ca autoradiographic study. *Neurol Res*, 2008. 30, p. 476–479.

Mori, T., X. Wang, T. Aoki, E.H. Lo, Downregulation of matrix metalloproteinase-9 and attenuation of edema via inhibition of ERK mitogen activated protein kinase in traumatic brain injury. *J Neurotrauma*, 2002. 19, p. 1411–1419.

Moro, N., S.S. Ghavim, D.A. Hovda, R.L. Sutton, Delayed sodium pyruvate treatment improves working memory following experimental traumatic brain injury. *Neurosci Lett*, 2011. 491, p. 158–162.

Mtchedlishvili, Z., E. Lepsveridze, H. Xu, E.A. Kharlamov, B. Lu, K.M. Kelly, Increase of GABAA receptor-mediated tonic inhibition in dentate granule cells after traumatic brain injury. *Neurobiol Dis*, 2010. 38, p. 464–475.

Mustafa, A.G., I.N. Singh, J. Wang, K.M. Carrico, E.D. Hall, Mitochondrial protection after traumatic brain injury by scavenging lipid peroxyl radicals. *J Neurochem*, 2010. 114, p. 271–280.

Myer, D.J., G.G. Gurkoff, S.M. Lee, D.A. Hovda, M.V. Sofroniew, Essential protective roles of reactive astrocytes in traumatic brain injury. *Brain*, 2006. 129, p. 2761–2772.

Mytar, J., K.K. Kibler, R.B. Easley, P. Smielewski, M. Czosnyka, D.B. Andropoulos, K.M. Brady, Static autoregulation is intact early after severe unilateral brain injury in a neonatal Swine model. *Neurosurgery*, 2012. 71, p. 138–145.

Natale, J.E., F. Ahmed, I. Cernak, B. Stoica, A.I. Faden, Gene expression profile changes are commonly modulated across models and species after traumatic brain injury. *J Neurotrauma*, 2003. 20, p. 907–927.

Nawashiro, H., K. Shima, H. Chigasaki, Immediate cerebrovascular responses to closed head injury in the rat. *J Neurotrauma*, 1995. 12, p. 189–197.

Nawashiro, H., K. Shima, H. Chigasaki, Selective vulnerability of hippocampal CA3 neurons to hypoxia after mild concussion in the rat. *Neurol Res*, 1995. 17, p. 455–460.

Neumann, M., Y. Wang, S. Kim, S.M. Hong, L. Jeng, M. Bilgen, J. Liu, Assessing gait impairment following experimental traumatic brain injury in mice. *J Neurosci Methods*, 2009. 176, p. 34–44.

Newcomb, J.K., A. Kampfl, R.M. Posmantur, X. Zhao, B.R. Pike, S.-J. Liu, G.L. Clifton, R.L. Hayes, Immunohistochemical study of calpain-mediated breakdown products to alpha-spectrin following controlled cortical impact injury in the rat. *J Neurotrauma*, 1997. 14, p. 369–383.

Newcomb, J.K., X. Zhao, B.R. Pike, R.L. Hayes, Temporal profile of apoptotic-like changes in neurons and astrocytes following controlled cortical impact injury in the rat. *Exp Neurol*, 1999. 158, p. 76–88.

O'Connor, C.A., I. Cernak, R. Vink, Interaction between anesthesia, gender, and functional outcome task following diffuse traumatic brain injury in rats. *J Neurotrauma*, 2003. 20, p. 533–541.

O'Connor, C.A., I. Cernak, F. Johnson, R. Vink, Effects of progesterone on neurologic and morphologic outcome following diffuse traumatic brain injury in rats. *Exp Neurol*, 2007. 205, p. 145–153.

Ohta, M., Y. Higashi, T. Yawata, M. Kitahara, A. Nobumoto, E. Ishida, M. Tsuda, Y. Fujimoto, K. Shimizu, Attenuation of axonal injury and oxidative stress by edaravone protects against cognitive impairments after traumatic brain injury. *Brain Res*, 2013. 1490, p. 184–192.

Okonkwo, D.O., J.T. Povlishock, An intrathecal bolus of cyclosporin A before injury preserves mitochondrial integrity and attenuates axonal disruption in traumatic brain injury. *J Cereb Blood Flow Metab*, 1999. 19, p. 443–451.

Olsen, A.S., C.N. Sozda, J.P. Cheng, A.N. Hoffman, A.E. Kline, Traumatic brain injury-induced cognitive and histological deficits are attenuated by delayed and chronic treatment with the 5-HT1A-receptor agonist buspirone. *J Neurotrauma*, 2012. 29, p. 1898–1907.

Onyszchuk, G., B. Al Hafez, Y.Y. He, M. Bilgen, N.E. Berman, W.M. Brooks, A mouse model of sensorimotor controlled cortical impact: Characterization using longitudinal magnetic resonance imaging, behavioral assessments and histology. *J Neurosci Methods*, 2007. 160, p. 187–196.

Opii, W.O., V.N. Nukala, R. Sultana, J.D. Pandya, K.M. Day, M.L. Merchant, J.B. Klein, P.G. Sullivan, D.A. Butterfield, Proteomic identification of oxidized mitochondrial proteins following experimental traumatic brain injury. *J Neurotrauma*, 2007. 24, p. 772–789.

Palmer, A.M., D.W. Marion, M.L. Botscheller, E.E. Redd, Therapeutic hypothermia is cytoprotective without attenuating the traumatic brain injury-induced elevations in interstitial concentrations of aspartate and glutamate. *J Neurotrauma*, 1993. 10, p. 363–372.

Pandya, J.D., J.R. Pauly, V.N. Nukala, A.H. Sebastian, K.M. Day, A.S. Korde, W.F. Maragos, E.D. Hall, P.G. Sullivan, Post-injury administration of mitochondrial uncouplers increases tissue sparing and improves behavioral outcome following traumatic brain injury in rodents. *J Neurotrauma*, 2007. 24, p. 798–811.

Pascual, J.M., J. Solivera, R. Prieto, L. Barrios, P. Lopez-Larrubia, S. Cerdan, J.M. Roda, Time course of early metabolic changes following diffuse traumatic brain injury in rats as detected by (1)H NMR spectroscopy. *J Neurotrauma*, 2007. 24, p. 944–959.

Pike, B.R., X. Zhao, J.K. Newcomb, R.M. Posmantur, K.K. Wang, R.L. Hayes, Regional calpain and caspase-3 proteolysis of alpha-spectrin after traumatic brain injury. *NeuroReport*, 1998. 9, p. 2437–2442.

Pleasant, J.M., S.W. Carlson, H. Mao, S.W. Scheff, K.H. Yang, K.E. Saatman, Rate of neurodegeneration in the mouse controlled cortical impact model is influenced by impactor tip shape: Implications for mechanistic and therapeutic studies. *J Neurotrauma*, 2011. 28, p. 2245–2262.

Portella, G., A. Beaumont, F. Corwin, P. Fatouros, A. Marmarou, Characterizing edema associated with cortical contusion and secondary insult using magnetic resonance spectroscopy. *Acta Neurochir Suppl*, 2000. 76, p. 273–275.

Posmantur, R.M., A. Kampfl, S.J. Liu, K. Heck, W.C. Taft, G.L. Clifton, R.L. Hayes, Cytoskeletal derangements of cortical neuronal processes three hours after traumatic brain injury in rats: An immunofluorescence study. *J Neuropathol Exp Neurol*, 1996. 55, p. 68–80.

Posmantur, R.M., J.K. Newcomb, A. Kampfl, R.L. Hayes, Light and confocal microscopic studies of evolutionary changes in neurofilament proteins following cortical impact injury in the rat. *Exp Neurol*, 2000. 161, p. 15–26.

Povlishock, J.T., A. Marmarou, T.K. McIntosh, J.Q. Trojanowski, J. Moroi, Impact acceleration injury in the rat: Evidence for focal axolemmal and related neurofilament sidearm alteration. *J Neuropathol Exp Neurol*, 1997. 56, p. 347–359.

Prasad, M.R., C. Ramaiah, T.K. McIntosh, R.J. Dempsey, S. Hipkens, D. Yurek, Regional levels of lactate and norepinephrine after experimental brain injury. *J Neurochem*, 1994. 63, p. 1086–1094.

Prieto, R., B. Tavazzi, K. Taya, L. Barrios, A.M. Amorini, V. Di, Pietro, J.M. Pascual, A. Marmarou, C.R. Marmarou, Brain energy depletion in a rodent model of diffuse traumatic brain injury is not prevented with administration of sodium lactate. *Brain Res*, 2011. 1404, 39–49.

Prins, M.L., D.A. Hovda, The effects of age and ketogenic diet on local cerebral metabolic rates of glucose after controlled cortical impact injury in rats. *J Neurotrauma*, 2009. 26, p. 1083–1093.

Rafols, J.A., R. Morgan, S. Kallakuri, C.W. Kreipke, Extent of nerve cell injury in Marmarou's model compared to other brain trauma models. *Neurol Res*, 2007. 29, p. 348–355.

Raghupathi, R., S.C. Fernandez, H. Murai, S.P. Trusko, R.W. Scott, W.K. Nishioka, T.K. McIntosh, BCL-2 overexpression attenuates cortical cell loss after traumatic brain injury in transgenic mice. *J Cereb Blood Flow Metab*, 1998. 18, p. 1259–1269.

Rancan, M., V.I. Otto, V.H. Hans, I. Gerlach, R. Jork, O. Trentz, T. Kossmann, M.C. Morganti-Kossmann, Upregulation of ICAM-1 and MCP-1 but not of MIP-2 and sensorimotor deficit in response to traumatic axonal injury in rats. *J Neurosci Res*, 2001. 63, p. 438–446.

Rao, V.L., M.K. Baskaya, A. Dogan, J.D. Rothstein, R.J. Dempsey, Traumatic brain injury down-regulates glial glutamate transporter (GLT-1 and GLAST) proteins in rat brain. *J Neurochem*, 1998. 70, p. 2020–2027.

Redell, J.B., A.N. Moore, R.J. Grill, D. Johnson, J. Zhao, Y. Liu, P.K. Dash, Analysis of functional pathways altered after mild traumatic brain injury. *J Neurotrauma*, 2013. 30, p. 752–764.

Rhodes, J.K., P.J. Andrews, M.C. Holmes, J.R. Seckl, Expression of interleukin-6 messenger RNA in a rat model of diffuse axonal injury. *Neurosci Lett*, 2002. 335, p. 1–4.

Robertson, C.L., M. Saraswati, G. Fiskum, Mitochondrial dysfunction early after traumatic brain injury in immature rats. *J Neurochem*, 2007. 101, p. 1248–1257.

Robertson, C.L., M. Saraswati, S. Scafidi, G. Fiskum, P. Casey, M.C. McKenna, Cerebral glucose metabolism in an immature rat model of pediatric traumatic brain injury. *J Neurotrauma*, 2013. 30, p. 1–7.

Roof, R.L., R. Duvdevani, D.G. Stein, Gender influences outcome of brain injury: Progesterone plays a protective role. *Brain Res*, 1993. 607, p. 333–336.

Roof, R.L., R. Duvdevani, J.W. Heyburn, D.G. Stein, Progesterone rapidly decreases brain edema: Treatment delayed up to 24 hours is still effective. *Exp Neurol*, 1996. 138, p. 246–251.

Rooker, S., S. Jander, J. Van Reempts, G. Stoll, P.G. Jorens, M. Borgers, J. Verlooy, Spatiotemporal pattern of neuroinflammation after impact-acceleration closed head injury in the rat. *Mediators Inflamm*, 2006. 2006, 90123.

Rose, M.E., M.B. Huerbin, J. Melick, D.W. Marion, A.M. Palmer, J.K. Schiding, P.M. Kochanek, S.H. Graham, Regulation of interstitial excitatory amino acid concentrations after cortical contusion injury. *Brain Res*, 2002. 943, p. 15–22.

Santos, A., N. Borges, A. Cerejo, A. Sarmento, I. Azevedo, Catalase activity and thiobarbituric acid reactive substances (TBARS) production in a rat model of diffuse axonal injury. Effect of gadolinium and amiloride. *Neurochem Res*, 2005. 30, p. 625–631.

Sauerbeck, A., J. Gao, R. Readnower, M. Liu, J.R. Pauly, G. Bing, P.G. Sullivan, Pioglitazone attenuates mitochondrial dysfunction, cognitive impairment, cortical tissue loss, and inflammation following traumatic brain injury. *Exp Neurol*, 2011. 227, p. 128–135.

Sawauchi, S., A. Marmarou, A. Beaumont, Y. Tomita, S. Fukui, A new rat model of diffuse brain injury associated with acute subdural hematoma: Assessment of varying hematoma volume, insult severity, and the presence of hypoxemia. *J Neurotrauma*, 2003. 20, p. 613–622.

Sawauchi, S., A. Marmarou, A. Beaumont, S. Signoretti, S. Fukui, Acute subdural hematoma associated with diffuse brain injury and hypoxemia in the rat: Effect of surgical evacuation of the hematoma. *J Neurotrauma*, 2004. 21, p. 563–573.

Scafidi, S., J. Racz, J. Hazelton, M.C. McKenna, G. Fiskum, Neuroprotection by acetyl-L-carnitine after traumatic injury to the immature rat brain. *Dev Neurosci*, 2010. 32, p. 480–487.

Scheff, S.W., S.A. Baldwin, R.W. Brown, P.J. Kraemer, Morris water maze deficits in rats following traumatic brain injury: Lateral controlled cortical impact. *J Neurotrauma*, 1997. 14, p. 615–627.

Schmidt, R.H., K.J. Scholten, P.H. Maughan, Cognitive impairment and synaptosomal choline uptake in rats following impact acceleration injury. *J Neurotrauma*, 2000. 17, p. 1129–1139.

Schuhmann, M.U., M. Mokhtarzadeh, D.O. Stichtenoth, M. Skardelly, P.M. Klinge, F.M. Gutzki, M. Samii, T. Brinker, Temporal profiles of cerebrospinal fluid leukotrienes, brain edema and inflammatory response following experimental brain injury. *Neurol Res*, 2003a. 25, p. 481–491.

Schuhmann, M.U., D. Stiller, M. Skardelly, J. Bernarding, P.M. Klinge, A. Samii, M. Samii, T. Brinker, Metabolic changes in the vicinity of brain contusions: A proton magnetic resonance spectroscopy and histology study. *J Neurotrauma*, 2003b. 20, p. 725–743.

Sharma, A.C., M. Misra, R. Prat, K. Alden, A.D. Sam, V.Z. Markiv, M. Dujovny, J.L. Ferguson, A differential response of diffuse brain injury on the concentrations of endothelin and nitric oxide in the plasma and brain regions in rats. *Neurol Res*, 1998. 20, p. 632–636.

Shelton, S.B., D.B. Pettigrew, A.D. Hermann, W. Zhou, P.M. Sullivan, K.A. Crutcher, K.I. Strauss, A simple, efficient tool for assessment of mice after unilateral cortex injury. *J Neurosci Methods*, 2008. 168, p. 431–442.

Shenaq, M., H. Kassem, C. Peng, S. Schafer, J.Y. Ding, V. Fredrickson, M. Guthikonda, C.W. Kreipke, J.A. Rafols, Y. Ding, Neuronal damage and functional deficits are ameliorated by inhibition of aquaporin and HIF1alpha after traumatic brain injury (TBI). *J Neurol Sci*, 2012. 323, p. 134–140.

Signoretti, S., A. Marmarou, B. Tavazzi, G. Lazzarino, A. Beaumont, R. Vignozzi, N-Acetylaspartate reduction as a measure of injury severity and mitochondrial dysfunction following diffuse traumatic brain injury. *J Neurotrauma*, 2001. 18, p. 977–991.

Singh, I.N., L.K. Gilmer, D.M. Miller, J.E. Cebak, J.A. Wang, E.D. Hall, Phenelzine mitochondrial functional preservation and neuroprotection after traumatic brain injury related to scavenging of the lipid peroxidation-derived aldehyde 4-hydroxy-2-nonenal. *J Cereb Blood Flow Metab*, 2013. 33, p. 593–599.

Singleton, R.H., J.R. Stone, D.O. Okonkwo, A.J. Pellicane, J.T. Povlishock, The immunophilin ligand FK506 attenuates axonal injury in an impact-acceleration model of traumatic brain injury. *J Neurotrauma*, 2001. 18, p. 607–614.

Smith, D.H., H.D. Soares, J.S. Pierce, K.G. Perlman, K.E. Saatman, D.F. Meaney, C.E. Dixon, T.K. McIntosh, A model of parasagittal controlled cortical impact in the mouse: Cognitive and histopathologic effects. *J Neurotrauma*, 1995. 12, p. 169–178.

Smith, S.L., P.K. Andrus, J.R. Zhang, E.D. Hall, Direct measurement of hydroxyl radicals, lipid peroxidation, and blood-brain barrier disruption following unilateral cortical impact head injury in the rat. *J Neurotrauma*, 1994. 11, p. 393–404.

Soblosky, J.S., M.A. Matthews, J.F. Davidson, S.L. Tabor, M.E. Carey, Traumatic brain injury of the forelimb and hindlimb sensorimotor areas in the rat: Physiological, histological and behavioral correlates. *Behav Brain Res*, 1996a. 79, p. 79–92.

Soblosky, J.S., S.L. Tabor, M.A. Matthews, J.F. Davidson, D.A. Chorney, M.E. Carey, Reference memory and allocentric spatial localization deficits after unilateral cortical brain injury in the rat. *Behav Brain Res*, 1996b. 80, p. 185–194.

Statler, K.D., K.L. Janesko, J.A. Melick, R.S. Clark, L.W. Jenkins, P.M. Kochanek, Hyperglycolysis is exacerbated after traumatic brain injury with fentanyl vs. isoflurane anesthesia in rats. *Brain Res*, 2003. 994, p. 37–43.

Stiefel, M.F., Y. Tomita, A. Marmarou, Secondary ischemia impairing the restoration of ion homeostasis following traumatic brain injury. *J Neurosurg*, 2005. 103, p. 707–714.

Stone, J.R., R.H. Singleton, J.T. Povlishock, Intra-axonal neurofilament compaction does not evoke local axonal swelling in all traumatically injured axons. *Exp Neurol*, 2001. 172, p. 320–331.

Stone, J.R., D.O. Okonkwo, A.O. Dialo, D.G. Rubin, L.K. Mutlu, J.T. Povlishock, G.A. Helm, Impaired axonal transport and altered axolemmal permeability occur in distinct populations of damaged axons following traumatic brain injury. *Exp Neurol*, 2004. 190, p. 59–69.

Stover, J.F., O.W. Sakowitz, S.N. Kroppenstedt, U.W. Thomale, O.S. Kempski, G. Flugge, A.W. Unterberg, Differential effects of prolonged isoflurane anesthesia on plasma, extracellular, and CSF glutamate, neuronal activity, 125 I-Mk801 NMDA receptor binding, and brain edema in traumatic brain-injured rats. *Acta Neurochir*, 2004. 146, p. 819–830.

Strauss, K.I., M.F. Barbe, R.M. Marshall, R. Raghupathi, S. Mehta, R.K. Narayan, Prolonged cyclooxygenase-2 induction in neurons and glia following traumatic brain injury in the rat. *J Neurotrauma*, 2000. 17, p. 695–711.

Strauss, K.I., R.K. Narayan, R. Raghupathi, Common patterns of bcl-2 family gene expression in two traumatic brain injury models. *Neurotox Res*, 2004. 6, p. 333–342.

Suehiro, E., Y. Ueda, E.P. Wei, H.A. Kontos, J.T. Povlishock, Posttraumatic hypothermia followed by slow rewarming protects the cerebral microcirculation. *J Neurotrauma*, 2003. 20, p. 381–390.

Sullivan, P.G., M. Thompson, S.W. Scheff, Continuous infusion of cyclosporin A postinjury significantly ameliorates cortical damage following traumatic brain injury. *Exp Neurol*, 2000. 161, p. 631–637.

Sullivan, P.G., J.N. Keller, W.L. Bussen, S.W. Scheff, Cytochrome c release and caspase activation after traumatic brain injury. *Brain Res*, 2002. 949, p. 88–96.

Sutton, R.L., L. Lescaudron, D.G. Stein, Unilateral cortical contusion injury in the rat: Vascular disruption and temporal development of cortical necrosis. *J Neurotrauma*, 1993. 10, p. 135–149.

Sutton, R.L., D.A. Hovda, P.D. Adelson, E.C. Benzel, D.P. Becker, Metabolic changes following cortical contusion: Relationships to edema and morphological changes. *Acta Neurochir Suppl (Wien)*, 1994. 60, p. 446–448.

Szmydynger-Chodobska, J., N. Strazielle, B.J. Zink, J.F. Ghersi-Egea, A. Chodobski, The role of the choroid plexus in neutrophil invasion after traumatic brain injury. *J Cereb Blood Flow Metab*, 2009. 29, p. 1503–1516.

Tavazzi, B., S. Signoretti, G. Lazzarino, A.M. Amorini, R. Delfini, M. Cimatti, A. Marmarou, R. Vagnozzi, Cerebral oxidative stress and depression of energy metabolism correlate with severity of diffuse brain injury in rats. *Neurosurgery*, 2005. 56, p. 582–589.

Tavazzi, B., R. Vagnozzi, S. Signoretti, A.M. Amorini, A. Belli, M. Cimatti, R. Delfini, V. Di Pietro, A. Finocchiaro, G. Lazzarino, Temporal window of metabolic brain vulnerability to concussions: Oxidative and nitrosative stresses—Part II. *Neurosurgery*, 2007. 61, p. 390–395.

Taya, K., C.R. Marmarou, K. Okuno, R. Prieto, A. Marmarou, Effect of secondary insults upon aquaporin-4 water channels following experimental cortical contusion in rats. *J Neurotrauma*, 2010. 27, p. 229–239.

Taylor, A.N., S.U. Rahman, N.C. Sanders, D.L. Tio, P. Prolo, R.L. Sutton, Injury severity differentially affects short- and long-term neuroendocrine outcomes of traumatic brain injury. *J Neurotrauma*, 2008. 25, p. 311–323.

Taylor, A.N., D.L. Tio, R.L. Sutton, Restoration of neuroendocrine stress response by glucocorticoid receptor or GABA(A) receptor antagonists after experimental traumatic brain injury. *J Neurotrauma*, 2013. 30, p. 1250–1256.

Tehranian, R. et al., Transgenic mice that overexpress the anti-apoptotic Bcl-2 protein have improved histological outcome but unchanged behavioral outcome after traumatic brain injury. *Brain Res*, 2006. 1101, p. 126–135.

Thomale, U.W., S.N. Kroppenstedt, T.F. Beyer, K.D. Schaser, A.W. Unterberg, J.F. Stover, Temporal profile of cortical perfusion and microcirculation after controlled cortical impact injury in rats. *J Neurotrauma*, 2002. 19, p. 403–413.

Thomale, U.W., M. Griebenow, A. Mautes, T.F. Beyer, N.K. Dohse, R. Stroop, O.W. Sakowitz, A.W. Unterberg, J.F. Stover, Heterogeneous regional and temporal energetic impairment following controlled cortical impact injury in rats. *Neurol Res*, 2007. 29, p. 594–603.

Thompson, S.N., T.R. Gibson, B.M. Thompson, Y. Deng, E.D. Hall, Relationship of calpain-mediated proteolysis to the expression of axonal and synaptic plasticity markers following traumatic brain injury in mice. *Exp Neurol*, 2006. 201, p. 253–265.

Thornton, E., R. Vink, P.C. Blumbergs, H.C. van den, Soluble amyloid precursor protein alpha reduces neuronal injury and improves functional outcome following diffuse traumatic brain injury in rats. *Brain Res*, 2006. 1094, p. 38–46.

Tuzgen, S., N. Tanriover, M. Uzan, E. Tureci, T. Tanriverdi, K. Gumustas, C. Kuday, Nitric oxide levels in rat cortex, hippocampus, cerebellum, and brainstem after impact acceleration head injury. *Neurol Res*, 2003. 25, p. 31–34.

Tyurin, V.A., Y.Y. Tyurina, G.G. Borisenko, T.V. Sokolova, V.B. Ritov, P.J. Quinn, M. Rose, P. Kochanek, S.H. Graham, V.E. Kagan, Oxidative stress following traumatic brain injury in rats: Quantitation of biomarkers and detection of free radical intermediates. *J Neurochem*, 2000. 75, p. 2178–2189.

Ucar, T., G. Tanriover, I. Gurer, M.Z. Onal, S. Kazan, Modified experimental mild traumatic brain injury model. *J Trauma*, 2006. 60, p. 558–565.

Ueda, Y., S.A. Walker, J.T. Povlishock, Perivascular nerve damage in the cerebral circulation following traumatic brain injury. *Acta Neuropathol*, 2006. 112, p. 85–94.

Ueno, M., T. Yamashita, Kinematic analyses reveal impaired locomotion following injury of the motor cortex in mice. *Exp Neurol*, 2011. 230, p. 280–290.

Vagnozzi, R., A. Marmarou, B. Tavazzi, S. Signoretti, D. Di Pierro, F. del Bolgia, A.M. Amorini, G. Fazzina, S. Sherkat, G. Lazzarino, Changes of cerebral energy metabolism and lipid peroxidation in rats leading to mitochondrial dysfunction after diffuse brain injury. *J Neurotrauma*, 1999. 16, p. 903–913.

Verrier, J.D., T.C. Jackson, R. Bansal, P.M. Kochanek, A.M. Puccio, D.O. Okonkwo, E.K. Jackson, The brain in vivo expresses the 2′,3′-cAMP-adenosine pathway. *J Neurochem*, 2012. 122, p. 115–125.

Vink, R., C.A. O'Connor, A.J. Nimmo, D.L. Heath, Magnesium attenuates persistent functional deficits following diffuse traumatic brain injury in rats. *Neurosci Lett*, 2003. 336, p. 41–44.

Wagner, A.K., B.A. Postal, S.D. Darrah, X. Chen, A.S. Khan, Deficits in novelty exploration after controlled cortical impact. *J Neurotrauma*, 2007. 24, p. 1308–1320.

Wakade, C., S. Sukumari-Ramesh, M.D. Laird, K.M. Dhandapani, J.R. Vender, Delayed reduction in hippocampal postsynaptic density protein-95 expression temporally correlates with cognitive dysfunction following controlled cortical impact in mice. *J Neurosurg*, 2010. 113, p. 1195–1201.

Wang, G., J. Zhang, X. Hu, L. Zhang, L. Mao, X. Jiang, A.K. Liou, R.K., Leak, Y. Gao, J. Chen, Microglia/macrophage polarization dynamics in white matter after traumatic brain injury. *J Cereb Blood Flow Metab*, 2013. 33, p. 1864–1874.

Wang, H.C., Z.X. Duan, F.F. Wu, L. Xie, H. Zhang, Y.B. Ma, A new rat model for diffuse axonal injury using a combination of linear acceleration and angular acceleration. *J Neurotrauma*, 2010. 27, p. 707–719.

Wang, Y., M. Neumann, K. Hansen, S.M. Hong, S. Kim, L.J. Noble-Haeusslein, J. Liu, Fluoxetine increases hippocampal neurogenesis and induces epigenetic factors but does not improve functional recovery after traumatic brain injury. *J Neurotrauma*, 2011. 28, p. 259–268.

Washington, P.M., P.A. Forcelli, T. Wilkins, D.N. Zapple, M. Parsadanian, M.P. Burns, The effect of injury severity on behavior: A phenotypic study of cognitive and emotional deficits after mild, moderate, and severe controlled cortical impact injury in mice. *J Neurotrauma*, 2012. 29, p. 2283–2296.

Whalen, M.J., T.M. Carlos, R.S.B. Clark, D.W. Marion, S.T. Dekosky, S. Heineman, J.K. Schiding, F. Memarzadeh, P.M. Kochanek, The effect of brain temperature on acute inflammation after traumatic brain injury in rats. *J Neurotrauma*, 1997. 14, p. 561–572.

Whalen, M.J. et al., Neutrophils do not mediate blood-brain barrier permeability early after controlled cortical impact in rats. *J Neurotrauma*, 1999. 16, p. 583–594.
Whishaw, I.Q., D.M. Piecharka, F. Zeeb, D.G. Stein, Unilateral frontal lobe contusion and forelimb function: Chronic quantitative and qualitative impairments in reflexive and skilled forelimb movements in rats. *J Neurotrauma*, 2004. 21, p. 1584–1600.
Xiong, Y., P.L. Peterson, B.H. Verweij, F.C. Vinas, J.P. Muizelaar, C.P. Lee, Mitochondrial dysfunction after experimental traumatic brain injury: Combined efficacy of SNX-111 and U-101033E. *J Neurotrauma*, 1998. 15, p. 531–544.
Xu, S., J. Zhuo, J. Racz, D. Shi, S. Roys, G. Fiskum, R. Gullapalli, Early microstructural and metabolic changes following controlled cortical impact injury in rat: A magnetic resonance imaging and spectroscopy study. *J Neurotrauma*, 2011. 28, p. 2091–2102.
Yamamoto, M., C.R. Marmarou, M.F. Stiefel, A. Beaumont, A. Marmarou, Neuroprotective effect of hypothermia on neuronal injury in diffuse traumatic brain injury coupled with hypoxia and hypotension. *J Neurotrauma*, 1999. 16, p. 487–500.
Yan, E.B., S.C. Hellewell, B.M. Bellander, D.A. Agyapomaa, M.C. Morganti-Kossmann, Post-traumatic hypoxia exacerbates neurological deficit, neuroinflammation and cerebral metabolism in rats with diffuse traumatic brain injury. *J Neuroinflammation*, 2011. 8, 147.
Yang, L., S. Afroz, H.B. Michelson, J.H. Goodman, H.A. Valsamis, D.S. Ling, Spontaneous epileptiform activity in rat neocortex after controlled cortical impact injury. *J Neurotrauma*, 2010. 27, p. 1541–1548.
Yang, S.Y., J.Z. Cui, Expression of the basic fibroblast growth factor gene in mild and more severe head injury in the rat. *J Neurosurg*, 1998. 89, p. 297–302.
Yu, S., Y. Kaneko, E. Bae, C.E. Stahl, Y. Wang, L.H. van, P.R. Sanberg, C.V. Borlongan, Severity of controlled cortical impact traumatic brain injury in rats and mice dictates degree of behavioral deficits. *Brain Res*, 2009. 1287, p. 157–163.
Zhang, Y., M. Chopp, Y. Meng, Z.G. Zhang, E. Doppler, A. Mahmood, Y. Xiong, Improvement in functional recovery with administration of Cerebrolysin after experimental closed head injury. *J Neurosurg*, 2013. 118, p. 1343–1355.
Zhao, Z., D.J. Loane, M.G. Murray, B.A. Stoica, A.I. Faden, Comparing the predictive value of multiple cognitive, affective, and motor tasks after rodent traumatic brain injury. *J Neurotrauma*, 2012. 29, p. 2475–2489.

Section 4.5: Fluid Percussion

Adams, J.H. et al., Neuropathological findings in disabled survivors of a head injury. *J Neurotrauma*, 2011. 28(5), p. 701–709.
Barkhoudarian, G., D.A. Hovda, C.C. Giza, The molecular pathophysiology of concussive brain injury. *Clin Sports Med*, 2011. 30(1), p. 33–48, vii–iii.
Behan, L.A. et al., Neuroendocrine disorders after traumatic brain injury. *J Neurol Neurosurg Psychiatry*, 2008. 79(7), p. 753–759.
Berger, R.P. et al., Serum concentrations of ubiquitin C-terminal hydrolase-L1 and alphaII-spectrin breakdown product 145 kDa correlate with outcome after pediatric TBI. *J Neurotrauma*, 2012. 29(1), p. 162–167.
Bergsneider, M. et al., Dissociation of cerebral glucose metabolism and level of consciousness during the period of metabolic depression following human traumatic brain injury. *J Neurotrauma*, 2000. 17(5), p. 389–401.
Chen, A.J., M. D'Esposito, Traumatic brain injury: From bench to bedside [corrected] to society. *Neuron*, 2010. 66(1), p. 11–14.
D'Ambrosio, R. et al., Post-traumatic epilepsy following fluid percussion injury in the rat. *Brain*, 2004. 127(Pt 2), p. 304–134.

Dixon, C.E., J.W. Lighthall, T.E. Anderson, Physiologic, histopathologic, and cineradiographic characterization of a new fluid-percussion model of experimental brain injury in the rat. *J Neurotrauma*, 1988. 5(2), p. 91–104.

Dusick, J.R. et al., Pathophysiology of hypopituitarism in the setting of brain injury. *Pituitary*, 2012. 15(1), p. 2–9.

Ekmark-Lewen, S. et al., Traumatic axonal injury in the mouse is accompanied by a dynamic inflammatory response, astroglial reactivity and complex behavioral changes. *J Neuroinflammation*, 2013. 10, p. 44.

Faul M, X.L., Wald MM, Coronado VG., *Traumatic Brain Injury in the United States: Emergency Department Visits, Hospitalizations and Deaths 2002–2006*, 2010. Atlanta, GA: Centers for Disease Control and Prevention, National Center for Injury Prevention and Control.

Floyd, C.L. et al., Craniectomy position affects morris water maze performance and hippocampal cell loss after parasagittal fluid percussion. *J Neurotrauma*, 2002. 19(3), p. 303–316.

Fujita, M., E.P. Wei, J.T. Povlishock, Effects of hypothermia on cerebral autoregulatory vascular responses in two rodent models of traumatic brain injury. *J Neurotrauma*, 2012. 29(7), p. 1491–1498.

Griesbach, G.S. et al., Differential effects of voluntary and forced exercise on stress responses after traumatic brain injury. *J Neurotrauma*, 2012. 29(7), p. 1426–1433.

Griesbach, G.S. et al., Recovery of stress response coincides with responsiveness to voluntary exercise after traumatic brain injury. *J Neurotrauma*, 2014. 31, p. 674–682.

Grundy, P.L. et al., The hypothalamo-pituitary-adrenal axis response to experimental traumatic brain injury. *J Neurotrauma*, 2001. 18(12), p. 1373–1381.

Hamm, R.J., Neurobehavioral assessment of outcome following traumatic brain injury in rats: An evaluation of selected measures. *J Neurotrauma*, 2001. 18(11), p. 1207–1216.

Hawkins, B.E. et al., Effects of trauma, hemorrhage and resuscitation in aged rats. *Brain Res*, 2013. 1496, p. 28–35.

Hillered, L., P.M. Vespa, D.A. Hovda, Translational neurochemical research in acute human brain injury: The current status and potential future for cerebral microdialysis. *J Neurotrauma*, 2005. 22(1), p. 3–41.

Hogg, S. et al., Neuroprotective effect of eliprodil: Attenuation of a conditioned freezing deficit induced by traumatic injury of the right parietal cortex in the rat. *J Neurotrauma*, 1998. 15(7), p. 545–553.

Hosseini, A.H., J. Lifshitz, Brain injury forces of moderate magnitude elicit the fencing response. *Med Sci Sports Exerc*, 2009. 41(9), p. 1687–1697.

Ingebrigtsen, T. et al., The clinical value of serum S-100 protein measurements in minor head injury: A Scandinavian multicentre study. *Brain Inj*, 2000. 14(12), p. 1047–1055.

Ip, E.Y. et al., Metabolic, neurochemical, and histologic responses to vibrissa motor cortex stimulation after traumatic brain injury. *J Cereb Blood Flow Metab*, 2003. 23(8), p. 900–910.

Jones, N.C. et al., Experimental traumatic brain injury induces a pervasive hyperanxious phenotype in rats. *J Neurotrauma*, 2008. 25(11), p. 1367–1374.

Kleindienst, A. et al., Assessment of cerebral S100B levels by proton magnetic resonance spectroscopy after lateral fluid-percussion injury in the rat. *J Neurosurg*, 2005. 102(6), p. 1115–1121.

Lafrenaye, A.D., M.J. McGinn, J.T. Povlishock, Increased intracranial pressure after diffuse traumatic brain injury exacerbates neuronal somatic membrane poration but not axonal injury: Evidence for primary intracranial pressure-induced neuronal perturbation. *J Cereb Blood Flow Metab*, 2012. 32(10), p. 1919–1932.

Lifshitz, J., Fluid percussion injury, in *Animal Models of Acute Neurological Injuries*, J. Chen et al., eds. 2008, Totowa, NJ: Humana Press.

McAllister, T.W., Neuropsychiatric sequelae of head injuries. *Psychiatr Clin North Am*, 1992. 15(2), p. 395–413.

McGinn, M.J. et al., Biochemical, structural, and biomarker evidence for calpain-mediated cytoskeletal change after diffuse brain injury uncomplicated by contusion. *J Neuropathol Exp Neurol*, 2009. 68(3), p. 241–249.

McIntosh, T.K. et al., Traumatic brain injury in the rat: Characterization of a lateral fluid-percussion model. *Neuroscience*, 1989. 28(1), p. 233–244.

McNamara, K.C., A.M. Lisembee, J. Lifshitz, The whisker nuisance task identifies a late-onset, persistent sensory sensitivity in diffuse brain-injured rats. *J Neurotrauma*, 2010. 27(4), p. 695–706.

Mez, J., R.A. Stern, A.C. McKee, Chronic traumatic encephalopathy: Where are we and where are we going? *Curr Neurol Neurosci Rep*, 2013. 13(12), p. 407.

Mondello, S. et al., Blood-based diagnostics of traumatic brain injuries. *Expert Rev Mol Diagn*, 2011. 11(1), p. 65–78.

Pierce, J.E. et al., Enduring cognitive, neurobehavioral and histopathological changes persist for up to one year following severe experimental brain injury in rats. *Neuroscience*, 1998. 87(2), p. 359–369.

Pitkanen, A., T. Bolkvadze, Head trauma and epilepsy, in *Jasper's Basic Mechanisms of the Epilepsies*, J.L. Noebels et al., eds. 2012, Bethesda, MD.

Pitkanen, A., T.K. McIntosh, Animal models of post-traumatic epilepsy. *J Neurotrauma*, 2006. 23(2), p. 241–261.

Povlishock, J.T., D.I. Katz, Update of neuropathology and neurological recovery after traumatic brain injury. *J Head Trauma Rehabil*, 2005. 20(1), p. 76–94.

Prins, M.L., D.A. Hovda, Developing experimental models to address traumatic brain injury in children. *J Neurotrauma*, 2003. 20(2), p. 123–137.

Raghupathi, R., T.K. McIntosh, D.H. Smith, Cellular responses to experimental brain injury. *Brain Pathol*, 1995. 5(4), p. 437–442.

Reeves, T.M. et al., Proteolysis of submembrane cytoskeletal proteins ankyrin-G and alphaII-spectrin following diffuse brain injury: A role in white matter vulnerability at Nodes of Ranvier. *Brain Pathol*, 2010. 20(6), p. 1055–1068.

Reger, M.L. et al., Concussive brain injury enhances fear learning and excitatory processes in the amygdala. *Biol Psychiatry*, 2012. 71(4), p. 335–343.

Riess, P. et al., Effects of chronic, post-injury Cyclosporin A administration on motor and sensorimotor function following severe, experimental traumatic brain injury. *Restor Neurol Neurosci*, 2001. 18(1), p. 1–8.

Rowe, R.K., J.L. Harrison, B.F. O'Hara, J. Lifshitz, Diffuse brain injury does not affect chronic sleep patterns in the mouse. *Brain Inj*, 2014. 28(4), p. 504–510.

Rowe, R.K., M. Striz, A.D. Bachstetter, L.J. Van Eldik, K.D. Donohue, B.F. O'Hara, J. Lifshitz, Diffuse brain injury induces acute post-traumatic sleep. *PLoS One*, 2014. 9(1), p. e82507.

Santiago, P.N. et al., A systematic review of PTSD prevalence and trajectories in DSM-5 defined trauma exposed populations: Intentional and non-intentional traumatic events. *PLoS One*, 2013. 8(4), p. e59236.

Schmidt, R.H., M.S. Grady, Regional patterns of blood-brain barrier breakdown following central and lateral fluid percussion injury in rodents. *J Neurotrauma*, 1993. 10(4), p. 415–430.

Shultz, S.R. et al., Repeated mild lateral fluid percussion brain injury in the rat causes cumulative long-term behavioral impairments, neuroinflammation, and cortical loss in an animal model of repeated concussion. *J Neurotrauma*, 2012. 29(2), p. 281–294.

Stein, D.G., Sex differences in brain damage and recovery of function: Experimental and clinical findings. *Prog Brain Res*, 2007. 161, p. 339–351.

Sviri, G.E. et al., Time course for autoregulation recovery following severe traumatic brain injury. *J Neurosurg*, 2009. 111(4), p. 695–700.

Thompson, H.J. et al., Lateral fluid percussion brain injury: A 15-year review and evaluation. *J Neurotrauma*, 2005. 22(1), p. 42–75.
Vink, R. et al., Small shifts in craniotomy position in the lateral fluid percussion injury model are associated with differential lesion development. *J Neurotrauma*, 2001. 18(8), p. 839–847.
Yan, X.X., A. Jeromin, Spectrin breakdown products (SBDPs) as potential biomarkers for neurodegenerative diseases. *Curr Transl Geriatr Exp Gerontol Rep*, 2012. 1(2), p. 85–93.
Yeh, C.C. et al., Risk of epilepsy after traumatic brain injury: A retrospective population-based cohort study. *J Neurol Neurosurg Psychiatry*, 2013. 84(4), p. 441–445.
Yoshino, A. et al., Dynamic changes in local cerebral glucose utilization following cerebral conclusion in rats: Evidence of a hyper- and subsequent hypometabolic state. *Brain Res*, 1991. 561(1), p. 106–119.

Section 4.6: Pediatric TBI

Adelson, P.D. et al., Guidelines for the acute medical management of severe traumatic brain injury in infants, children, and adolescents. *Pediatr Crit Care Med*, 2003. 4, p. S1–S75.
Adelson, P.D., C.E. Dixon, P.M. Kochanek, Long-term dysfunction following diffuse traumatic brain injury in the immature rat. *J Neurotrauma*, 2000. 17, p. 273–282.
Adelson, P.D., C.E. Dixon, P. Robichaud, P.M. Kochanek, Motor and cognitive functional deficits following diffuse traumatic brain injury in the immature rat. *J Neurotrauma*, 1997. 14, p. 99–108.
Adelson, P.D., L.W. Jenkins, R.L. Hamilton, P. Robichaud, M.P. Tran, P.M. Kochanek, Histopathologic response of the immature rat to diffuse traumatic brain injury. *J Neurotrauma*, 2001. 18, p. 967–976.
Ajao D.O. et al., Traumatic brain injury in young rats leads to progressive behavioral deficits coincident with altered tissue properties in adulthood. *J Neurotrauma*, 2012. 29(11), p. 2060–2074.
Aldrich, E.F. et al., Diffuse brain swelling in severely head-injured children. A report from the NIH Traumatic Coma Data Bank. *J Neurosurg*, 1992. 76, p. 450–454.
Anderson, V., C. Catroppa, S. Morse, F. Haritou, J. Rosenfeld, Functional plasticity or vulnerability after early brain injury? *Pediatrics*, 2005. 116, p. 1374–1382.
Bittigau, P. et al., Apoptotic neurodegeneration following trauma is markedly enhanced in the immature brain. *Ann Neurol*, 1999. 45, p. 724–735.
Bruce, D.A., A. Alavi, L. Bilaniuk, C. Dolinskas, W. Obrist, B. Uzzell, Diffuse cerebral swelling following head injuries in children: The syndrome of "malignant brain edema." *J Neurosurg*, 1981. 54, p. 170–178.
Cernak, I. et al., Pathophysiological response to experimental diffuse brain trauma differs as a function of developmental age. *Dev Neurosci*, 2010. 32, p. 442–453.
Clause, C.P. et al., Age is a determinant of leukocyte infiltration and loss of cortical volume after traumatic brain injury. *Dev Neurosci*, 2011. 32, p. 454–465.
Coronado, V.G., L. Xu, S.V. Basavaraju, L.C. McGuire, M.M. Wald, M.D. Faul, B.R. Guzman, J.D. Hemphill,; Centers for Disease Control and Prevention (CDC), Surveillance for traumatic brain injury-related deaths—United States, 1997–2007. *MMWR Surveill Summ*, 2011. 60(5), p. 1–32.
Dikranian, K. et al., Mild traumatic brain injury to the infant mouse causes robust white matter axonal degeneration which precedes apoptotic death of cortical and thalamic neurons. *Exp Neurol*, 2008. 211, p. 551–560.
DiLeonardi, A.M., J.W. Huh, R. Raghupathi, Differential effects of FK506 on structural and functional axonal deficits after diffuse brain injury in the immature rat. *J Neuropathol Exp Neurol*, 2012. 71(11), p. 959–972.

Duhaime, A.C., S. Durham, Traumatic brain injury in infants: The phenomenon of subdural hemorrhage with hemispheric hypodensity ("Big Black Brain"). *Prog Brain Res*, 2007. 161, p. 293–302.

Duhaime, A.-C. et al., Magnetic resonance imaging studies of age-dependent responses to scaled focal brain injury in the piglet. *J Neurosurg*, 2003. 99, p. 542–548.

Duhaime, A.C. et al., Maturation-dependent response of the piglet brain to scaled cortical impact. *J Neurosurg*, 2000. 93, p. 455–462.

Fineman, I., C.C. Giza, B.V. Nahed, S.M. Lee, D.A. Hovda, Inhibition of neocortical plasticity during development by a moderate concussive brain injury. *J Neurotrauma*, 2000. 17, p. 739–749.

Friess, S.H. et al., Neurobehavioral functional deficits following closed head injury in the neonatal pig. *Exp Neurol*, 2007. 204, p. 234–243.

Giza, C.C., N.S. Maria, D.A. Hovda, N-methyl-D-aspartate receptor subunit changes after traumatic injury to the developing brain. *J Neurotrauma*, 2006. 23, p. 950–961.

Gurkoff, G.G., C.C. Giza, D.A. Hovda, Lateral fluid percussion injury in the developing rat causes an acute, mild behavioral dysfunction in the absence of significant cell death. *Brain Res*, 2006. 1077, p. 24–36.

Huh, J.W., R. Raghupathi, Chronic cognitive deficits and long-term histopathological alterations following contusive brain injury in the immature rat. *J Neurotrauma*, 2007. 24, p. 1460–1474.

Huh, J.W., M.A. Franklin, A.G. Widing, R. Raghupathi, Regionally distinct patterns of calpain activation and traumatic axonal injury following contusive brain injury in immature rats. *Dev Neurosci*, 2006. 28, p. 466–476.

Huh, J.W., A.G. Widing, R. Raghupathi, Midline brain injury in the immature rat induces sustained cognitive deficits, bihemispheric axonal injury and neurodegeneration. *Exp Neurol*, 2008. 213, p. 84–92.

Huh, J.W., A.G. Widing, R. Raghupathi, Differential effects of injury severity on cognition and cellular pathology after contusive brain trauma in the immature rat. *J Neurotrauma*, 2011. 28, p. 245–257.

Ibrahim, N.G., J. Ralston, C. Smith, S.S. Margulies, Physiological and pathological responses to head rotations in toddler piglets. *J Neurotrauma*, 2010. 27, p. 1021–1035.

Ip, E.Y.Y., C.C. Giza, G.S. Griesbach, D.A. Hovda, Effects of enriched environment and fluid percussion injury on dendritic arborization within the cerebral cortex of the developing rat. *J Neurotrauma*, 2002. 19, p. 573–585.

Kochanek, P.M. et al., Guidelines for the acute medical management of severe traumatic brain injury in infants, children, and adolescents, 2nd ed. *Pediatr Crit Care Med*, 2012. 13 Suppl 1, p. S1–S82.

Levin, H.S. et al., Severe head injury in children: Experience of the Traumatic Coma Data Bank. *Neurosurg*, 1992. 31, p. 435–443.

Marmarou, C.R., S.A. Walker, C.L. Davis, J.T. Povlishock, Quantitative analysis of the relationship between intra-axonal neurofilament compaction and impaired axonal transport following diffuse traumatic brain injury. *J Neurotrauma*, 2005. 22, p. 1066–1080.

Monyer, H., N. Burnashev, D.J. Laurie, Sa kmann B., P.H. Seeburg, Developmental and regional expression in the rat brain and functional properties of four NMDA receptors. *Neuron*, 1994. 12, p. 529–540.

Osteen, C.L., C.C. Giza, D.A. Hovda, Injury-induced alterations in N-methyl-D-aspartate receptor subunit composition contribute to prolonged 45 calcium accumulation following lateral fluid percussion. *Neuroscience*, 2004. 128, p. 305–322.

Ozdemir, D. et al., Protective effect of melatonin against head trauma-induced hippocampal damage and spatial memory deficits in immature rats. *Neurosci Lett*, 2005. 385, p. 234–239.

Prins, M.L., D.A. Hovda, Traumatic brain injury in the developing rat: Effects of maturation on Morris water maze acquisition. *J Neurotrauma*, 1998. 15, p. 799–811.
Pullela, R. et al., Traumatic injury to the immature brain results in progressive neuronal loss, hyperactivity and delayed cognitive impairments. *Dev Neurosci*, 2006. 28, p. 396–409.
Raghupathi, R.R., Cell death mechanisms following traumatic brain injury. *Brain Pathol*, 2004. 14(2), p. 215–222.
Raghupathi, R., J.W. Huh, Diffuse brain injury in the immature rat: Evidence for an age-at-injury effect on cognitive function and histopathologic damage. *J Neurotrauma*, 2007. 24, p. 1596–1608.
Raghupathi, R., S.S. Margulies, Traumatic axonal injury after closed head injury in the neonatal pig. *J Neurotrauma*, 2002. 19, p. 843–853.
Reeves, T.M., L.L. Phillips, J.T. Povlishock, Myelinated and unmyelinated axons of the corpus callosum differ in vulnerability and functional recovery following traumatic brain injury. *Exp Neurol*, 2005. 196(1), p. 126–137.
Semple, B.D., S.A. Canchola, L.J. Noble-Haeusslein, Deficits in social behavior emerge during development after pediatric traumatic brain injury in mice. *J Neurotrauma*, 2012. 29(17), p. 2672–2683.
Smith, D.H. et al., Immediate coma following inertial brain injury dependent on axonal damage in the brainstem. *J Neurosurg*, 2000. 93, p. 315–322.
Tovar, K.R., G.L. Westbrook, The incorporation of NMDA receptors with a distinct subunit composition at nascent hippocampal synapses in vitro. *J Neurosci*, 1999. 19, p. 4180–4188.

Section 4.7: Explosive/Blast TBI

Overview

Armed Forces Health Surveillance Center, DoD TBI statistics 2000–2013, in *DoD Defense*, 2013. Washington, DC: U.S. Government Printing Office, p. 1–5.
Armonda, R.A. et al., Wartime traumatic cerebral vasospasm: Recent review of combat casualties. *Neurosurgery*, 2006. 59(6), p. 1215–1225; discussion 1225.
Baker, W.E., *Explosions in Air*, 1973. Austin, TX: University of Texas Press.
Camm, A.J., C.J. Garratt, Adenosine and supraventricular tachycardia [see comments]. *N Engl J Med*, 1991. 325(23), p. 1621–1629.
Cernak, I. et al., Blast injury from explosive munitions. *J Trauma*, 1999. 47(1), p. 96–103; discussion 103–104.
Cernak, I. et al., Cognitive deficits following blast injury-induced neurotrauma: Possible involvement of nitric oxide. *Brain Inj*, 2001. 15(7), p. 593–612.
Cernak, I. et al., The pathobiology of blast injuries and blast-induced neurotrauma as identified using a new experimental model of injury in mice. *Neurobiol Dis*, 2011. 41(2), p. 538–551.
Cernak, I. et al., Ultrastructural and functional characteristics of blast injury-induced neurotrauma. *J Trauma*, 2001. 50(4), p. 695–706.
Courtney, A.C., M.W. Courtney, A thoracic mechanism of mild traumatic brain injury due to blast pressure waves. *Med Hypotheses*, 2009. 72(1), p. 76–83.
Cullis, I.G., Blast waves and how they interact with structures. *J R Army Med Corps*, 2001. 147(1), p. 16–26.
Defense and Veterans Brain Injury Center (DVBIC), Military Acute Concussion Evaluation (MACE). *DVBIC Brainwaves*, 2008. Summer, p. 1–2.
Defense and Veterans Brain Injury Center (DVBIC), *Consensus Conference on the acute management of mild traumatic brain injury in military operational settings: Clinical practice guidelines and recommendations*. 2006 [cited April 2, 2013].

Defense and Veterans Brain Injury Center (DVBIC), *Military Acute Concussion Evaluation*. 2007 July, 2007 [cited December 27, 2011]; Available from: http://www.pdhealth.mil/downloads/MACE.pdf.

de Lanerolle, N.C. et al., Characteristics of an explosive blast-induced brain injury in an experimental model. *J Neuropathol Exp Neurol*, 2011. 70(11), p. 1046–1057.

Desmoulin, G.T., J.P. Dionne, Blast-induced neurotrauma: Surrogate use, loading mechanisms, and cellular responses. *J Trauma*, 2009. 67(5), p. 1113–1122.

Desnoyers, M.R. et al., Pulmonary hemorrhage in lupus erythematosus without evidence of an immunologic cause. *Arch Intern Med*, 1984. 144(7), p. 1398–1400.

Giza, C.C. et al., Summary of evidence-based guideline update: Evaluation and management of concussion in sports: Report of the Guideline Development Subcommittee of the American Academy of Neurology. *Neurology*, 2013. 80(24), p. 2250–2257.

Ling, G. et al., Explosive blast neurotrauma. *J Neurotrauma*, 2009. 26(6), p. 815–825.

Ling, G.S., J.M. Ecklund, Traumatic brain injury in modern war. *Curr Opin Anaesthesiol*, 2011. 24(2), p. 124–130.

Lu, J. et al., Effect of blast exposure on the brain structure and cognition in macaca fascicularis. *J Neurotrauma*, 2012. 29, p. 1434–1454.

Magnuson, J., F. Leonessa, G.S. Ling, Neuropathology of explosive blast traumatic brain injury. *Curr Neurol Neurosci Rep*, 2012. 12(5), p. 570–579.

Masel, B.E. et al., Galveston Brain Injury Conference 2010: Clinical and experimental aspects of blast injury. *J Neurotrauma*, 2012. 29(12), p. 2143–2171.

Phillips, Y.Y., Primary blast injuries. *Ann Emerg Med*, 1986. 15(12), p. 1446–1450.

The Brain Trauma Foundation. The American Association of Neurological Surgeons. Guidelines for the management of severe traumatic brain injury. *J Neurotrauma*, 2007. 24(Suppl 1), p. S1–106.

VA/DoD Clinical Practice Guideline for Management of Concussion/Mild Traumatic Brain Injury. *J Rehabil Res Dev*, 2009. 46(6): p. CP1–CP68.

Wilk, J.E. et al., Mild traumatic brain injury (concussion) during combat: Lack of association of blast mechanism with persistent postconcussive symptoms. *J Head Trauma Rehabil*, 2010. 25(1), p. 9–14.

Experimental Blast TBI

Axelsson, H., H. Hjelmqvist, A. Medin, J.K. Persson, A. Suneson, 2000. Physiological changes in pigs exposed to a blast wave from a detonating high-explosive charge. *Mil Med*, 165, p. 119–126.

Bass, C.R., M.B. Panzer, K.A. Rafaels, G. Wood, J. Shridharani, B. Capehart, Brain injuries from blast. *Ann Biomed Eng*, 2012. 40(1), 185–202.

Bauman, R.A. et al., An introductory characterization of a combat-casualty-care relevant swine model of closed head injury resulting from exposure to explosive blast. *J Neurotrauma*, 2009 26(6), 841–60.

Benzinger, T.L., D. Brody, S. Cardin, K.C. Curley, M.A. Mintun, S.K. Mun, K.H. Wong, J.R. Wrathall, Blast-related brain injury: Imaging for clinical and research applications: Report of the 2008 St. Louis workshop. *J Neurotrauma*, 2009. 26, p. 2127–2144.

Bochicchio, G.V., K. Lumpkins, J. O'Connor, M. Simard, S. Schaub, A. Conway, K. Bochicchio, T.M. Scalea, Blast injury in a civilian trauma setting is associated with a delay in diagnosis of traumatic brain injury. *Am Surg*, 2008. 74, p. 267–270.

Bolander, R., B. Mathie, C. Bir, D. Ritzel, P. VandeVord, Skull flexure as a contributing factor in the mechanism of injury in the rat when exposed to a shock wave. *Ann Biomed Eng*, 2011. 39(10), p. 2550–2559.

Bowen, I.G., A. Holladay, E.R. Fletcher, D.R. Richmond, C.S. White. *A fluid-mechanical model of the thoracoabdominal system with applications to blast biology*. DASA-1675. Albuquerque, NM: Lovelace Foundation for Medical Education and Research, 1965.

Bowen, I.G., E.R. Fletcher, D.R. Richmond. *Estimate of man's tolerance to the direct effects of air blast*. DASA-2113. Albuquerque, NM: Lovelace Foundation for Medical Education and Research, 1968.

Celander, H., C.J. Clemedson, U.A. Ericsonn, H.I. Hultman. The use of a compressed air operated shock tube for physiological blast research. *Acta Physiol Scand*, 1955. 33(1), p. 6–13.

Cernak, I., D. Ignjatovic, G. Andelic, J. Savic, [Metabolic changes as part of the general response of the body to the effect of blast waves]. *Vojnosanit Pregl*, 1991. 48, 515–522.

Cernak, I., L.J. Noble-Haeusslein, Traumatic brain injury: An overview of pathobiology with emphasis on military populations. *J Cereb Blood Flow Metab*, 2010. 30, 255–266.

Cernak, I., Z. Wang, J. Jiang, X. Bian, J. Savic, Ultrastructural and functional characteristics of blast injury-induced neurotrauma. *J Trauma*, 2001. 50(4), 695–706.

Champion, H.R., J.B. Holcomb, L.A. Young, 2009. Injuries from explosions: Physics, biophysics, pathology, and required research focus. *J Trauma*, 66, 1468–1477, discussion 77.

Chu, J.J., J.G. Beckwith, D.S. Leonard, C.M. Paye, R.M. Greenwald, Development of a multimodal blast sensor for measurement of head impact and over-pressurization exposure. *Ann Biomed Eng*, 2012. 40(1), 203–212.

Clemedson, C.J., C.O. Criborn. A detonation chamber for physiological blast research. *J Aviat Med*, 1955. 26(5), 373–381.

Clemedson, C.J. Shock wave transmission to the central nervous system. *Acta Physiol Scand*, 1956. 37(2–3), 204–214.

Clemedson, C.J., H. Pettersson. Genesis of respiratory and circulatory changes in blast injury. *Am J Physiol*, 174, 316–320, 1953.

Clemedson, C.J., H. Pettersson. Propagation of a high explosive air shock wave through different parts of an animal body. *Am J Physiol*, 1956. 184(1), p. 119–126.

Cooper, G.J., R.L. Maynard, N L. Cross, J.F. Hill. Casualties from terrorist bombings. *J Trauma*, 1983. 23, p. 955–967.

DMDC. Table 501. Military Personnel on Active Duty by Rank or Grade: 1990 to 2006. http://siadapp.dmdc.osd.mil/personnel/CASUALTY/castop.htm.

Engin, A.E., The axisymmetric response of a fluid-filled spherical shell to local radial impulse—A model for head injury. *J Biomechanics*, 1969. 2, p. 325–341.

Ganpule, S., A. Alai, E. Plougonven, N. Chandra, Mechanics of blast loading on the head models in the study of traumatic brain injury using experimental and computational approaches. *Biomech Model Mechanobiol*, 2013. 12(3), p. 511–531.

Goeller, J., A. Wardlaw, D. Treichler, J. O'Bruba, G. Weiss, Investigation of cavitation as a possible damage mechanism in blast-induced traumatic brain injury. *J Neurotrauma*, 2012. 29(10), p. 1970–1981.

Goldstein, L.E. et al., Chronic traumatic encephalopathy in blast-exposed military veterans and a blast neurotrauma mouse model. *Sci Transl Med*, 2012. 4(134), p. 134ra60.

Hicks, R.R., S.J. Fertig, R.E. Desrocher, W.J. Koroshetz, J.J. Pancrazio. Review. Neurological effects of blast injury. *J Trauma*, 2010. 68(5), p. 1257–1263.

Hoge, C.W., H.M. Goldberg, C.A. Castro. Care of war veterans with mild traumatic brain injury—Flawed perspectives. *N Engl J Med*, 2009. 360(16), p. 1588–1591.

Hoge, C.W., D. McGurk, J.L. Thomas, A.L. Cox, C.C. Engel, C.A. Castro. Mild traumatic brain injury in U.S. Soldiers returning from Iraq. *N Engl J Med*, 2008. 358(5), p. 453–463.

Institute of Medicine. 2009. *Gulf War and Health, Volume 7: Long-Term Consequences of Traumatic Brain Injury*. Washington, DC: National Academies Press.

Lakis, N., R.J. Corona, G. Toshkezi, L.S. Chin. Chronic traumatic encephalopathy—Neuropathology in athletes and war veterans. *Neurol Res*, 2013. 35(3), p. 290–299.

Leibovici, D., O.N. Gofrit, M. Stein, S.C. Shapira, Y. Noga, R.J. Heruti, J. Shemer. Blast injuries: Bus versus open-air bombings—A comparative study of injuries in survivors of open-air versus confined-space explosions. *J Trauma*. 1996. 41: p. 1030–1035.

Leung, L.Y., P.J. VandeVord, A.L. Dal Cengio, C. Bir, K.H. Yang, A.I. King. Blast related neurotrauma: A review of cellular injury. *Mol Cell Biomech*, 2008. 5(3), p. 155–168. Review.

Ling, G., F. Bandak, R. Armonda, G. Grant, J. Ecklund. 2009. Explosive blast neurotrauma. *J Neurotrauma*, 26, p. 815–825.

Long, J.B. et al., Blast-induced traumatic brain injury: Using a shock tube to recreate a battlefield injury in the laboratory. *IFMBE Proc*, 2010. 32: p. 26–30.

Lu, J. et al., Effect of blast exposure on the brain structure and cognition in Macaca fascicularis. *J Neurotrauma*, 2012. 29(7), p. 1434–1454.

Lucci, E.B., Civilian preparedness and counter-terrorism: Conventional weapons. *Surg Clin N Am*. 2006. 86, 579–600.

Mediavilla Varas, J., M. Philippens, S.R. Meijer, A.C. van den Berg, P.C. Sibma, J.L. van Bree, D.V. de Vries, Physics of IED blast shock tube simulations for mTBI Research. *Front Neurol*, 2011. 2, p. 58.

Military Health System (MHS). http://www.health.mil/Pages/Page.aspx?ID=49.

Nakagawa, A., G.T. Manley, A.D. Gean, K. Ohtani, R. Armonda, A. Tsukamoto, H. Yamamoto, K. Takayama, T. Tominaga, Mechanisms of primary blast-induced traumatic brain injury: Insights from shock-wave research. *J Neurotrauma*, 2011. 28(6), p. 1101–1119.

Petrie, E.C. et al., Neuroimaging, behavioral, and psychological sequelae of repetitive combined blast/impact mild traumatic brain injury in Iraq and Afghanistan war veterans. *J Neurotrauma*, 2013. Epub ahead of print.

Richmond, D.R., Blast criteria for open spaces and enclosures. *Scand Audiol Suppl*, 1991. 34, p. 49–76.

Rigby, P., Wong, J., Juhas, B., Eslami, P., Rapo, M., Baumer, T. Using helmet sensors in predicting head kinematics, in *A Survey of Blast Injury across the Full Landscape of Military Science–NATOHFM-207*, 2011. Halifax.

Ritzel, D.V., S.A. Parks, J. Roseveare, G. Rude, T. Sawyer, Experimental blast simulation for injury studies, *NATO/RTO HFM-207 Symposium*, Halifax, Canada, 3–5. 2011.

Rosenfeld, J.V., N.L. Ford. Bomb blast, mild traumatic brain injury and psychiatric morbidity: A review. *Injury*, 41(5), p. 437–443, 2010.

Rubovitch, V. et al., A mouse model of blast-induced mild traumatic brain injury. *Exp Neurol*, 2011. 232(2), p. 280–289.

Saljo, A., A. Hamberger, Intracranial sound pressure levels during impulse noise exposure, in *7th International Neurotrauma Symposium*. 2004, Medimond, Monduzzi Editore: Adelaide, Australia.

Saljo, A., F. Arrhén, H. Bolouri, M. Mayorga, A. Hamberger, Neuropathology and pressure in the pig brain resulting from low-impulse noise exposure. *J Neurotrauma*, 2008. 25(12), p. 1397–1406.

Stein, M.B., T.W. McAllister. Exploring the convergence of posttraumatic stress disorder and mild traumatic brain injury. *Am J Psychiatry*, 2009. 166(7), p. 768–776.

Tanielian, T., L.H. Jaycox, eds. 2008. *Invisible Wounds of War: Psychological and Cognitive Injuries, Their Consequences, and Services to Assist Recovery*. RAND Corporation: Santa Monica, CA.

Turégano-Fuentes, F. et al., Injury patterns from major urban terrorist bombings in trains: The Madrid experience. *World J Surg*, 2008. 32, p. 1168–1175.

U.S. Department of Defense, Structures to Resist the Effects of Accidental Explosions, Unified Facilities Criteria, UFC 3-340-02, 2008.

Weiner, M.W. et al., Military risk factors for Alzheimer's disease. *Alzheimers Dement.*, 2013. 9(4), p. 445–451.

Wightman, J.M., S.L. Gladish, 2001. Explosions and blast injuries. *Ann Emerg Med*, 37, p. 664–678.

Wilk, J.E., J.L. Thomas, D.M. McGurk, L.A. Riviere, C.A. Castro, C.W. Hoge. Mild traumatic brain injury (concussion) during combat: Lack of association of blast mechanism with persistent postconcussive symptoms. *J Head Trauma Rehabil*, 2010. 25(1), p. 9–14.

5 Behavioral Testing after Experimental Traumatic Brain Injury

Bridgette D. Semple and
Linda J. Noble-Haeusslein

CONTENTS

5.1 INTRODUCTION

Traumatic brain injury (TBI) is a devastating condition that results in significant mortality and morbidity worldwide. Defined as any head injury with traumatic etiology, including blunt or penetrating trauma and nonaccidental injury, TBI commonly results from motor vehicle accidents, falls, and sports-related injuries (Myburgh et al., 2008). Affecting over 1.5 million individuals annually in the United States, survivors of severe TBI can experience sustained and debilitating physical, psychological, and cognitive deficits (Langlois, 2000; Faul et al., 2010). The lifetime cost of medical costs, rehabilitation, and lost productivity for survivors of TBI can be substantial, placing a major economic burden on families and society at large (Humphreys et al., 2013). Although historically much focus has been on brain injuries in adults, there is now an increasing understanding that those injured at a young age are more likely to develop symptoms such as attention deficit and hyperactivity disorders, anxiety, depression, motor problems, and learning deficits, all of which considerably impact long-term quality of life (Ewing-Cobbs et al., 1998; Jorge, 2005; Donders and Warschausky, 2007).

Given the significant medical and societal importance of this condition, there is substantial need for the appropriate modeling of TBI in the experimental setting. Models aid in understanding the mechanisms and pathophysiology of injury, as well as allow for the critical evaluation of potential therapeutic targets. Here, we focus on behavioral testing in experimental TBI models in rodents, with an emphasis on paradigms of sensorimotor, cognitive, and psychosocial outcome measures. We address the relationship between structural integrity and functionality, and consider translation from rodent models into the clinical setting. Last, we offer recommendations to optimize the reproducibility and translatability of behavioral read-outs in experimental TBI models.

5.2 CLINICAL FINDINGS: LONG-TERM DISABILITY AFTER TRAUMATIC BRAIN INJURY

Survivors of TBI may present with functional impairments that persist for many years. Long-term recovery and disability are influenced by both the severity of the brain injury and the age-at-insult, while preinjury factors such as socioeconomic status and alcohol abuse have also been identified as predictors of poorer outcomes (Spitz et al., 2012; Ponsford et al., 2013). Functional outcomes may be broadly classified as cognitive dysfunction, neurobehavioral disorders, somatosensory disruptions, or somatic symptoms (Halbauer et al., 2009) (Figure 5.1). Although motor and physical symptoms tend to stabilize or diminish over time, functional impairments including cognitive and emotional disturbances reportedly cause the greatest long-term distress (Yeates et al., 2004; Catroppa et al., 2008; Chapman et al., 2010; Catroppa et al., 2012). A recently completed longitudinal comparison of functional outcomes over 10 years after moderate-to-severe TBI found that fatigue and balance problems were the most common of the neurological symptoms (Ponsford et al., 2014). Changes in a broad range of cognitive functions, particularly in the domains of memory, attention, and cognitive fatigue, are more common than physical changes and fail to resolve over time (Ponsford et al., 2014).

Understanding long-term outcomes is particularly important as TBI most commonly impacts young adults or children who have many productive years ahead of them. A TBI sustained during childhood may cause difficulties in a wide range of physical, neurological, cognitive, social, and functional domains (Beauchamp et al., 2013). Further, the young brain shows particular vulnerability to early life disruption, such that deficits may become apparent over time after injury as TBI survivors fail to meet developmentally appropriate milestones (Anderson and Moore, 1995; Koskiniemi et al., 1995; Anderson et al., 2005; Donders and Warschausky, 2007; Ryan et al., 2014). Unfortunately, the ability of health professionals/researchers to accurately predict the outcome of pediatric TBI remains limited, and a better understanding of contributing factors and underlying mechanisms is urgently needed.

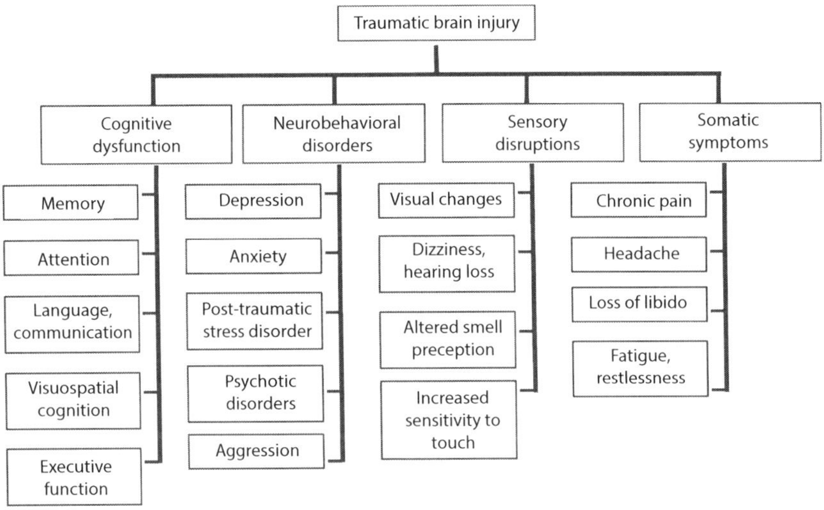

FIGURE 5.1 Functional outcomes may be classified as cognitive dysfunction, neurobehavioral disorders, somatosensory disruptions, or somatic symptoms. (Adapted from Halbauer JD et al., 2009. *J Rehabil Res Dev* 46:757–796.)

A range of measures are used by clinicians/researchers to measure long-term outcome, quantify the burden of TBI, and the degree of functional impairment in survivors (Wilde et al., 2010; Nichol et al., 2011). The Glasgow Outcome Scale (GOS) is a commonly used measure of functional limitations. The extended GOS (GOSe), typically conducted at 6 months after injury, is currently the most reliable and validated functional outcome measurement scale in randomized controlled trials of patients following TBI (Nichol et al., 2011). Other measures are designed to specifically assess activity and independence, participation and community integration, neuropsychological status, or quality of life (Shukla et al., 2011).

5.3 EXPERIMENTAL TBI IN RODENTS

Animal models of TBI are used to elucidate mechanisms underlying the sequelae of human head injury, with the ultimate goal of identifying potential neuroprotective therapies and developing rehabilitation strategies to support favorable recovery. To achieve this, an experimental model must replicate some aspects of the pathobiology of the human condition, such as the mechanical, biochemical, cellular, and behavior consequences of TBI. Given the heterogeneity and complexity of human TBI, there is no single ideal model in rodents; instead, several have been developed and are well-characterized across a spectrum of outcome measures, and the choice of experimental model typically depends on the research goal and underlying objectives (Cernak, 2005). Rodents are the most common species used for modeling human TBI, as their small size and relatively low cost make them a practical choice for quantification of multiple parameters or repetitive measures (Cernak, 2005). The availability of genetically modified rodents also allows for the investigation of specific molecular pathways.

TABLE 5.1
Features of the Most Common Models of Traumatic Brain Injury in Rodents

Experimental Model	Focal Contusion	Axonal Injury	Skull Fracture	Strengths	Weaknesses
Controlled cortical impact	+++	+	–	High reproducibility	Requires craniotomy
Fluid percussion injury	+++	++	–	High reproducibility within a laboratory	Requires craniotomy
Weight drop	+	+	+/–	Injury mechanism close to human TBI; easy and quick to perform	High variability, risk of skull fractures
Impact acceleration	+/–	+++	+/–	Model of commonly seen diffuse axonal injury	High mortality due to apnea, requires early respiratory support
Blast	–	++	–	Models a specific injury mechanism	Requires specific apparatus

Experimental TBI models involve a primary insult that is either focal or diffuse in nature, or a combination of both, and the mechanism may be either penetrating or closed. Table 5.1 illustrates the key features of the most common TBI models, as described briefly below: fluid percussion, controlled cortical impact, weight-drop, impact acceleration, and blast injury. For a detailed description of experimental TBI models, there are several thorough reviews available (Cernak, 2005; Morales et al., 2005; O'Connor et al., 2011; Xiong et al., 2013).

Fluid percussion injury (FPI) is one of the most frequently used models of direct brain deformation in rodents (McIntosh et al., 1987; Dixon et al., 1988; Carbonell et al., 1998). Replicating a clinical contusion in the absence of skull fracture, this model is widely used in neurotrauma research for both mechanistic studies and drug discovery (Morales et al., 2005). The insult is inflicted by the release of a pendulum, which causes a pressure pulse of fluid to rapidly strike the intact dura surface through a craniotomy located either laterally or at the midline. Injury severity is determined by the pressure pulse and injury location (Thompson et al., 2005). Rodents subjected to midline or lateral FPI show both cognitive and neurological deficits (Morales et al., 2005; Thompson et al., 2005). Motor deficits have also been reported, typically within several days after injury (Dixon et al., 1987; Hamm et al., 1994). Injury location is a key determinant of injury outcomes, as small variations in the craniotomy position can affect the behavioral consequences and associated neuropathology (Floyd et al., 2002).

Controlled cortical impact (CCI) is the other most commonly used model, first characterized in the ferret and later modified for application in both rats (Dixon et al., 1991) and mice (Smith et al., 1995). Injury is induced by a compressed air-driven mechanical piston that directly impacts the intact dura through a craniotomy, causing deformation of the underlying cortex. Compared to most other models, CCI affords better control over impact parameters, allowing for fine adjustments of the impact velocity, angle, depth of penetration, and dwell time. In general, CCI produces a more focal injury compared to FPI, which may influence the subsequent pathology and resulting behavioral consequences (Dunn-Meynell and Levin, 1997). Cognitive deficits including memory dysfunction are commonly seen after CCI (Scheff et al., 1997; Fox et al., 1998), and these may persist up to 1 year after injury (Lindner et al., 1998; Shear et al., 2004). Sensorimotor deficits are typically more transient and may involve gross motor function and/or fine motor coordination dysfunction (Morales et al., 2005).

A third commonly used injury model involves a free-falling, guided weight impacting the exposed skull (Feeney et al., 1981; Chen et al., 1996). In comparison to the FPI and CCI models that require a craniotomy, this *weight-drop model* impacts a closed, intact skull surface. Injury severity is rapidly and easily controlled through adjustment of the height and mass of the weight used for impact (Flierl et al., 2009). Although this model arguably reflects injury mechanisms in humans more accurately, the trade-off is an increase in interanimal variability resulting from skull fractures and possible rebound impacts. Acute impairments of motor function and cognition have been observed after weight-drop injury (Chen et al., 1996; Tsenter et al., 2008), although weight drop typically results in fewer and more transient impairments compared to lateral fluid percussion injury (LFPI) at an equivalent injury severity (Hallam et al., 2004). Of note, however, there is a paucity of studies evaluating behavioral dysfunction across a chronic time course in this model.

In contrast to the FPI, CCI, and standard weight-drop models, the *impact-acceleration model* generates a more diffuse axonal injury (Marmarou et al., 1994). Impact is delivered by a brass weight impacting a steel disc that has previously been rigidly attached to the animal's exposed skull, usually centrally. The dropped weight is guided through a Plexiglass tube by gravity, and injury severity is determined by the mass and height from which the weight is released (Piper et al., 1996). The impact force needs to be widely distributed across the skull in order to produce a predominantly diffuse brain injury—this is achieved by positioning the rodent on a foam bed such that the head accelerates upon impact. As with the above-described focal injury models, both motor and cognitive impairments have been reported after impact-acceleration TBI, particularly acutely after injury (Beaumont et al., 1999; Vink et al., 2003).

Most recently, models of *blast-induced TBI* have been established. Blast injuries have become the predominant cause of brain injury in military populations, yet the etiology is poorly understood, prompting the need for experimental model development. Traumatic brain injury resulting from blasts may be caused by particles propelled by the blast-force, acceleration and deceleration forces, and/or the blast wave itself (Cernak and Noble-Haeusslein, 2010). To model this phenomenon, rodents are restrained and exposed to blast waves caused by the detonation of either explosives (Cheng et al., 2010) or compressed air in a shock tube (Kuehn et al., 2011). Systemic injuries may be avoided by shielding of the body, and injury severity is controlled by altering the duration of exposure or blast pressure. Although studies examining functional outcomes are limited to date, cognitive difficulties have been reported up to 30 days after experimental blast injury (Cernak et al., 2001; Rubovitch et al., 2011).

Last, most of the above-mentioned experimental models may be adapted to represent mild or concussive brain injury, defined as a short-lived impairment of neurologic function induced by biomechanical forces to the brain (Prigatano and Gale,

2011; McCrory et al., 2013). This has been an area of intense research in recent years, in light of intense media attention highlighting the risks of concussion in both professional and amateur athletes. Several studies to date have demonstrated acute neurobehavioral consequences after experimental concussions in rodents, in line with what is seen in concussed individuals. Further, there is accumulating evidence that repeated injuries may result in poorer sensorimotor and cognitive outcomes, which persist across a prolonged time course and may be associated with chronic neurodegeneration (Meehan et al., 2012; Hylin et al., 2013).

Behavioral assays have been developed to profile emerging deficits and/or recovery after experimental TBI, and are often key to interpreting the relationship between pathogenesis and recovery over time. Although each of these experimental models may have a unique "signature" in terms of behavior characteristics, they also produce some similar findings across different models in terms of sensorimotor, cognitive, and affective behaviors.

5.4 SENSORIMOTOR ASSESSMENTS

Functional deficits and recovery after TBI can be assessed by a range of different outcome measures in rodents. The following sections will briefly describe the most commonly used tasks that have been applied to experimental TBI research, as summarized in Table 5.2. First, global assessments of neurological outcome may be employed, consisting of several brief tasks for somatosensory and motor function. These measurements are used to determine injury severity, as inclusion/exclusion criteria for assignment to treatment groups, or as stand-alone endpoint measures (Gold et al., 2013). The *Neurological Severity Score* (NSS) or an extended, modified version (mNSS), evaluates mice in a series of motor, sensory, balance, and reflex tasks (Flierl et al., 2009). A higher score indicates greater dysfunction, and scores have been correlated with the degree of pathology evident by neuroimaging and histology methods (Tsenter et al., 2008). An alternative global assessment is the *Composite Neuroscore* (McIntosh et al., 1987). Components of the test vary between laboratories but typically include evaluations of gross motor function, reflexes, and general activity. This score has been proposed to correspond to components of the GCS in the clinical setting (Gold et al., 2013), and can detect deficits up to 1 year after lateral FPI (Pierce et al., 1998). Third, the simple neuroassessment of asymmetric impairment (SNAP) test has been proposed as a rapid assessment of visible, proprioception, motor strength, and posture, which may be useful to evaluate injury severity and recovery over time (Shelton et al., 2008). Acutely postinjury, time to regain the righting reflex has also been used as a measure of brain injury severity and loss of consciousness (Kane et al., 2012; Mouzon et al., 2012).

Although aggregate tests provide an overall behavioral profile, individual assays offer insight into sensorimotor function that may be specifically affected by the injury. Sensorimotor tasks may assess either gross or fine performance, or a combination of both. Several tests also incorporate vestibular function by evaluating balance, and most tasks include a component of sensory feedback and proprioception. The *balance beam* is one example—this task involves an elevated narrow beam that rodents are required to balance on, and the latency to fall off the beam is recorded (Singleton et al., 2010).

TABLE 5.2
Modeling of Human Symptoms in Experimental TBI in Rodents

	Rodent Behavior Test	Outcome Measure	Selected References
SENSORIMOTOR	Composite neuroscore	General neurological function	McIntosh et al. (1989); Murai et al. (1998)
	Neurological severity score (NSS)	General neurological function	Chen et al. (1996); Flierl et al. (2009)
	Foot fault, grid walk	Balance, motor function	Hylin et al. (2013); Baskin et al. (2003)
	Rotarod	Motor function, coordination, balance	Hamm et al. (1994, 2001); Monville et al. (2006)
	Balance beam	Vestibulomotor function	Hamm (2001); Hylin et al (2013)
	Ledged beam test	Balance, motor function	Klint et al. (2003); Semple et al. (2010)
	Cylinder test	Spontaneous forelimb use	Woodlee et al. (2005); Carballosa Gonzalez et al. (2013)
	Inclined plane test	Vestibulomotor function	Hallam et al. (2004); Chang et al. (2008)
	Automated gait analyses	Gait during locomotion	Russell et al. (2011); Mountney et al. (2013)
	Rotating pole	Motor coordination	Piot-Grosjean et al. (2001); Hoover et al. (2004)
	Grip strength	Fine motor function	Long et al. (1996); Whalen et al. (1999)
	Tactile adhesion removal	Fine motor skills, sensory impairments	Schallert et al. (1982); Riess et al. (2001)
	Acoustic startle response	Auditory function, startle response	Washington et al. (2012); Bijlsma et al. (2010)
COGNITIVE	Water maze—latency, distance	Spatial learning	Hoover et al. (2004); Griesbach et al. (2004)
	Water maze—probe trials	Spatial memory retention	Hicks et al. (1993); Smith et al. (1991)
	Water maze—reversal	Learning flexibility	Vorhees and Williams (2006)
	Radial arm maze	Spatial learning and memory	Lyeth et al. (1990; 2001)
	Barnes maze	Spatial learning and memory	Fox et al. (1998, 1999)
	Novel object recognition	Recognition memory	Bevins et al. (2006); Prins et al. (2010)
	Y-maze, T-maze	Spatial memory, exploratory behavior	Schultz et al. (2012); Rubovich et al. (2011)
	Passive avoidance	Non-spatial learning and memory	Hamm et al. (1994); Yamaguchi et al. (1996)

(Continued)

TABLE 5.2 (*Continued*)
Modeling of Human Symptoms in Experimental TBI in Rodents

	Rodent Behavior Test	Outcome Measure	Selected References
PSYCHOSOCIAL	Open field	General activity, anxiety, exploration	Pullela et al. (2006); Washington et al. (2012)
	Elevated plus/zero maze	Anxiety	Pullela et al. (2006); Walf et al. (2007)
	Marble burying	Anxiety; repetitive/compulsive behaviors	Pandey et al. (2009)
	Fear-conditioning (contextual, cued)	Anxiety, fear-associated response/memory	Meyer et al. (2012); Klemenhagen et al. (2013)
	Forced swim test	Behavioral despair (depressive-like)	Slattery et al. (2012); Washington et al. (2012)
	Tail suspension test	Behavioral despair (depressive-like)	Ando et al. (2011)
	Reaction to aversive stimuli	Hyper-emotionality	Pandey et al. (2009)
	Sucrose preference test	Anhedonia	Klemenhagen et al. (2013); Strekalova and Steinbusch (2010)
	Resident-intruder, social interactions	Social interactions, aggression	Semple et al. (2012); Schultz et al. (2012)
	Socio-sexual interactions	Sexual anhedonia	Pandey et al. (2009)
	Tube-dominance	Aggressive tendencies	Semple et al. (2012)
	Three-chamber social approach	Sociability, social recognition and memory	Semple et al. (2012)

Source: Adapted from Thompson HJ et al. 2005. *J Neurotrauma* 22:42–75; Fujimoto et al. 2004. *Neurosci Biobehav Rev* Jul; 28(4):365–378. Malkesman O et al. 2013. *Front Neurol* 4:157.

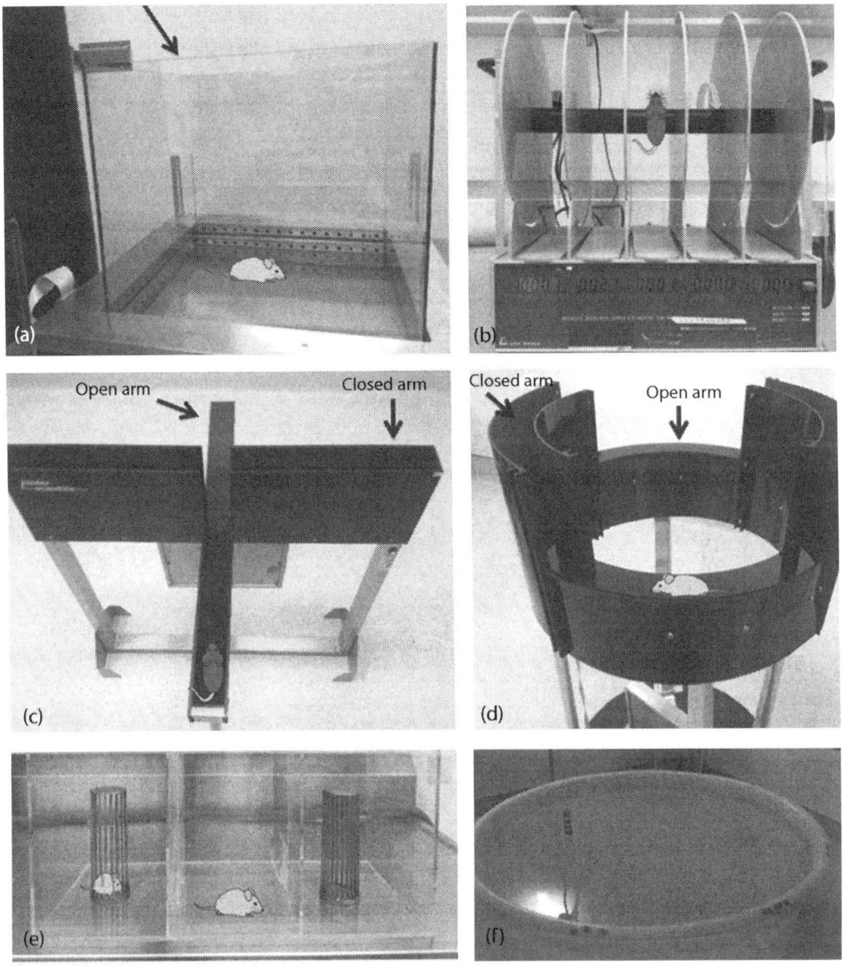

FIGURE 5.2 A range of behavioral assays may be used to evaluate functional deficits after experimental TBI in rodents, including (a) open field, (b) rotarod, (c) elevated plus maze, (d) elevated zero maze, (e) the three-chamber social approach, and (f) Morris water maze. (Images courtesy of the Neurobehavioral Core for Rehabilitation Research at the University of California–San Francisco.)

The *beam walk* task uses similar apparatus; however, in this task rodents must traverse the elevated beam, and the number of foot faults or slips is recorded, as well as latency to reach the goal side (Hallam et al., 2004). This task can be made more challenging by tapering the beam across its length (Semple et al., 2010).

The *rotarod* is another commonly used test for general motor performance—the latency to fall from a rotating rod is recorded, often with acceleration of the rod speed to increase difficulty over time (Figure 5.2b). This task is considered sensitive to brain injury which may impair integration of motor control required to maintain balance and grip, and actively move in a forward direction (Hamm et al., 1994), including damage to cortical motor areas or the cerebellum. Repeated testing over multiple days can provide a measure of motor learning (Chen et al., 2013). A more general task for motor performance, which also incorporates a measure of anxiety, exploration, and general activity, is the *open field* paradigm (Figure 5.2a). Animals are allowed to freely explore an open field arena, typically a square or rounded open-topped box, and movements are recorded either manually or with tracking software. Brain-injured animals may show hyper- or hypoactivity depending on the injury severity and age at the time of insult (Pullela et al., 2006; Zhao et al., 2012).

Several behavioral tests are particularly useful for detecting unilateral sensorimotor deficits, contralateral to the injury location. One such task is the *tactile adhesion removal* or *sticky tape test*, which has a strong somatosensory component. This task involves placing small pieces of adhesive tape onto the rodent's forepaws, and the time taken for removal is recorded (Barth and Stanfield, 1990). Although this task may be vulnerable to a degree of handler-dependent variability, it is sensitive enough to detect treatment efficacy in experimental TBI (Riess et al., 2001). The *cylinder test* (or *spontaneous forelimb use test*) is another that assesses asymmetries in forelimb function. A rodent is placed within a clear open-topped cylinder for a predetermined time limit, and the number of times it places each forepaw during rearing is quantified. Brain-injured animals will often avoid using the limb contralateral to the injury, and favor the ipsilateral paw (Baskin et al., 2003; Woodlee et al., 2005). The *grid walk test* also provides a measure of unilateral deficits, including those affecting fine motor control of the digits. This task challenges the rodent to explore and traverse a wire grid, during which time individual paw slips through the grid are counted. An increase in foot faults is evident up to 4 weeks after CCI in mice (Baskin et al., 2003; Xiong et al., 2008).

Apparatus for automated gait analyses are also available, such as the CatWalk™ (Noldus Information Technology) or Digigait™ (MouseSpecifics, Inc) systems. These consist of a long plexiglass chamber often on a motorized treadmill to encourage forward locomotion, and paw placements are viewed and captured via a ventrally positioned camera (Russell et al., 2011). A large body of data are produced for a single animal, including information about paw placement, posture, stride length and width, and paw pressure. By allowing for individual paws to be distinguished, these systems may be particularly useful to detect subtle, clinically relevant changes in motor deficits (Mountney et al., 2013)

5.5 COGNITIVE OUTCOMES

Among the most debilitating consequences of TBI are changes in cognition, including confusion, memory impairments, and deficits in executive function (Horneman and Emanuelson, 2009). Several aspects of cognitive function can be quantified in rodents, and deficits are typically associated with damage to the hippocampus as well as other brain regions involved in learning and memory (Gold et al., 2013). Although cognitive testing of brain-injured patients includes both verbal and visual recognition tasks, in the laboratory setting, researchers typically rely on performance in spatial and nonspatial mazes (Thompson et al., 2005).

The *Morris water maze* (MWM) is the most commonly used assessment of spatial learning and memory in rodents (Figure 5.2f). Although labor intensive, this task is versatile and can be adapted to investigate specific questions using distinct training and testing protocols. Mice are placed in a large water-filled pool and tasked to locate a platform hidden below the water surface. Visual features of the testing room provide spatial cues, and spatial learning is evaluated across repeated trials beginning from different starting positions around the pool edge. Latency or distance traveled to the platform is recorded (D'Hooge and De Deyn, 2001). Reference memory may be subsequently tested, by removing the platform completely and quantifying the time spent searching the space where the platform was previously located. This "probe trial" may be performed at different times after MWM training, to test both short- and long-term memory. By relocating the target platform to a new location, researchers can also evaluate reversal learning, or the ability to extinguish and replace the initial learning. Deficits in this task in the context of experimental TBI are often associated with hippocampal damage (Skelton, 1998; Hallam et al., 2004).

The *Barnes maze* is also a test of spatial reference memory. In this task, the rodent is placed onto a brightly lit, flat circular board raised above the ground, with holes lining the perimeter. Only one of these holes contains a dark, escape compartment, and animals are trained over several days to learn the location of this box. By moving the location of the escape box during training, investigators can also detect injury-related deficits in working memory performance (Vink et al., 2003). The *radial arm maze* is another alternative to the Morris water maze. Instead of using water immersion as the motivating factor, rodents are food restricted and then required to navigate a 6–12 arm maze using spatial cues to seek a food reward (Sebastian et al., 2013). Recently, a combined approach has been developed, appropriately dubbed the *radial arm water maze*. By inserting maze arms into a water pool, this design limits the spatial complexity of the task to accelerate task learning (Alamed et al., 2006; Luo et al., 2014).

The *Y maze* and *T maze* are two similar tasks that also challenge hippocampal-based working memory. After being placed on the apparatus, consisting of three arms of equal length, a rodent is expected to explore all arms equally during spontaneous exploration. In a two-trial design, one of the arms is blocked during the first trial and then all arms are accessible during the second trial (Gold et al., 2013). A preference for the newly unveiled maze arm indicates memory retention (Rubovitch et al., 2011). Although the Morris water maze is the most commonly used of the spatial memory tasks, others such as the Y maze may be used preferentially due to time efficiency.

In contrast, the *novel object recognition (NOR) task* evaluates nonspatial hippocampal-mediated declarative memory. In this test, animals are allowed to freely explore two identical objects in an arena for a predetermined amount of time (several minutes). After an intertrial interval (minutes to hours), animals are placed back into the same arena with one familiar object and one novel object. The time spent with the objects is recorded, and control mice typically show a preference for novelty (Gold et al., 2013). Brain injury, particularly that which affects the hippocampal dentate gyrus and perirhinal cortex, may interfere with object discrimination (Zhao et al., 2012). The *passive avoidance task* is also used to assess simple nonspatial learning and memory. Here, animals are placed into a brightly lit chamber with a door opening to an adjoining dark chamber. If the animal enters the dark chamber, the door closes and the animal receives a mild foot shock. When the same animal is tested the following day, latency to enter the dark chamber is recorded, and a delayed response indicates that learning has occurred (Hamm et al., 1993; Zhao et al., 2012).

Historically, behavioral tasks to assess more complex cognitive abilities in rodents have been relatively scarce. However, a surge of publications over recent years have evaluated new cognitive tasks for their application in experimental TBI models, such as the *attentional set-shifting task* (AST). In this test for executive function and behavioral flexibility, rodents are presented with a series of increasingly difficult perceptual discrimination choices to obtain a food reward, requiring them to form and maintain directional attention (Young et al., 2010). Difficulties with this task indicate cognitive inflexibility and perseverance, and are associated with dysfunction of the prefrontal cortex (Bondi et al., 2014). Frontal injury may also disrupt decision-making behaviors, which can be evaluated in rodents using a *scent-discrimination test*, during which rodents are exposed to different samples of scented sand across several training sessions and days (Martens et al., 2012).

5.6 PSYCHIATRIC, SOCIAL, AND EMOTIONAL OUTCOMES

In contrast to sensorimotor and cognitive assessments, behavioral tests to evaluate neuropsychiatric symptoms are less frequently employed in animal models of TBI. As a result, findings to date are fairly inconsistent and require further replication and validation across different laboratory settings (Malkesman et al., 2013). Here, we briefly describe tests used in rodents to model affective behaviors, including depression, anxiety, social dysfunction, and emotional disturbances after injury.

In rodents, *fear-conditioning* models may be used to evaluate the emotional conditioning processes that are known to become dysfunctional in posttraumatic stress disorder (PTSD) (Meyer et al., 2012). A negative stimulus (e.g., a mild foot shock) is associated with either a location ("contextual fear") or prompt ("cued fear"), and the startle or freezing response time is heightened if conditioning has occurred. Continued presentation of the stimulus typically produces a gradual extinction of the conditioned fear, and time to extinction is another quantifiable measure (Klemenhagen et al., 2013; Luo et al., 2014).

Anxiety behavior can be assessed using the *elevated plus maze* (Figure 5.2c). This cross-shaped maze consists of two open arms and two closed arms. Rodents, with a natural preference for dark places, will usually spend most of their time in the closed arms. Changes in the percentage time spent in the open versus closed can be evaluated to gauge hypo- or hyperanxiety after injury relative to control animals (Pullela et al., 2006; Bao et al., 2012). The *elevated zero maze* is an alternative task based on the same principles, except that the maze is a continuous elevated circle composed of alternating closed and open segments (Siopi et al., 2012) (Figure 5.2d).

Depression is one of the most common symptoms reported after mild TBI in patients (Malkesman et al., 2013). In the *forced swim test*, which involves placing an animal in a nonescapable cylinder of water for several minutes, immobility is quantified as a measure of behavioral despair, an indicator of depressive-like behavior (Washington et al., 2012). The *tail suspension* test similarly assesses behavioral despair by quantifying immobility—in this task, rodents are hung by their tails in an isolated chamber, and the time until the animal ceases to right itself is recorded (Ando et al., 2011). Anhedonia is another key component of depression, and this can be measured in rodents by their preference to consume a sweetened solution in the *sucrose-preference test*. During this test, animals are given a free choice between two identical bottles, one containing water and the other containing a sucrose solution (0.5%–2%). Weighing of the bottles before and after exposure provides an estimation of fluid consumption, and a preference for the sucrose can be calculated. Typically, a decrease of sucrose preference below 65% is considered as criteria for anhedonia (Strekalova and Steinbusch, 2010; Klemenhagen et al., 2013). Only one study to date has examined *hyperemotionality* after weight-drop injury in rats, by calculating an aggregate score determined by the animal's response to various provocative stimuli. Here, the authors related an observed increase in the hyperemotionality score after injury with an increase in anxiety-like behaviors (Pandey et al., 2009).

Changes in social behaviors seen in many patients after TBI, such as social isolation, poor social skills, and poor social communication deficits, can also be modeled in rodents. Social investigation or interest is quantified as the time spent sniffing different body regions, or distance between two unfamiliar rodents during free exploration or a *resident-intruder paradigm* (Pandey et al., 2009; Semple et al., 2012; Shultz et al., 2012). Alternatively, a social choice task can be employed, in which animals are presented with a choice between spending time in the proximity of other mice or remaining alone. The *three-chamber social approach task*, developed and validated by the Crawley laboratory to detect social deficits in models of autism spectrum disorders (Crawley, 2007; Moy et al., 2008), is a useful paradigm to examine sociability in mice (Figure 5.2e). This task can also assess a preference for social novelty by presenting test mice with a choice between familiar and unfamiliar stimulus mice, thus enabling detection of potential injury-related deficits in social recognition and memory (Semple et al., 2012).

5.7 MODELING CLINICALLY RELEVANT BEHAVIOR—WHAT ARE WE MISSING?

Comorbidities are common in TBI patients but often neglected in preclinical investigations. After evaluation of a potential therapeutic in young, healthy male animals, additional studies should be conducted taking into account gender, age, and preexisting medical conditions (Lapchak et al., 2013). For example, age at the time of injury can significantly affect behavioral outcomes. Traumatic injury to the developing brain modified subsequent maturation and may interfere with the acquisition of age-appropriate skills over time (Anderson et al., 2005; Donders and Warschausky, 2007). At the other end of the spectrum, the elderly is another population with particular vulnerability to TBI. A comparable CCI injury in adult and aging rodents results in significantly poorer outcomes for the older age group in terms of edema, neurodegeneration, and functional deficits (Onyszchuk et al., 2008). In addition to the clinical importance of better understanding these differences, such findings have implications for experimental TBI research, in that the selection of behavioral measures must be age appropriate and may be complicated by developmental changes across the life span.

Another important factor to consider when modeling behavioral outcomes after TBI is gender. Experimental models of TBI have historically focused on males; however, it is becoming clear that gender may be a key determinant of performance in many behavioral tasks particularly memory (Roof and Stein, 1999; Wagner et al., 2007). These differences may result from variation in hormone levels or be present even at the chromosomal level, and may be reflected by differing pathophysiology after injury. Clinically, there is disparity in both incidence and outcomes after TBI in males compared to females; however, the effect of gender on injury mechanisms remains controversial. There remains a paucity of experimental studies to address this issue, particularly in the context of TBI to the developing brain.

Last, we consider comorbidities and symptoms resulting from TBI that are currently afforded little attention in experimental TBI models. Chronic pain, or pain that persists for more than 3 months after onset, is a common comorbidity that frequently manifests as headaches or migraines. This is particularly prevalent in the military population, with up to 50% of male veterans reporting that they experience regular pain (Otis et al., 2011; Bosco et al., 2013). Although often coexisting with other psychosocial disorders including PTSD and depression, chronic pain can occur independently even after a mild injury (Nampiaparampil, 2008). In rodents, pain syndromes can be modeled by examining evoked withdrawal responses to mechanical or thermal stimuli (Berge, 2014). A recent study using the CCI model in both rats and mice demonstrated heightened cutaneous sensitivity to mechanical stimuli applied to the paws or periorbital regions after injury, and the authors propose that the latter region may be a useful model of headache-related pain (Macolino et al., 2014). Importantly, experiencing pain may confound performance in other behavioral assays, for example, by suppression of general activity. Understanding whether pain is coexisting with functional deficits in an experimental model may therefore be quite important to the interpretation of the overall phenotype.

Coexisting injuries, another comorbidity, have recently gained some attention in experimental models. Clinically, head injuries rarely occur in isolation and may be complicated by additional injuries or peripheral responses, as well as systemic complications that may arise following injury, such as sepsis (Venturi et al., 2009). Combined models of TBI and spinal cord injury may be useful to elucidate some of the interactions between the central nervous system (CNS) and peripheral neural networks, and how this affects remodeling and subsequent functional outcomes (Inoue et al., 2013).

Finally, sleep disturbances and resulting fatigue syndromes are highly prevalent after TBI—more than 70% of TBI patients report persistent problems up to 3 years after injury (Orff et al., 2009; Ponsford et al., 2012). Symptoms include reduced sleep efficiency, increased sleep onset latency, and increased time awake after sleep onset (Ponsford et al., 2012). To better understand these symptoms, several studies have recently begun to model sleep behavior in rodents after FPI or CCI and demonstrated disturbances by electroencephalography/electromyography and video monitoring systems (Willie et al., 2012; Lim et al., 2013).

5.8 LINKING BEHAVIORAL OUTCOMES WITH HISTOPATHOLOGY

Ideally, one might expect that outcomes relating to brain structure and function after experimental TBI should correlate, such that a treatment-induced reduction in tissue damage corresponds with improved behavioral findings, and vice versa. Damage to specific neuroanatomical regions is associated with deficits in specific behavioral tasks, particularly after moderate-to-severe TBI; for example, spatial memory deficits in the MWM are frequently related to hippocampal integrity (Floyd et al., 2002; Broadbent et al., 2004; Budde et al., 2013). Sensorimotor deficits may also correlate directly with the degree of damage to the sensorimotor cortex and associated brain regions (Hamm et al., 1994).

However, the link between structure and function is less established with injuries of a milder nature. Neuropsychiatric dysfunction may be present across the spectrum of injury severities, as seen in both human and experimental TBI models (Washington et al., 2012; Malkesman et al., 2013). Behavioral difficulties are commonly reported up to 10 years after a TBI acquired during childhood, regardless of injury severity (Catroppa et al., 2012). Further, cognitive and neuropsychiatric difficulties may persist despite a lack of detectable anatomical changes in brain structures related to cognitive functioning (Heffernan et al., 2013). Imaging approaches including quantitative high-resolution fMRI and susceptibility-weighted imaging show promise for better detection of structural damage that may be more predictive of functional outcomes in both experimental and clinical settings (Beauchamp et al., 2013; Budde et al., 2013). However, given the complexity of the mammalian brain and its extensive network of connectivity, as well as the potential for neuroplasticity contributing to remodeling and compensation, perhaps it is naïve to necessarily expect an obvious and direct correlation between structure and function. Social behavior is one example of a function that requires many brain regions to be working synergistically, and it has been proposed that impaired social function after TBI may result from changes in global connectivity rather than being attributed to any particular region (Ryan et al., 2014).

5.9 APPLICATION OF BEHAVIORAL ASSESSMENTS IN THERAPEUTICS DISCOVERY

Once established, behavioral assessments in experimental TBI models are an invaluable tool to evaluate potential new interventions and treatments. Candidate pharmacotherapies for TBI should ideally have broad-spectrum action in several preclinical models of TBI, across multiple species, and be reproducible by multiple laboratories. Further, robust neuroprotection should be reflected by improvements across a spectrum of neurological functions (Diaz-Arrastia et al., 2014).

Progesterone is one compound with potential for therapeutic benefit in the context of TBI. A neurosteroid produced in the brain, progesterone has pleiotropic effects in the injured brain, and multiple preclinical models of TBI have demonstrated neuroprotective properties of progesterone including the prevention of cerebral edema, apoptosis, inflammation, and neuronal cell death (Wei and Xiao, 2013). Progesterone may also enhance myelination and neurogenesis, and its metabolite allopregnanolone also exhibits significant neuroprotective effects (Wei and Xiao, 2013; Diaz-Arrastia et al., 2014).

Of note, preclinical studies in multiple laboratories have also demonstrated that progesterone enhances functional recovery across a spectrum of behavioral outcomes, including spatial learning and memory in the MWM and Barnes maze, improved locomotor activity on the rotarod, and reduced anxiety-like behaviors in the elevated plus maze (Gibson et al., 2008; Diaz-Arrastia et al., 2014). Two completed phase II randomized controlled trials (RCTs) using progesterone after moderate-to-severe TBI have demonstrated good tolerance and potential efficacy. Two multicenter phase III RCTs are now underway, named the SyNAPSe (Clinicaltrials.gov Identifier: NCT00822900) and ProTECT III (Clinicaltrials.gov Identifier: NCT01143064) studies; however, the latter trial was recently halted due to futility (Kabadi and Faden, 2014). Although disappointing, recent preclinical data suggest that progesterone in combination with other agents such as vitamin D may potentiate its neuroprotective effects (Aminmansour et al., 2012; Tang et al., 2013), an approach that warrants further exploration. Of note, the optimal dosage and duration for treatment with progesterone have not yet been determined, and pharmacokinetics may also be an important factor here. A meta-analysis examining progesterone treatment in experimental TBI also highlighted a low standard of methodological quality and evidence of potential statistical biases in many of the preclinical studies (Gibson et al., 2008).

5.10 LOST IN TRANSLATION? RECOMMENDATIONS TO IMPROVE TRANSLATION OF FUNCTIONAL OUTCOMES FROM EXPERIMENTAL MODELS TO THE CLINIC

To date, although there have been more than 45 phase II or III clinical trials in the United States for a drug or intervention to treat TBI, most have failed to demonstrate any significant improvement in outcomes, and there are still no U.S. Food and Drug Administration (FDA)–approved therapies (Narayan et al., 2002). These failures highlight the need to reevaluate both clinical translation strategies and preclinical models (Kabadi and Faden, 2014). In part, failure in translation may result from heterogeneity of the human TBI population, and the mismatch between broad inclusion criteria in clinical trials compared to the use of highly homogeneous animal models (Maas and Menon, 2012).

As no single animal model accurately mimics all of the features of human TBI, individual laboratories employ different experimental approaches deemed to best fit their specific research goals. The signature neuroanatomical and neurobehavioral characteristics of these distinct models may be capitalized on during therapeutic testing. Assessment of functional outcomes is one aspect of preclinical research that can be optimized; below, we have compiled several recommendations to potentially enhance translation from the bench to the bedside.

5.10.1 Recommendation #1: Optimize Study Design and Standardization

Improving general scientific rigor within both experimental and clinical TBI research communities is paramount (Landis et al., 2012). Progress has been made in this regard particularly in the ischemia field, with a set of basic standardized criteria now in circulation as guidelines for clinical study design and analysis (STAIR, 1999; Lapchak et al., 2013) (http://www.nih.gov/research-training/rigor-reproducibility/principles-guidelines-reporting-preclinical-research). The interagency TBI Outcomes Workgroup has generated a set of Common Data Elements (CDE), a series of outcome measures for TBI research to address primary clinical research objectives. Three tiers of CDE have been recommended: core, supplemental, and emerging (McCauley et al., 2012). Core measures are those that have demonstrated validity and robustness, and show wide applicability, including measures of global function level, neuropsychological impairment, psychological status, TBI-related symptoms, executive functions, cognitive and physical activity limitations, social role participation, and perceived health-related quality of life (Wilde et al., 2010; McCauley et al., 2012). Application of a similar set of guidelines for designing and reporting preclinical studies would only improve the quality of TBI research and minimize the potential for false-positive results (Landis et al., 2012). Preclinical studies should apply the same level of rigor required for clinical trial design, including randomization into treatment conditions, blinded assessment, predetermined outcome measures, and application of appropriate statistical methods.

The standardization of existing behavioral protocols and introduction of new models when scientifically necessary are also required to enhance translatability of preclinical findings. The novel object recognition test is one behavioral test for which the procedure varies widely in terms of implementation, design, and timing, limiting comparisons between studies (Antunes and Biala, 2012). Standardization should therefore be encouraged to increase potential for reproducibility across different laboratories.

5.10.2 Recommendation #2: Choose the Right Outcome Measures and Evaluation Period

The choice of an appropriate behavioral assessment will depend on the study design (e.g., is repeated testing required), the manipulation of interest, the injury severity, the animal's age, and the time following injury (acute, subacute, or chronic). We recommend a battery of tests across the spectrum of sensorimotor, cognitive, and psychosocial behaviors, to provide a comprehensive understanding of the animal's phenotype. Endpoint measurements with relevance to quality of life, such as tests for depression and anxiety, should be included given their strong clinical importance and parallel use in clinical trials. Functional outcome measures employed in TBI patients should also be reliable, reproducible, and sensitive to injury-induced disabilities (Nichol et al., 2011).

Many of the behavioral measures used in preclinical TBI models have not been validated for long-term studies (Gold et al., 2013), or may not be sufficiently sensitive to long-term behavioral and cognitive deficits. As the goal of TBI research is to improve the lives of TBI patients long term, studies that can discriminate injury severity beyond several weeks after injury are needed. The timeframe for outcome is particularly important for translation to clinical trials, in which a 6-month endpoint is common (Narayan et al., 2002).

5.10.3 Recommendation #3: Consider a Multimodal, Clinically Relevant Approach

Although experimental TBI models typically examine cognitive, motor, and somatic deficits through behavior tests, it has been noted that the clinical evaluation of TBI also relies heavily on the assessment of neuroimaging modalities such as fMRI to predict dysfunction and recovery (Heffernan et al., 2013). There is often a mismatch between the primary outcome measures used in animal models (e.g., histopathology) compared to clinical trials (e.g., eGOS score). A greater emphasis on clinically relevant behavioral measures and long-term follow-up may reduce this disconnect (Narayan et al., 2002). Both preclinical and clinical research in TBI could benefit from adopting a more multidimensional approach to diagnosis and assessment, such as the incorporation of advanced neuroimaging measures into experimental TBI studies.

CONCLUSION

Assessing the performance of animals in a behavioral assay provides an understanding as to the functional consequences of TBI. Successful translation of potential novel therapeutics in the neurotrauma field requires evidence of reproducible behavioral improvements across different experimental models, and ideally, across the spectrum of functionality. There are now a wide range of behavioral assays available for use in rodents spanning neurocognitive, neurological, and psychosocial domains. Moving forward, we recommend increased communication between clinicians, clinician/scientists, and researchers to optimize study design, standardization, and general scientific rigor when assessing behavioral outcomes after TBI, and consider inclusion of multimodel approaches to improve translatability from the bench to the bedside. Increased application of standardized guidelines for scientific rigor will aid in the achievement of this goal. Preclinical trials would benefit from the expansion into longer-term studies and the evaluation of behavioral assays that better overlap with functional outcome assessments in TBI patients. Often-neglected factors that may have strong implications for functional outcomes, including gender, age, and comorbidities, also require further investigation.

REFERENCES

Alamed J, Wilcock DM, Diamond DM, Gordon MN, Morgan D. 2006. Two-day radial-arm water maze learning and memory task; robust resolution of amyloid-related memory deficits in transgenic mice. *Nat Protoc* 1:1671–1679.

Aminmansour B, Nikbakht H, Ghorbani A, Rezvani M, Rahmani P, Torkashvand M, Nourian M, Moradi M. 2012. Comparison of the administration of progesterone versus progesterone and vitamin D in improvement of outcomes in patients with traumatic brain injury: A randomized clinical trial with placebo group. *Adv Biomed Res* 1:58.

Anderson V, Moore C. 1995. Age at injury as a predictor of outcome following pediatric head injury: A longitudinal perspective. *Child Neuropsychol* 1:187–202.

Anderson V, Catroppa C, Morse S, Haritou F, Rosenfeld J. 2005. Functional plasticity or vulnerability after early brain injury? *Pediatrics* 116:1374–1382.

Ando T, Xuan W, Xu T, Dai T, Shama SK, Kharkwal GBea. 2011. Comparison of therapeutic effects between pulsed and continuous wave 810-nm wavelength laser irradiation for traumatic brain injury in mice. *PLoS One*. 6:e26212.

Antunes M, Biala G. 2012. The novel object recognition memory: Neurobiology, test procedure, and its modifications. *Cogn Process* 13:93–110.

Bao F, Shultz SR, Hepburn JD, Omana V, Weaver LC, Cain DP, Brown A. 2012. A CD11d monoclonal antibody treatment reduces tissue injury and improves neurological outcome after fluid percussion brain injury in rats. *J Neurotrauma* 29:2375–2392.

Barth TM, Stanfield BB. 1990. The recovery of forelimb-placing behavior in rats with neonatal unilateral cortical damage involves the remaining hemisphere. *J Neurosci* 10:3449–3459.

Baskin YK, Dietrich WD, Green EJ. 2003. Two effective behavioral tasks for evaluating sensorimotor dysfunction following traumatic brain injury in mice. *J Neurosci Methods* 129:87–93.

Beauchamp MH, Beare R, Ditchfield M, Coleman L, Babl FE, Kean M, Crossley L, Catroppa C, Yeates KO, Anderson V. 2013. Susceptibility weighted imaging and its relationship to outcome after pediatric traumatic brain injury. *Cortex* 49:591–598.

Beaumont A, Marmarou A, Czigner A, Yamamoto M, Demetriadou K, Shirotani T, Marmarou C, Dunbar J. 1999. The impact-acceleration model of head injury: Injury severity predicts motor and cognitive performance after trauma. *Neurol Res* 21:742–754.

Berge OG. 2014. Behavioral pharmacology of pain. *Curr Topics Behav Neurosci* 20:33–56.

Bondi CO, Cheng JP, Tennant HM, Monaco CM, Kline AE. 2014. Old dog, new tricks: The attentional set-shifting test as a novel cognitive behavioral task after controlled cortical impact injury. *J Neurotrauma* 31(10):926–937.

Bosco MA, Murphy JL, Clark ME. 2013. Chronic pain and traumatic brain injury in OEF/OIF service members and Veterans. *Headache* 53:1518–1522.

Broadbent NJ, Squire LR, Clark RE. 2004. Spatial memory, recognition memory, and the hippocampus. *Proc Natl Acad Sci* 101:14515–14520.

Budde MD, Shah A, McCrea M, Cullinan WE, Pintar FA, Stemper BD. 2013. Primary blast traumatic brain injury in the rat: Relating diffusion tensor imaging and behavior. *Front Neurol* 4:154.

Carbonell WS, Maris DO, McCall T, Grady MS. 1998. Adaption of the fluid percussion injury model to the mouse. *J Neurotrauma* 15:217–229.

Catroppa C, Anderson V, Morse S, Haritou F, Rosenfeld J. 2008. Outcome and predictors of functional recovery 5 years following pediatric traumatic brain injury (TBI). *J Pediatr Psychol* 33:707–718.

Catroppa C, Godfrey C, Rosenfeld JV, Hearps SS, Anderson VA. 2012. Functional recovery ten years after pediatric traumatic brain injury: Outcomes and predictors. *J Neurotrauma* 29:2539–2547.

Cernak I. 2005. Animal models of head trauma. *NeuroRx* 2:410–422.

Cernak I, Noble-Haeusslein LJ. 2010. Traumatic brain injury: An overview of pathobiology with emphasis on military populations. *J Cereb Blood Flow Metab* 30:255–266.

Cernak I, Wang Z, Jiang J, Bian X, Savic J. 2001. Ultrastructural and functional characteristics of blast injury-induced neurotrauma. *J Trauma* 50:695–706.

Chapman LA, Wade SL, Walz NC, Taylor HG, Stancin T, Yeates KO. 2010. Clinically significant behavior problems during the initial 18 months following early childhood traumatic brain injury. *Rehabil Psychol* 55:48–57.

Chen CY, Noble-Haeusslein LJ, Ferriero D, Semple BD. 2013. Traumatic injury to the immature frontal lobe: A new murine model of long-term motor impairment in the absence of psychosocial or cognitive deficits. *Dev Neurosci* 35:474–490.

Chen Y, Constantini S, Trembovler V, Weinstock M, Shohami E. 1996. An experimental model of closed head injury in mice: Pathophysiology, histopathology, and cognitive deficits. *J Neurotrauma* 13:557–568.

Cheng J, Gu J, Ma Y, Yang T, Kuang Y, Li B, Kang J. 2010. Development of a rat model for studying blast-induced traumatic brain injury. *J Neurol Sci* 294:23–28.

Crawley JN. 2007. Mouse behavioral assays relevant to the symptoms of autism. *Brain Pathol* 17:448–459.

D'Hooge R, De Deyn PP. 2001. Applications of the Morris water maze in the study of learning and memory. *Brain Res Rev* 36:60–90.

Diaz-Arrastia R, Kochanek PM, Bergold P, Kenney K, Marx CE, Grimes JB, Loh Y. 2014. Pharmacotherapy of traumatic brain injury: State of the science and the road forward report of the Department of Defense Neurotrauma Pharmacology Workgroup. *J Neurotrauma* 31(2):135–158.

Dixon CE, Lighthall JW, Anderson TE. 1988. Physiologic, histopathologic, and cineradiographic characterization of a new fluid-percussion model of experimental brain injury in the rat. *J Neurotrauma* 5:91–104.

Dixon CE, Clifton GL, Lighthall JW, Yaghmai AA, Hayes RL. 1991. A controlled cortical impact model of traumatic brain injury in the rat. *J Neurosci Methods* 39:253–262.

Dixon CE, Lyeth BG, Povlishock JT, Findling RL, Hamm RJ, Marmarou A, Young HF, Hayes RL. 1987. A fluid percussion model of experimental brain injury in the rat. *J Neurosurg* 67:110–119.

Donders J, Warschausky S. 2007. Neurobehavioral outcomes after early versus late childhood traumatic brain injury. *J Head Trauma Rehabil* 22:296–302.

Dunn-Meynell AA, Levin BE. 1997. Histological markers of neuronal, axonal and astrocytic changes after lateral rigid impact traumatic brain injury. *Brain Res* 761:25–41.

Ewing-Cobbs L, Fletcher JM, Levin HS, Iovino I, Miner ME. 1998. Academic achievement and academic placement following traumatic brain injury in children and adolescents: A two-year longitudinal study. *J Clin Exp Neuropsychol* 20:769–781.

Faul M, Xu L, Wald MM, Coronado VG. 2010. *Traumatic Brain Injury in the United States: Emergency Department Visits, Hospitalizations and Deaths 2002–2006*. Atlanta, GA: Centers for Disease Control and Prevention, National Center for Injury Prevention and Control.

Feeney DM, Boyeson MG, Linn RT, Murray HM, Dail WG. 1981. Responses to cortical injury: I. Methodology and local effects of contusions in the rat. *Brain Res* 211:67–77.

Flierl MA, Stahel PF, Beauchamp KM, Morgan SJ, Smith WR, Shohami E. 2009. Mouse closed head injury model induced by a weight-drop device. *Nat Protoc* 4:1328–1337.

Floyd CL, Golden KM, Black RT, Hamm RJ, Lyeth BG. 2002. Craniectomy position affects morris water maze performance and hippocampal cell loss after parasagittal fluid percussion. *J Neurotrauma* 19:303–316.

Fox GB, Fan L, Levasseur RA, Faden AI. 1998. Sustained sensory/motor and cognitive deficits with neuronal apoptosis following controlled cortical impact brain injury in the mouse. *J Neurotrauma* 15:599–614.

Fujimoto ST, Longhi L, Saatman KE, Conte V, Stocchetti N, McIntosh TK. 2004. Motor and cognitive function evaluation following experimental traumatic brain injury. *Neurosci Biobehav Rev* Jul; 28(4):365–378.

Gibson CL, Gray LJ, Bath PM, Murphy SP. 2008. Progesterone for the treatment of experimental brain injury; a systematic review. *Brain* 131:318–328.

Gold EM, Su D, López-Velázquez L, Haus DL, Perez H, Lacuesta GA, Anderson AJ, Cummings BJ. 2013. Functional assessment of long-term deficits in rodent models of traumatic brain injury. *Regen Med* 8:483–516.

Halbauer JD, Ashford JW, Zeitzer JM, Adamson MM, Lew HL, Yesavage JA. 2009. Neuropsychiatric diagnosis and management of chronic sequelae of war-related mild to moderate traumatic brain injury. *J Rehabil Res Dev* 46:757–796.

Hallam TM, Floyd CL, Folkerts MM, Lee LL, Gong QZ, Lyeth BG, Muizelaar JP, Berman RF. 2004. Comparison of behavioral deficits and acute neuronal degeneration in rat lateral fluid percussion and weight-drop brain injury models. *J Neurotrauma* 21:521–539.

Hamm RJ, Lyeth BG, Jenkins LW, O'Dell DM, Pike BR. 1993. Selective cognitive impairment following traumatic brain injury in rats. *Behav Brain Res* 59:169–173.

Hamm RJ, Pike BR, O'Dell DM, Lyeth BG, Jenkins LW. 1994. The rotarod test: An evaluation of its effectiveness in assessing motor deficits following traumatic brain injury. *J Neurotrauma* 11:187–196.

Heffernan ME, Huang W, Sicard KM, Bratane BT, Sikoglu EM, Zhang N, Fisher M, King JA. 2013. Multi-modal approach for investigating brain and behavior changes in an animal model of traumatic brain injury. *J Neurotrauma* 30:1007–1012.

Horneman G, Emanuelson I. 2009. Cognitive outcome in children and young adults who sustained severe and moderate traumatic brain injury 10 years earlier. *Brain Injury* 23:907–914.

Humphreys I, Wood RL, Phillips CJ, Macey S. 2013. The costs of traumatic brain injury: A literature review. *ClinicoEcon Outcomes Res* 5:281–287.

Hylin MJ, Osri SA, Rozas NS, Hill JL, Zhao J, Redell JB, Moore NM, Dash PK. 2013. Repeated mild closed head injury impairs short-term visuospatial memory and complex learning. *J Neurotrauma* 30:716–726.

Inoue T, Lin A, Ma X, McKenna SL, Creasey GH, Manley GT, Ferguson AR, Bresnahan JC, Beattie MS. 2013. Combined SCI and TBI: recovery of forelimb function after unilateral cervical spinal cord injury (SCI) is retarded by contralateral traumatic brain injury (TBI), and ipsilateral TBI balances the effects of SCI on paw placement. *Exp Neurol* 248:136–147.

Jorge RE. 2005. Neuropsychiatric consequences of traumatic brain injury: A review of recent findings. *Cur Opin Psychiatry* 18:289–299.

Kabadi SV, Faden AI. 2014. Neuroprotective strategies for traumatic brain injury: Improving clinical translation. *Int J Mol Sci* 15:1216–1236.

Kane MJ, Angoa-Pérez M, Briggs DI, Viano DC, Kreipke CW, Kuhn DM. 2012. A mouse model of human repetitive mild traumatic brain injury. *J Neurosci Methods* 203:41–49.

Klemenhagen KC, O'Brien SP, Brody DL. 2013. Repetitive concussive traumatic brain injury interacts with post-injury foot shock stress to worsen social and depression-like behavior in mice. *PLoS One* 8:e74510.

Koskiniemi M, Kyykkä T, Nybo T, Jarho L. 1995. Long-term outcome after severe brain injury in preschoolers is worse than expected. *Arch Pediatr Adolesc Med* 149:249–254.

Kuehn R, Simard PF, Driscoll I, Keledjian K, Ivanova S, Tosun C, Williams A, Bochicchio G, Gerzanich V, Simard JM. 2011. Rodent model of direct cranial blast injury. *J Neurotrauma* 28:2155–2169.

Landis SC, Amara SG, Asadullah K, Austin CP, Blumenstein R, Bradley EW, Crystal RG, Darnell RB et al. 2012. A call for transparent reporting to optimize the predictive value of preclinical research. *Nature* 490:187–191.

Langlois JA. 2000. *Traumatic Brain Injury in the United States: Assessing Outcomes in Children: Summary and Recommendations from the Expert Working Group, October 26–27.* Atlanta, GA: Division of Acute Care, Rehabilitation Research and Disability Prevention, National Center for Injury Prevention and Control, Centers for Disease Control and Prevention, Department of Health and Human Services.

Lapchak PA, Zhang JH, Noble-Haeusslein LJ. 2013. RIGOR Guidelines: Escalating STAIR and STEPS for effective translational research. *Transl Stroke Res* 4:279–285.

Lim MM, Elkind J, Xiong G, Galante R, Zhu J, Zhang L, Lian J, Rodin J, Kuzma NN, Pack AI, Cohen AS. 2013 Dietary therapy mitigates persistent wake deficits caused by mild traumatic brain injury. *Sci Transl Med* 5:215ra173.

Lindner MD, Plone MA, Cain CK, Frydel B, Francis JM, Emerich DF, Sutton RL. 1998. Dissociable long-term cognitive deficits after frontal versus sensorimotor cortical contusions. *J Neurotrauma* 15:199–216.

Luo J, Nguyen A, Villeda S, Zhang H, Ding Z, Lindsey D, Bieri G, Castellano JM, Beaupre GS, Wyss-Coray T. 2014. Long-term cognitive impairments and pathological alterations in a mouse model of repetitive mild traumatic brain injury. *Front Neurol* 5(5):2–34.

Maas AI, Menon DK. 2012. Traumatic brain injury: Rethinking ideas and approaches. *Lancet Neurol* 11:12–13.

Macolino CM, Daiutolo BV, Alberston BK, Elliott MB. 2014. Mechanical alloydnia induced by traumatic brain injury is independent of restraint stress. *J Neurosci Methods* 15:139–146.

Malkesman O, Tucker LB, Ozl J, McCabe JT. 2013. Traumatic brain injury—Modeling neuropsychiatric symptoms in rodents. *Front Neurol* 4:157.

Marmarou A, Foda MA, van den Brink W, Campbell J, Kita H, Demetriadou K. 1994. A new model of diffuse brain injury in rats. Part I: Pathophysiology and biomechanics. *J Neurosurg* 80:291–300.

Martens KM, Vonder Haar C, Hutsell BA, Hoane MR. 2012. A discrimination task used as a novel method of testing decision-making behavior following traumatic brain injury. *J Neurotrauma* 29:2505–2512.

McCauley SR, Wilde EA, Anderson VA, Bedell G, Beers SR, Campbell TF, Chapman SB et al. 2012. Recommendations for the use of common outcome measures in pediatric traumatic brain injury research. *J Neurotrauma* 29:678–705.

McCrory P, Meeuwisse W, Aubry M, Cantu B, Dvorak J, Echemendia RJ, Engebretsen Lea. 2013. Consensus statement on concussion in sport-the 4th International Conference on Concussion in Sport Held in Zurich, November 2012. *Clinical Journal of Sport Medicine* 23:89–117.

McIntosh TK, Noble LJ, Andrews B, Faden AI. 1987. Traumatic brain injury in the rat: characterization of a midline fluid-percussion model. *Cent Nerv Syst Trauma* 4:119–134.

Meehan WPr, Zhang, J., Mannix, R., Whalen, M.J. 2012. Increasing recovery time between injuries improves cognitive outcome after repetitive mild concussive brain injuries in mice. *Neurosurgery* 71:885–891.

Meyer DL, Davies DR, Barr JL, Manzerra P, Frster GL. 2012. Mild traumatic brain injury in the rat alters neuronal number in the limbic system and increases conditioned fear and anxiety-like behaviors. *Exp Neurol* 235:574–587.

Morales DM, Marklund N, Lebold D, Thompson HJ, Pitkanen A, Maxwell WL, Longhi L et al. 2005. Experimental models of traumatic brain injury: Do we really need to build a better mousetrap? *Neuroscience* 136:971–989.

Mountney A, Leung LY, Pedersen R, Shear D, Tortella F. 2013. Longitudinal assessment of gait abnormalities following penetrating ballistic-like brain injury in rats. *J Neurosci Methods* 212:1–16.

Mouzon B, Chaytow H, Crynen G, Bachmeier C, Stewart J, Mullan M, Stewart W, Crawford F. 2012. Repetitive mild traumatic brain injury in a mouse model produces learning and memory deficits accompanied by histological changes. *J Neurotrauma* 29:2761–2773.

Moy SS, Nadler JJ, Young NB, Nonneman RJ, Segall SK, Andrade GM, Crawley JN, Magnuson TR. 2008. Social approach and repetitive behavior in eleven inbred mouse strains. *Behav Brain Res* 191:118–129.

Myburgh JA, Cooper DJ, Finfer SR, Venkatesh B, Jones D, Higgins A, Bishop N, Higlett T, (ATBIS) TATBIS, Group IftANZICSCT. 2008. Epidemiology and 12-month outcomes from traumatic brain injury in Australia and New Zealand. *J Trauma* 64:854–862.

Nampiaparampil DE. 2008. Prevalence of chronic pain after traumatic brain injury: A systematic review. *JAMA* 300:711–719.

Narayan RK, Michel ME, Group TCTiHIS. 2002. Clinical trials in head injury. *J Neurotrauma* 19:503–557.

Nichol AD, Higgins AM, Gabbe BJ, Murray LJ, Cooper DJ, Cameron PA. 2011. Measuring functional and quality of life outcomes following major head injury: Common scales and checklists. *Injury, Int J Care Injured* 42:281–287.

O'Connor WT, Smyth A, Gilchrist MD. 2011. Animal models of traumatic brain injury: A critical evaluation. *Pharmacol Ther* 130:106–113.

Onyszchuk G, He YY, Berman NE, Brooks WM. 2008. Detrimental effects of aging on outcome from traumatic brain injury: A behavioral, magnetic resonance imaging, and histological study in mice. *J Neurotrauma* 25:153–171.

Orff HJ, Avalon L, Drummond SP. 2009. Traumatic brain injury and sleep disturbance: A review of current research. *J Head Trauma Rehabil* 24:155–165.

Otis JD, McGlinchey R, Vasterling JJ, Kerns RD. 2011. Complicating factors associated with mild traumatic brain injury: Impact on pain and posttraumatic stress disorder treatment. *J Clin Psychol Med Settings* 18:145–154.

Pandey DK, Yadav SK, Mahesh R, Rajkumar R. 2009. Depression-like and anxiety-like behavioural aftermaths of impact accelerated traumatic brain injury in rats: A model of comorbid depression and anxiety? *Behav Brain Res* 205:436–442.

Pierce JE, Smith DH, Trojanowski JQ, McIntosh TK. 1998. Enduring cognitive, neurobehavioral and histopathological changes persist for up to one year following severe experimental brain injury in rats. *Neuroscience* 87:359–369.

Piper IR, Thomson D, Miller JD. 1996. Monitoring weight drop velocity and foam stiffness as an aid to quality control of a rodent model of impact acceleration neurotrauma. *J Neurosci Methods* 69:171–174.

Ponsford J, Tweedly L, Taffe J. 2013. The relationship between alcohol and cognitive functioning following traumatic brain injury. *J Clin Exp Neuropsychol* 35:103–112.

Ponsford J, Downing M, Olver J, Ponsford M, Acher R, Carty M, Spitz G. 2014. Longitudinal follow-up of patients with traumatic brain injury: Outcome at 2, 5, and 10-years post-injury. *J Neurotrauma* 31(1):64–77.

Ponsford JL, Ziino C, Parcell DL, Shekleton JA, Roper M, Redman JR, Phipps-Nelson J, Rajaratnam SM. 2012. Fatigue and sleep disturbance following traumatic brain injury—Their nature, causes, and potential treatments. *J Head Trauma Rehabil* 27:224–233.

Prigatano GP, Gale SD. 2011. The current state of postconcussive syndrome. *Curr Opin Psychiatry* 24:243–250.

Pullela R, Raber J, Pfankuch T, Ferriero DM, Claus CP, Koh S-E, Yamauchi T, Rola R, Fike JR, Noble-Haeusslein LJ. 2006. Traumatic injury to the immature brain results in progressive neuronal loss, hyperactivity and delayed cognitive impairments. *Dev Neurosci* 28:396–409.

Riess P, Bareyre FM, Saatman KE, Cheney JA, Lifshitz J, Raghupathi R, Grady MS, Neugebauer E, McIntosh TK. 2001. Effects of chronic, post-injury Cyclosporin A administration on motor and sensorimotor function following severe, experimental traumatic brain injury. *Restor Neurol Neurosci* 18:1–8.

Roof RL, Stein DG. 1999. Gender differences in Morris water maze performance depend on task parameters. *Physiol Behav* 68:81–86.

Rubovitch V, Ten-Bosch M, Zohar O, Harrison CR, Tempel-Brami C, Stein E, Hoffer BJ. et al. 2011. A mouse model of blast-induced mild traumatic brain injury. *Exp Neurol* 232:280–289.

Russell KL, Kutchko KM, Fowler SC, Berman NEJ, Levant B. 2011. Sensorimotor behavioral tests for use in a juvenile rat model of traumatic brain injury: Assessment of sex differences. *J Neurosci Methods* 199:214–222.

Ryan NP, Anderson V, Godfrey C, Beauchamp MH, Coleman L, Eren S, Rosema S, Taylor K, Catroppa C. 2014. Predictors of very long-term socio-cognitive function after pediatric traumatic brain injury (TBI): Support for the vulnerability of the immature "social brain" [published online December 20, 2013]. *J Neurotrauma* 31:649–657.

Scheff SW, Baldwin SA, Brown RW, Kraemer PJ. 1997. Morris water maze deficits in rats following traumatic brain injury: Lateral controlled cortical impact. *J Neurotrauma* 14:615–627.

Sebastian V, Diallo A, Ling DS, Serrano PA. 2013. Robust training attenuates TBI-induced deficits in reference and working memory on the radial 8-arm maze. *Front Behav Neurosci* 7:38.

Semple BD, Canchola SA, Noble-Haeusslein L. 2012. Deficits in social behavior emerge during development after pediatric traumatic brain injury in mice. *J Neurotrauma* 29:2672–2683.

Semple BD, Bye N, Rancan M, Ziebell JM, Morganti-Kossmann MC. 2010. Role of CCL2 (MCP-1) in traumatic brain injury (TBI): Evidence from severe TBI patients and CCL2-/- mice. *J Cereb Blood Flow Metab* 30:769–782.

Shear DA, Tate MC, Archer DR, Hoffman SW, Hulce VD, Laplaca MC, Stein DG. 2004. Neural progenitor cell transplants promote long-term functional recovery after traumatic brain injury. *Brain Res* 1026:11–22.

Shelton SB, Pettigrew DB, Hermann AD, Zhou W, Sullivan PM, Crutcher KA, Strauss KI. 2008. A simple, efficient tool for assessment of mice after unilateral cortex injury. *J Neurosci Methods* 168:431–442.

Shukla D, Devi BI, Agrawal A. 2011. Outcome measures for traumatic brain injury. *Clin Neurol Neurosurg* 113(6):435–441.

Shultz SR, Bao F, Omana V, Chiu C, Brown A, Cain DP. 2012. Repeated mild lateral fluid percussion brain injury in the rat causes cumulative long-term behavioral impairments, neuroinflammation, and cortical loss in an animal model of repeated concussion. *J Neurotrauma* 29:281–294.

Singleton RH, Yan HQ, Fellows-Mayle W, Dixon CE. 2010. Resveratrol attenuates behavioral impairments and reduces cortical and hippocampal loss in a rat controlled cortical impact model of traumatic brain injury. *J Neurotrauma* 27:1091–1099.

Siopi E, Llufriu-Dabén G, Fanucchi F, Plotkine M, Marchand-Leroux C, Jafarian-Tehrani M. 2012. Evaluation of late cognitive impairment and anxiety states following traumatic brain injury in mice: The effect of minocycline. *Neurosci Lett* 511:110–115.

Skelton RW. 1998. Modelling recovery of cognitive function after traumatic brain injury: Spatial navigation in the Morris water maze after complete or partial transections of the perforant path in rats. *Behav Brain Res* 96:13–35.

Smith DH, Soares HD, Pierce JS, Perlman KG, Saatman KE, Meaney DF, Dixon CE, McIntosh TK. 1995. A model of parasagittal controlled cortical impact in the mouse: Cognitive and histopathologic effects. *J Neurotrauma* 12:169–178.

Spitz G, Ponsford JL, Rudzki D, Maller JJ. 2012. Association between cognitive performance and functional outcome following traumatic brain injury: A longitudinal multilevel examination. *Neuropsychology* 26:604–612.

STAIR. 1999. Recommendations for standards regarding preclinical neuroprotective and restorative drug development. *Stroke* 30:2752–2758.

Strekalova T, Steinbusch HW. 2010. Measuring behavior in mice with chronic stress depression paradigm. *Prog Neuropsychopharmacol Biol Psychiatry* 34:348–361.

Tang H, Hua F, Wang J, Sayeed I, Wang X, Chen Z, Yousuf S, Atif F, Stein DG. 2013. Progesterone and vitamin D: Improvement after traumatic brain injury in middle-aged rats. *Horm Behav* 64:527–538.

Thompson HJ, Lifshitz J, Marklund N, Grady MS, Graham DI, Hovda DA, McIntosh TK. 2005. Lateral fluid percussion brain injury: A 15-year review and evaluation. *J Neurotrauma* 22:42–75.

Tsenter J, Beni-Adani L, Assaf Y, Alexandrovich AG, Trembovler V, Shohami E. 2008. Dynamic changes in the recovery after traumatic brain injury in mice: Effect of injury severity on T2-weighted MRI abnormalities, and motor and cognitive functions. *J Neurotrauma* 25:324–333.

Venturi L, Miranda M, Selmi V, Vitali L, Tani A, Margheri M, De Gaudio AR, Adembri C. 2009. Systemic sepsis exacerbates mild post-traumatic brain injury in the rat. *J Neurotrauma* 26:1547–1556.

Vink R, O'Connor CA, Nimmo AJ, Heath DL. 2003. Magnesium attenuates persistent functional deficits following diffuse traumatic brain injury in rats. *Neurosci Lett* 336:41–44.

Wagner AK, Kline AE, Ren D, Willard LA, Wenger MK, Zafonte RD, Dixon CE. 2007. Gender associations with chronic methylphenidate treatment and behavioral performance following experimental traumatic brain injury. *Behav Brain Res* 181:200–209.

Washington PM, Forcelli PA, Wilkins T, Zapple DN, Parsadanian M, Burns MP. 2012. The effect of injury severity on behavior: A phenotypic study of cognitive and emotional deficits after mild, moderate, and severe controlled cortical impact injury in mice. *J Neurotrauma* 29:2283–2296.

Wei J, Xiao GM. 2013. The neuroprotective effects of progesterone on traumatic brain injury: Current status and future prospects. *Acta Pharmacologic Sinica* 34:1485–1490.

Wilde EA, Whiteneck GG, Bogner J, Bushnik T, Cifu DX, Dikmen S, French L. 2010. Recommendations for the use of common outcome measures in traumatic brain injury research. *Arch Phys Med Rehabil* 91:1650–1660.

Willie JT, Lim MM, Bennett RE, Azarion AA, Schwetye KE, Brody DL. 2012. Controlled cortical impact traumatic brain injury acutely disrupts wakefulness and extracellular orexin dynamics as determined by intracerebral microdialysis in mice. *J Neurotrauma* 29:1908–1921.

Woodlee MT, Asseo-García AM, Zhao X, Liu SJ, Jones TA, Schallert T. 2005. Testing forelimb placing "across the midline" reveals distinct, lesion-dependent patterns of recovery in rats. *Exp Neurol* 191:310–317.

Xiong Y, Lu D, Qu C, Goussev A, Schallert T, Mahmood A, Chopp M. 2008. Effects of erythropoietin on reducing brain damage and improving functional outcome after traumatic brain injury in mice. *J Neurosurg* 109:510–521.

Xiong Y, Mahmood A, Chopp M. 2013. Animal models of traumatic brain injury. *Nat Rev Neurosci* 14:128–142.

Yeates KO, Swift E, Taylor HG, Wade SL, Drotar D, Stancin T, Minich N. 2004. Short- and long-term social outcomes following pediatric traumatic brain injury. *J Int Neuropsychol Soc* 10:412–426.

Young JW, Powell SB, Geyer MA, Jeste DV, Risbrough VB. 2010. The mouse attentional-set-shifting task: A method for assaying successful cognitive aging? *Cogn Affect Behav Neurosci* 10:243–251.

Zhao Z, Loane DJ, Murray MGI, Stoica BA, Faden AI. 2012. Comparing the predictive value of multiple cognitive, affective, and motor tasks after rodent traumatic brain injury. *J Neurotrauma* 29:2475–2489.

6 Translation Biomarkers for Traumatic Brain Injury

Pramod K. Dash, Michael Hylin, and Anthony N. Moore

CONTENTS

6.1 INTRODUCTION

Traumatic brain injury (TBI) results from transduction of energy into the brain from an outside source. Approximately 1.4 million people sustain a TBI each year in the United States (Zohar et al. 2011), of which 40,000–50,000 die as a result of the injury. Traumatic brain injury has its greatest impact on young men and women, and it poses a tremendous burden to families and society in terms of years of lost productivity, increased demands on the health care system, and reduction in quality of life. It has been estimated that 5.3 million people in the United States are living with disabilities resulting from TBI, underscoring the necessity for proper diagnosis and intervention. Unfortunately, numerous therapies and treatments have failed in their ability to translate from basic science research to a clinically useful intervention.

Although the human brain is protected by the skull, its large size and relatively fragile composition make it highly vulnerable to trauma. Depending on the severity and location of the injury, brain trauma can result in deficits ranging from difficulties in fine motor control to lasting impairments in learning and memory or even death. Most commonly, trauma to the brain occurs as either the result of sudden change in acceleration/deceleration (especially angular acceleration/deceleration), impact with an object, penetration of the brain by a projectile, or as the result of exposure to blast overpressure. As the underlying pathology of TBI can differ based on the injury type, there is a need to identify these pathologies in individual patients in order to formulate effective mechanism-based treatments. Although computed tomography (CT) scans are useful for the detection of focal lesions, bleeding/hematomas, and brain herniation, the resolution is not sufficient to detect all types of brain pathologies, especially in persons whose injuries can be classified as mild. Although studies have indicated that magnetic resonance imaging (MRI) is more sensitive than CT for detecting mild brain damage, it is not routinely performed and can be expensive. Thus, great gain can come from validated biomarkers that can be used to supplement clinical and neuropsychological evaluation for diagnosing brain injury, for determining the different types of brain pathology, and for objectively assessing the effectiveness of a potential therapy.

6.2 DEFINITION OF A BIOMARKER

A biomarker is a molecular entity (e.g., protein, lipid, RNA) whose abundance and/or modification state can be used to indicate a specific biological or disease state, including injury. Two of the most commonly reported measures of a biomarker's usefulness are sensitivity and specificity. Sensitivity is the measure of the probability of a positive test to identify the condition when it is actually present (i.e., no false negatives). Specificity is defined as the probability of a negative test to correctly identify healthy individuals (i.e., no false positives). At a given threshold, each biomarker will have a sensitivity and a specificity that determine the diagnostic utility of the biomarker for diagnosing/predicting the disease state. For example, serum levels of prostate-specific antigen (PSA) of 4 ng/mL have been reported to have high specificity for the detection of prostate cancer (only 6% false positives), but low sensitivity (80% of cancers missed) (Ankerst and Thompson 2006). Lowering the threshold allows for more sensitive detection of prostate cancer but increases the false-positive rate. The diagnostic accuracy of a biomarker is therefore determined by examining the relationship between sensitivity and specificity using a receiver operator characteristic (ROC) curve, a graph of (sensitivity) versus (1-specificity). Area under the curve (AUC) values of 0.98–1.0 are considered excellent; values ranging from 0.9 to 0.98 are very good, 0.8 to 0.9 are good; 0.7 to 0.8 are fair; and 0.5 to 0.7 are poor indicators of the disease.

6.3 UTILITY OF TRAUMATIC BRAIN INJURY BIOMARKERS

Identical cellular and molecular mechanisms may not underlie the pathophysiology for the entire spectrum of TBI that can range from mild to severe. For instance, axonal injury is a dominant pathology underlying the dysfunction associated with mild TBI, whereas a combination of pathologies including axonal injury, contusion, neuronal death, and hemorrhage are commonly detected in patients who have sustained a severe TBI. Although the Glasgow Coma Scale (GCS) is widely used to assess injury severity, this test has limitations including its dependency on a verbal score, its insensitivity to subtle impairments of consciousness, and its sensitivity to alcohol and drugs of abuse. Further, GCS may not have a linear relationship with outcome. Thus, biomarkers may help clinicians in diagnosing and managing TBI patients in several ways:

1. Validated biomarkers can supplement clinical evaluations and imaging results to assess injury severity.
2. Identification of pathology-specific biomarkers can aid clinicians in formulating appropriate treatment strategies.
3. Identification of biomarkers to predict ensuing clinical complications such as hypoxia or elevated intracranial pressure (ICP) would be invaluable for initiating early treatment.
4. Biomarkers that can predict long-term outcome can allow caregivers to better prepare for the future needs of the patient.
5. Biomarkers can reduce the need for repeated CT scans, thereby avoiding unnecessary radiation exposure, especially to the developing brain.
6. Biomarker measurements can be performed relatively quickly and can be inexpensive, thus allowing for on-site evaluation.
7. Readily accessible bodily fluids such as blood, saliva, or urine can be used to measure a number of TBI biomarkers, reducing risk to the patient.

6.4 CELLULAR SOURCES OF TBI BIOMARKERS

Astrocytes, neurons, oligodendrocytes, and endothelial cells are the major cell types within the brain, with glia being the most abundant. In addition, oligodendrocytes and microglia are widely present. These cells not only express common housekeeping proteins, but also specific proteins that are critical for their structure and function. Therefore, damage to these cells as a result of trauma can cause the release of these proteins into the cerebrospinal fluid (CSF) and blood where they can be

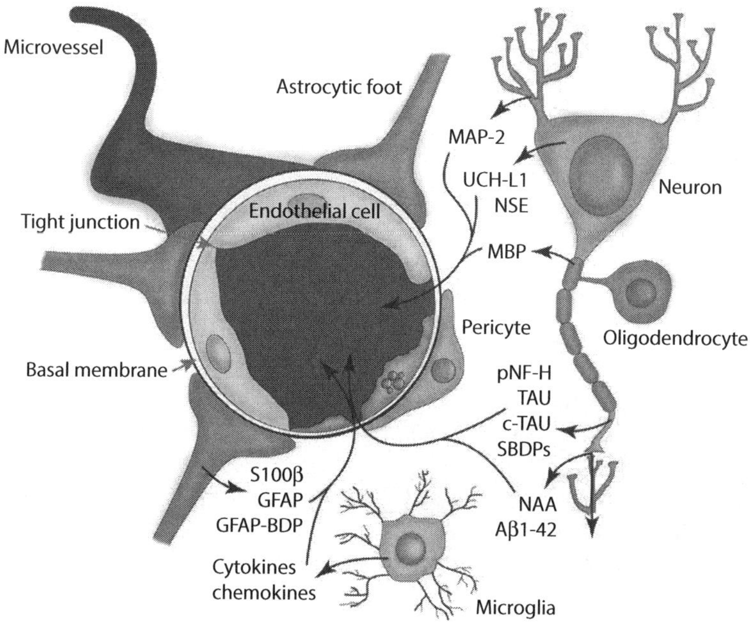

FIGURE 6.1 Injured neurovascular unit and cellular source of TBI biomarkers. Biomarkers can be released from damaged astrocytes, neurons, oligodendrocytes, and activated microglia. These markers can gain access to CSF and/or blood as a result of breakdown of CSF-brain and blood-brain barriers, respectively. Specific transporters (e.g., for Aβ1-42) can also assist in ferrying some of these markers across the CSF-brain and blood-brain barrier.

detected and used as injury biomarkers (Figure 6.1). Although none are currently approved for the diagnosis of brain trauma, several clinical studies have been carried out to test the diagnostic accuracy of these putative biomarkers.

6.5 MARKERS RELEASED FROM ASTROCYTES

S100β is a low molecular weight (10.5 kDa) calcium-binding protein that is primarily expressed and secreted by astrocytes. It is an abundantly expressed protein, constituting 1–1.5 μg/mg of all soluble protein in the brain. However, S100β is present at very low levels in normal human CSF and serum. Following brain injury, however, S100β is released into the CSF where it can pass through the compromised blood-brain barrier (BBB) and/or arachnoid villi and enter the bloodstream. Several studies have shown that TBI increases the levels of this protein (Ingebrigtsen et al. 1999; Ingebrigtsen et al. 2000; Herrmann et al. 2001; Savola et al. 2004), and that its levels correlate with the severity of the injury (Savola et al. 2004; Goyal et al. 2013). The serum half-life of S100β is less than 60 min, and it can be readily detected for days after injury (Jonsson et al. 2000). However, enthusiasm for use of S100β as a TBI biomarker has been somewhat tempered as its levels have been found to be increased in response to extracranial trauma (Savola et al. 2004) and burns (Anderson et al. 2001). More recent reports have found that the timing of the detected increase is a critical consideration, as persons with multiple organ injury without TBI were found to have S100β levels that had normalized by 6 hours following injury. This lack of overt extracranial contribution at later time points supports the use of circulating S100β as an indicator of brain injury (Raabe et al. 2003; Biberthaler et al. 2006; Muller et al. 2007). Furthermore, it has been proposed that if serum S100β levels are normal in a patient with suspected TBI, it is unlikely that any brain injury has occurred (Savola et al. 2004). If this diagnosis is further verified, CT scans (current standard of care) for these patients may not be needed, thereby minimizing unnecessary radiation exposure, especially in pediatric patients. Although it has been demonstrated that normal S100β levels are indicative of the absence of a central nervous system (CNS) injury, acute increases in this biomarker in symptomatic patients with a GCS = 15 are able to serve as an indicator of intracranial lesions (Egea-Guerrero et al. 2012). Furthermore, a recent study comparing TBI patients with diffused axonal injury to focal injuries has found that serum S100β levels are significantly higher in patients with focal injuries (Matek et al. 2012). Together with GFAP and phosphorylated neurofilament-H (discussed below), these markers may be useful in segregating patients with diffused or focal injuries.

6.5.1 S100β and Outcome

Increased levels of S100β have been proposed as a biomarker of poor outcome. For example, Vos et al. demonstrated that in patients with severe TBI, serum S100β concentrations greater than 1.13 ng/mL were associated with increased mortality (100% sensitivity; 41% specificity) and morbidity (88% sensitivity; 43% specificity) (Vos et al. 2004). Further, elevated levels of S100β in the CSF of TBI patients have been found to correlate with an increased mortality and worse disability at 6 months after injury, an effect not observed in nontrauma intensive care unit (ICU) patients (Bloomfield et al. 2007; Goyal et al. 2013; Macedo et al. 2013). Further, ROC analysis has found that S100β levels 24 hours after admission are a good predictor of brain death for TBI patients with an admission GCS less than 8. Postmortem analysis of CSF and serum S100β levels revealed that the levels of this marker were significantly elevated in subjects who had died from a TBI as compared to non-TBI death (Ondruschka et al. 2013). In mild TBI patients, increased serum S100β levels were detected in 38% of patients with postconcussive symptoms as assessed using the Rivermead Post-concussive Symptoms Questionnaire (RPSQ) administered 3 months after injury. Similarly, logistic regression analysis revealed a significant association between initial S100β levels and long-term disability (Stalnacke et al. 2005). However, Lima et al. has reported that high initial serum S100β levels do not correlate with reduced quality of life, anxiety, and depression when tested 18 months after injury (Lima et al. 2008). As a prognostic marker for pediatric patients, conflicting reports have been published regarding the usefulness of this marker in predicting outcome after pediatric TBI (Berger et al. 2002; Piazza et al. 2007; Geyer et al. 2009).

6.5.1.1 Glial Fibrillary Acidic Protein

Glial fibrillary acidic protein (GFAP) is a monomeric intermediate filament protein expressed by astrocytes. Pelinka et al. were the first to report that after TBI,

GFAP is released from the injured brain and that its circulating levels correlate with the degree of brain injury (Pelinka et al. 2004b). Consistent with its CNS-specific expression, GFAP levels were found to be normal in polytrauma patients that did not have brain injury. Based on this observation, it has been suggested that GFAP may be an excellent marker for the diagnosis of mild TBI. Consistent with this, it has been reported that serum GFAP levels are significantly higher in mild TBI patients with intracranial lesions (as detected using CT scans) compared to CT-negative mild TBI patients (Papa et al. 2012a). Compared to S100β, GFAP has been found to have higher diagnostic accuracy for predicting brain lesions in patients with isolated mild TBI (GFAP: AUC = 0.84; S100β: AUC = 0.78) and in those with multiple-organ injuries (GFAP: AUC = 0.93; S100β: AUC = 0.75) (Papa et al. 2014). Furthermore, it has been reported that serum GFAP levels are significantly increased in patients with focal injuries compared to persons with diffused injuries (Pelinka et al. 2004a; Mondello et al. 2011). In addition to GFAP, it has been demonstrated that breakdown products of GFAP (GFAP-BDP), can be found in the serum of patients within 1 hour of a mild or moderate TBI and are able to differentiate patients with mild TBI from uninjured controls (AUC = 0.88) (Papa et al. 2012a; Okonkwo et al. 2013).

6.5.2 GFAP and Outcome

Increased GFAP levels have been shown to be predictive of elevated ICP, reduced mean arterial pressure, low cerebral perfusion pressure, poor Glasgow Outcome Score (GOS), and increased mortality (Pelinka et al. 2004b; Honda et al. 2010). For example, GFAP serum concentrations above 1.5 ng/mL were found to be predictive of death (85% sensitivity, 52% specificity) and poor outcome (GOS at 6 months; 80% sensitivity and 59% specificity). High maximal GFAP serum levels have been associated with death and poor outcome in both adult severe TBI patients, as well as in pediatric TBI patients (Nylen et al. 2006; Zurek and Fedora 2012). When GFAP-BDP was evaluated for its ability to predict a 6-month outcome, a modest predictive value (AUC = 0.65) was determined (Okonkwo et al. 2013).

6.6 MARKERS RELEASED FROM DAMAGED NEURONS

6.6.1 Spectrin Breakdown Products

One of the key pathological features of TBI is the unregulated breakdown of cellular proteins through activation of protease families such as calpain and caspase. Alpha II-spectrin (αII-spectrin) is a structural protein that is abundant in central nervous system neurons. Following TBI, αII-spectrin is proteolyzed by calpain and caspase-3 to generate spectrin breakdown products (SBDPs) indicative of either necrosis or apoptosis, respectively (Pike et al. 1998). In the CSF of patients with severe TBI, calpain- and caspase-3-mediated SBDP levels have been found to be increased at several time points after injury. However, the time course of calpain-generated SBDP145 (i.e., spectrin breakdown product of 145 kDa) and SBDP150 differs from that of caspase-3-generated SBDP120 (Pineda et al. 2007), suggesting different time courses for necrosis and apoptosis. Although both calpain- and caspase-generated SBDPs can be detected in the CSF within hours of TBI, the levels of SBDP145, but not SBDP120, have been reported to correlate with injury severity (Mondello et al. 2010). These breakdown products have also been used to examine injury type, with the serum levels of SBDP150 being significantly higher in diffused injury as compared to focal injury at the 48-hour time point (Pineda et al. 2007).

6.6.1.1 Spectrin Breakdown Products and Outcome

Mondello et al. analyzed mortality in relation to SBDP levels in the CSF and found that both SBDP145 (>6 ng/mL) and SBDP120 (>17.55 ng/mL) strongly predicted death (Mondello et al. 2010). In patients with better outcome (measured by the Glasgow Outcome Scale 6 months after injury), a significant decrease in SBDP levels was found to occur over the first 4 days after injury. Correspondingly, patients whose SBDP levels remained elevated had a worse outcome (Cardali and Maugeri 2006).

6.6.2 Ubiquitin C-Terminal Hydrolase-L1

Ubiquitin carboxyl-terminal hydrolase-L1 (UCH-L1) is a small (~25 kDa) cysteine protease that hydrolyzes C-terminal adducts of ubiquitin precursors and ubiquitinated proteins to generate ubiquitin monomers (Setsuie and Wada 2007). This enzyme comprises approximately 1%–2% of total soluble protein in the brain, where it is expressed predominately in neurons, with very low levels also expressed in some neuroendocrine cells. The UCH-L1 can be detected in the serum of TBI patients within 4 hours of injury (Berger et al. 2012). It has been reported that the serum levels of UCH-L1 can be used to stratify mild and moderate TBI with intracranial lesions (Papa et al. 2012b). Furthermore, the levels of UCH-L1 have been shown to be significantly higher in mild TBI patients with a positive CT scan versus mild TBI patients with no CT findings (Papa et al. 2012b). Serum levels of UCH-L1 have been reported to correlate with blood-brain barrier permeability as assessed by measuring the albumin quotient Q_A, defined as $Q_A = \text{albumin}_{CSF}/\text{albumin}_{serum}$. At 12 hours after injury, serum UCH-L1 and Q_A had a significant relationship with an area under the curve (AUC) of 0.76 (Blyth et al. 2011). Although this suggests that UCH-L1 can be used to indicate a disruption of the BBB, its inability to cross the intact barrier may limit its usefulness in milder forms of injury. Consistent with this, no detectable increases in serum UCH-L1 levels were found in pediatric patients with mTBI (Berger et al. 2012). It has been reported that the GFAP:UCHL1 ratio, when examined <12 hours after injury, can be used to distinguish between diffused and focal injuries, with a higher ratio observed in patients with focal injuries (Mondello et al. 2012b).

6.6.2.1 UCHL1 and Outcome

A prospective study to determine the CSF levels of UCH-L1 in severe brain injury patients and uninjured controls (Papa et al. 2010) found that UCH-L1 levels were significantly increased in the CSF of brain-injured patients, and that the magnitude of this increase correlated with mortality, postinjury complications, and outcome tested

6 months after discharge (Brophy et al. 2011; Papa et al. 2010). Furthermore, the GFAP:UCH-L1 ratio has been reported to be significantly higher in TBI patients who die from their injuries, although this ratio did not have predictive value (Mondello et al. 2012b).

6.6.3 Neuron-Specific Enolase

Neuron-specific enolase (NSE) is one of the five isozymes of the glycolytic enzyme, enolase. Initially found to be expressed in neurons, NSE has subsequently been identified in neuroendocrine cells, oligodendrocytes, thrombocytes, and erythrocytes. The serum half-life of NSE is approximately 24 hours and can be detected within 6 hours of injury. The serum level for NSE (normally <12.5 ng/mL) has been reported to increase after TBI, the levels of which have been correlated with the severity of the injury (Skogseid et al. 1992; Ross et al. 1996; Herrmann et al. 2000). Further, the serum levels of NSE after TBI have been found to significantly correlate with the volume of cortical contusion (Herrmann et al. 2000).

As a biomarker of TBI, NSE was thought to have the opposite problem of S100β, in that NSE appears to have high specificity but is not very sensitive (de, Jr. et al. 2001; Ingebrigtsen and Romner 2003). However, recent clinical studies have challenged the specificity of NSE, especially when evaluated in patients with documented extracranial injuries (Pelinka et al. 2005). Further complicating its utility as a TBI biomarker in the context of other organ injuries is the observation that elevated serum NSE levels can be detected in patients with endocrine tumors of the gut, liver, and lung (Tapia et al. 1981), and have been reported to be increased in experimental models of liver and kidney ischemia (Pelinka et al. 2005). Compared to S100β, NSE has been found to be a better indicator of neuroinflammation in patients with severe traumatic brain injury, whereas S100β better reflects the extent of injury and outcome (Pleines et al. 2001).

6.6.3.1 NSE and Outcome

The levels of NSE on admission have been shown to be associated with an unfavorable outcome at discharge and at a follow-up 6 months after injury (Herrmann et al. 2000; Vos et al. 2004; Chabok et al. 2012). For example, Vos Verbeek et al. found that at serum levels >21.7 ng/mL, NSE is a sensitive indicator of mortality (85% sensitivity) and poor outcome (80% sensitivity) (Vos et al. 2004). A recent meta-analysis study that evaluated the overall diagnostic accuracy of NSE in 16 clinical studies found that the AUC for NSE levels to predict unfavorable outcome was 0.73, whereas it had a AUC of 0.76 for predicting mortality. NSE has also been associated with β-amyloid levels, as TBI patients with poor outcomes (determined by the Glasgow Outcome Scale) demonstrate a significant correlation between increased CSF levels of β-amyloid oligomers and increased CSF levels of NSE (Gatson et al. 2013).

6.6.4 TAU

Brain injury results in damage to cell soma, dendrites, and axons which, in turn, results in disruption of neuronal cytoskeletal elements. TAU is a microtubule-binding protein that is abundantly expressed in neurons, and at low levels in astrocytes and oligodendrocytes. TAU is not expressed in dendrites but is enriched in thin, nonmyelinated axons of cortical interneurons (Binder et al. 1985; Kosik and Finch 1987; Trojanowski et al. 1989). As such, TAU has been proposed to be a biomarker of axonal injury. Consistent with this, the concentrations of TAU in the CSF from boxers suspected of having a diffuse axonal injury (DAI) were found to be enhanced for up to 10 days after a fight (Zetterberg et al. 2006; Neselius et al. 2012; Neselius et al. 2013). The levels of CSF TAU have been reported to correlate with lesion size (Franz et al. 2003).

6.6.4.1 TAU and Outcome

In a study examining the relationship between TAU in the CSF (on days 2 and 3 after injury) and 1-year outcome after severe TBI, Ost et al. have reported that a TAU level of >2.1 ng/mL has a sensitivity of 100% and a specificity of 81% in predicting mortality, whereas a level of >702 pg/mL was found to be indicative of poor outcome (Ost et al. 2006). Further, Liliang et al. have reported that TAU has an 88% sensitivity and a 94% specificity in predicting poor outcome (Liliang et al. 2010). In addition to examining total TAU, several studies have examined the utility of cleaved TAU to predict outcome after TBI. Patients after TBI have been observed to have a more than 1000 times more cleaved TAU (c-TAU) in their CSF than non-TBI patients (Zemlan et al. 1999). When proteolyzed by calpain, TAU releases a 17 kDa fragment, whereas its breakdown by caspase yields a 50kDa fragment. This suggests that the pattern of fragment generation may be indicative of the time courses/levels of necrosis and apoptosis (Park et al. 2007). In patients with severe brain trauma, initial postinjury CSF c-TAU levels correlated with clinical outcome (sensitivity 92% and specificity 94%) and could predict the occurrence of elevated intracranial pressure (Zemlan et al. 2002). However, in patients with mild brain injury, c-TAU levels have been shown to be a poor predictor of postconcussive syndrome (PCS) (Bazarian et al. 2006; Bulut et al. 2006; Ma et al. 2008). For example, when the levels of c-TAU were evaluated for their ability to predict PCS, it was found that mTBI patients with long-term symptomology did not have any higher initial levels of c-TAU than did those without PCS (Ma et al. 2008). Consistent with this, a review of 11 prospective studies revealed that S100β, NSE, and c-TAU were not consistently able to predict PCS in mTBI patients (Begaz et al. 2006).

6.6.5 Neurofilaments

Neurofilaments are abundantly expressed in neurons where they provide structural support to axons and dendrites. Three major subtypes of neurofilaments: 68–70 kDa light (NF-L), 145–160 kDa medium (NF-M), and 200–220 heavy (NF-H) have been identified. The NF-H contains multiple serine phosphorylation sites, and phosphorylated NF-H has been found in axons but not in dendrites. As increased levels of phosphorylated NF-H have been detected in the CSF and serum of TBI patients (Siman et al. 2009; Gatson et al. 2014), this suggests that this marker may serve as a surrogate marker of axonal damage. Consistent with this, it has been reported that the serum levels of phosphorylated NF-H are inversely correlated to GCS.

As a biomarker for the diagnosis of mild TBI, serum levels of phosphorylated NF-H have been reported to be significantly increased by 24 hours after injury (Gatson et al. 2014). Likewise, in boxers, the CSF levels of phosphorylated NF-H have been shown to be significantly elevated between 1 and 6 days after a fight (Neselius et al. 2013). When measured 24 hours after mild TBI, serum phosphorylated NF-H was found to have a sensitivity of 87.5% and specificity of 70% in identifying mild TBI patients who have clinically relevant findings on a CT scan.

6.6.5.1 Neurofilaments and Outcome

Although limited outcome studies have been performed for phosphorylated NF-H, studies examining severe TBI in pediatric patients have reported that its serum levels are predictive of mortality and poor outcome (Zurek et al. 2011, 2012). Two studies have examined the utility of phosphorylated NF-H to diagnose injury type and have found that its levels are significantly higher in patients with diffused axonal injury than in patients with more focal injuries (Zurek et al. 2011; Vajtr et al. 2012). If confirmed, the serum levels of phosphorylated NF-H may have clinical utility in detecting diffuse axonal injuries and may help direct appropriate therapies.

6.6.6 Microtubule-Associated Protein-2

Microtubule-associated protein-2 (MAP2) is exclusively expressed by neurons where it binds to tubulin and stabilizes the microtubules. As opposed to TAU which is found predominantly in axons, MAP2 is predominately dendritic with its release being suggestive of dendritic damage. In a prospective study examining 16 adult TBI patients, Mondello et al. found that serum MAP-2 levels were significantly higher 6 months after injury when compared to persons without TBI (Mondello et al. 2012a). There was a significant relationship between the magnitude of MAP-2 elevations and degree of cognitive function as assessed using the Los Amigos Levels of Cognitive Functioning Scale (LALCFS) and the Glasgow Outcome Scale (GOS). Although the sample size was small, the authors observed that patients with better outcome had higher levels of serum MAP-2 (Mondello et al. 2012a). Further studies, including examining its breakdown products, may help to determine if MAP-2 release/breakdown is indicative of dendritic damage and provide insights into the underlying mechanisms of dendritic disruptions after TBI. These mechanisms could be potentially targeted to reduce dendritic damage.

6.6.7 N-Acetylaspartate

A derivative of aspartate, N-acetylaspartate (NAA) is the second most abundant chemical in the brain. N-Acetylaspartate is synthesized by neurons and is thought to be involved in a variety of processes including fluid balance in the brain, energy production, and myelin synthesis. It also serves as the precursor for the generation of the neurotransmitter NAAG (N-acetylaspartylglutamate). Following concussion, alterations of NAA levels have been coupled with levels of creatine and choline (Vagnozzi et al. 2013). There is an increase in the NAA-to-creatine ratio shortly after injury, whereas there is a decrease in the NAA to choline ratio that normalizes by 1.5 months after injury.

6.6.7.1 N-Acetylaspartate and Outcome

Global decreases in the levels of NAA, as detected by proton magnetic resonance spectroscopy, have been associated with poor outcome following mild TBI (Govind et al. 2010). Furthermore, the postinjury ratios of NAA to creatine in various brain regions have been demonstrated to predict aspects of neuropsychological dysfunction (Yeo et al. 2006). These authors have measured the ratios of NAA/Cre and Cho/Cre in the anterior and posterior halves of supraventricular region of the brain in pediatric TBI patients using magnetic resonance spectroscopy. Their findings revealed that each ratio predicts aspects of neuropsychological deficits, indicating the utility of measuring region/structure-specific biomarker measurements.

6.7 MARKERS RELEASED FROM DAMAGED OLIGODENDROCYTES

6.7.1 Myelin Basic Protein

Myelin basic protein (MBP) is expressed by oligodendrocytes and is the major protein component of myelin, which is critical for propagation of neuronal action potentials. A MBP test is often used to assess the levels of this protein in the CSF in conditions where demyelination is suspected. Sudden acceleration/deceleration of the brain as a result of trauma causes shearing of white matter, leading to DAI and the release of MBP into the CSF and serum. For example, it has been found that MBP levels remain elevated for up to 2 weeks after TBI (Berger et al. 2010). Interestingly, MBP can cause opening of the blood-brain barrier, thereby facilitating its own entry (and possibly other CNS-derived biomarkers) into the blood. Normally, MBP levels are less than 4 ng/mL in the CSF, with levels exceeding 9 ng/mL indicative of active myelin degradation. Serum levels in pediatric TBI have a peak myelin basic protein concentration resulting in a ROC curve predicting intracranial hemorrhage with an AUC of 0.69 with 44% sensitivity and 96% specificity (Berger et al. 2005). Interestingly, it has been reported that initial serum levels of MBP are higher in children with inflicted TBI than in noninflicted TBI (Berger et al. 2005).

6.8 INFLAMMATORY MARKERS

Inflammatory markers can be released by multiple cell types resident to the brain, as well as by circulating cells (e.g., neutrophils) that infiltrate the injured brain. Although the importance of inflammation on the progression of TBI-associated pathologies has been the focus of numerous studies (Morganti-Kossmann et al. 2001; Allan and Rothwell 2001; Kadhim et al. 2008; Whitney et al. 2009), their utility as markers of injury has only recently been appreciated. Acute-phase proteins (e.g., C-reactive protein, amyloid A), proinflammatory cytokines (e.g., IL-1, TNF-α, IL-6), anti-inflammatory cytokines (IL-10, TGF-β), and chemokines (e.g., ICAM-1, MIP-1, MIP-2) have all been reported to change in response to TBI. The CSF and/or serum levels of a number of these markers have been correlated with injury, and in some cases with outcome. For example, it has been reported that the acute-phase reactant proteins C-reactive protein and serum amyloid A are rapidly increased in the serum of brain trauma patients (Hergenroeder et al. 2008; Su et al. 2014). Using a multiplex approach to simultaneously evaluate a panel of chemokines and cytokines, Buttram et al. reported that severe TBI in children was associated with significant increases in the CSF levels of IL-1ß, IL-6, IL-12p70, IL-10, IL-8, and MIP-1α (Buttram et al. 2007). Because injury to other organs can increase the serum level of these markers, they do not have high specificity. However, the temporal changes in the serum levels of these cytokines may be different following TBI and injury to other organs. Furthermore, in certain situations where selective brain injury is suspected (e.g., inflicted TBI), these markers can be helpful in diagnosis (Berger et al. 2006). A recent study has used the PET ligand (R)PK11195 (PK) to examine neuroinflammation after TBI. This ligand binds to the translocator protein expressed by the mitochondria of activated microglia. The results from this study indicate that microglial activation remains elevated for up to 17 years after the initial insult (Ramlackhansingh et al. 2011). This in vivo imaging technique may be useful as a biomarker to assess the effectiveness of therapies designed to reduce TBI-associated inflammation.

6.9 BIOMARKERS OF NEURODEGENERATIVE DISEASES AND TBI

It has been reported that TBI can increase the risk of developing Alzheimer's disease (AD) by as much as 4.5-fold (Mortimer et al. 1991; Plassman et al. 2000; Johnson et al. 2010). Alzheimer's disease is a neurodegenerative disease that is characterized by progressive neurocognitive impairments. Deposition of extracellular amyloid Aβ plaques and neurofibrillary tangles (NFTs) consisting of hyperphosphorylated TAU are the most prominent pathologies. Aβ peptides (Aβ1-40 and Aβ1-42) are generated from its precursor protein (i.e., amyloid precursor protein [APP]) by proteolytic cleavage. Aβ1-42, which is a minor product, is hydrophobic, aggregates, and is a major constituent of amyloid plaques. In severe TBI patients, the CSF levels of Aß1-42 in the CSF are markedly increased (>1000-fold) by day 5 following injury (Olsson et al. 2004). However, imaging and biomarker analyses of familial AD have shown that approximately 15 years prior to the manifestation of clinical symptoms, a decrease in CSF Aß1-42 is observed that coincides with Aß deposition (detected by PET imaging using the ligand PIB8) (Bateman et al. 2012). Thus, it is unclear if the acute increase in CSF Aß1-42 levels observed in TBI patients contributes to plaque formation or if it represents a protective mechanism. Longitudinal PIB8 imaging (or a combination of CSF biomarkers, e.g., total TAU, Aß1-42, imaging, and neuropsychological testing) in TBI patients would be valuable to determine the relationship between elevated extracellular Aß-42 and plaque formation (Weiner et al. 2013).

Athletes with a history of concussions have been observed to have postconcussive symptoms that exhibit a more protracted rate of recovery (Collins et al. 2002). Chronic traumatic encephalopathy (CTE) was first recognized in boxers and was termed *dementia pugilistica* and is now being recognized in athletes involved in a variety of contact sports (Corsellis et al. 1973). Chronic traumatic encephalopathy is thought to result from TAU aggregation and deposition, resulting in memory dysfunction, difficulties with balance, behavioral changes, and loss of intellect that result in symptoms similar to those seen in Parkinson's disease. Symptoms usually appear several years after the injuries with the appearance of neurofibrillary tangles as are seen in Alzheimer's disease. However, the main difference between CTE and Alzheimer's disease is the anatomical location of these tangles (McKee et al. 1991). Although CTE is typically diagnosed posthumously, a newly developed PET ligand (T807) has been shown to bind to NFTs and may be useful to identify patients at risk for developing CTE after repeated concussion (Chien et al. 2013).

6.10 BIOMARKERS FOR REPEATED CONCUSSIONS AND RETURN-TO-PLAY/-WORK

Significant interest has been generated on the diagnosis and management of sport-related concussions, especially with respect to return-to-play. Evidence indicates that younger athletes are more susceptible to concussion than older athletes, take longer time to recover from concussion, and have more significant neurocognitive deficits as a result of their injury (Field et al. 2003; Purcell 2009). High school athletes who had sustained three or more concussions were found to be 9.3 times more likely than athletes with no previous history of head injury to have abnormal (more severe) markers of concussion following an on-field incident (Collins et al. 2002). These markers include loss of consciousness, anterograde amnesia, and confusion. Due to these cumulative effects, it has been postulated that the brain may have a period of vulnerability during which a second (or third) impact exacerbates ongoing pathology (Doolan et al. 2012). Thus, an appreciation of the underlying pathology and the identification of markers that can be used to delineate the period of vulnerability are required to minimize the chances of long-term disabilities caused by repeated head injury.

A recent study to examine altered metabolism using magnetic resonance spectroscopy (MRS) in older athletes indicates that the NAA:Cr ratio of injured patients was diminished by 18.5% at 3 days after concussion and was back to normal by 30 days (Vagnozzi et al. 2008, 2010). These changes were observed despite patients' assertion of symptom resolution, indicating that recovery from symptomology may not be sufficient for return-to-play decisions (Vagnozzi et al. 2010). Athletes who sustained repeated concussion within 3 to 15 days of the initial injury had similar initial decreases in NAA:Cr ratio, but this change did not normalize until 45 days following injury. This sustained decrease was associated with exacerbated concussion symptoms, suggesting that brain NAA:Cr ratio may be a good biomarker to define return-to-play. However, this finding may be related to the age of the patient, as Maugans et al. found no significant changes in NAA or the NAA:Cr ratio in children with sports-related concussion versus controls (Maugans et al. 2012). Another report using diffusion tensor imaging (DTI) examined diffusion changes in long association white matter tracts and found that patients with poor outcome had higher mean diffusivity values in several fiber tracts (Messe et al. 2011). Although yet to be measured in athletes who have sustained repeated concussions, these changes have the potential to serve as indicators of lasting brain pathologies that need to resolve before return-to-play.

6.11 USE OF TBI BIOMARKERS TO ASSESS EFFICACY OF A THERAPEUTIC AGENT

The TBI biomarkers, in conjunction with other outcome measures, have the potential to objectively assess the efficacy of a therapeutic agent. A recent study evaluated the serum levels of S100β to assess the efficacy of hypertonic saline and mannitol as resuscitation fluids after moderate-severe TBI. Although hyperosmolar treatment significantly increased GCS, serum S100β levels did not significantly decrease. However, a significant inverse correlation between serum S100β levels and GCS was found.

Several experimental and clinical studies have examined the benefits of therapeutic hypothermia and have begun to use biomarkers to evaluate treatment effectiveness. Using a rodent model of TBI, Yokobori et al. reported that hypothermia reduced the extracellular levels of both UCHL1 and GFAP (Yokobori et al. 2013). However, the utility of these markers as surrogate markers after treatment in human TBI has yielded mixed results. For example, MBP concentration has been reported to be significantly elevated after pediatric TBI, but its levels were found to be unaffected by hypothermia. Likewise, neither CSF nor serum levels of GFAP were found to be differentially altered in pediatric TBI patients receiving hypothermia versus normothermic treatment (Stewart et al. 2011).

6.12 SUMMARY

Traumatic brain injury has emerged as a major health problem for all age groups. Although the diagnosis of mild TBI remains a challenge, there is currently no proven method to predict the long-term prognosis of these patients. The pathophysiology of TBI is not well delineated. Even less understood is how TBI acts as a trigger for the development of other neurodegenerative disorders such as Alzheimer's disease and chronic traumatic encephalopathy. The biomarkers discussed herein, including GFAP, UCHL1, NF-H, and PET imaging ligands, have shown promise in assessing injury severity, assessing injury type, and predicting outcome (Table 6.1). Furthermore, increased levels of some of the markers (e.g., cleaved TAU, αII spectrin breakdown products) may provide insights into the underlying mechanisms activated by TBI that can be targeted for therapeutic intervention. Although not examined extensively, some of these markers can be used in conjunction with clinical, imaging, and neuropsychological data to evaluate the effectiveness of therapies in clinical settings. Additional studies to further test the diagnostic utility of these markers using easily accessible samples such as blood and urine, as well as the identification of novel biomarkers capable of predicting acute and long-term outcome, are still required.

TABLE 6.1
Summary Table for Translational TBI Biomarkers

Marker	Major Cellular Source	Sampling Fluid	Diagnostic Utility	Outcome Prediction
S100β	Astrocytes/ peripheral cells	Serum	Negative result excludes brain injury	High initial levels predictive of increased mortality and poor outcome
GFAP/GFAP-BDP	Astrocytes	Serum	Sensitive indicator of brain injury	High initial levels predictive of increased mortality and poor outcome
SDBPs	Neurons	CSF	Indicators of cell death mechanism	Prolonged elevation predictive of increased mortality and poor outcome
UCH-L1	Neurons	CSF	Indicator of compromised blood-brain barrier	High levels predictive of death
NSE	Neurons	Serum	Indicator of injury severity	High initial levels predictive of increased mortality and poor outcome
TAU/c-TAU	Neurons (axons)	CSF	Indicator of axonal injury	High TAU levels predictive of poor outcome; high c-TAU levels predictive of elevated ICP
Phospho-NF-H	Neurons (axons)	Serum	Can detect mild injury; differentiates between DAI and focal injuries	High initial levels predictive of mortality and poor outcome
MAP-2	Neurons (dendrites)	Serum	Indicator of dendritic damage	High levels in chronic phase associated with improved outcome
NAA	Neurons	Tissue	Increases indicative of injury	High global levels predictive of poor outcome following mild TBI
MBP	Oligodendrocytes	CSF/ Serum	Indicator of myelin degradation; differentiates between inflicted and noninflicted pediatric TBI	Fair predictor of intracranial hemorrhage

(Continued)

TABLE 6.1 (*Continued*)
Summary Table for Translational TBI Biomarkers

Marker	Major Cellular Source	Sampling Fluid	Diagnostic Utility	Outcome Prediction
Aß1-42	Neurons	CSF	Acute increase in TBI	Unknown
[116](R)PK11195	PET ligand		Indicates activated microglia	Indicates chronic inflammation
PIB8	PET ligand		Indicates Aß deposition	Can be used to examine link between TBI and Alzheimer's disease
T807	PET ligand		Indicates NFT formation	Can be used to examine link between TBI and CTE

REFERENCES

Allan, S.M. and Rothwell, N.J. 2001. Cytokines and acute neurodegeneration. *Nat. Rev. Neurosci.* 2, 734–744.

Anderson, R.E., Hansson, L.O., Nilsson, O., jlai-Merzoug, R., and Settergren, G. 2001. High serum S100B levels for trauma patients without head injuries. *Neurosurgery* 48, 1255–1258.

Ankerst, D.P. and Thompson, I.M. 2006. Sensitivity and specificity of prostate-specific antigen for prostate cancer detection with high rates of biopsy verification. *Arch. Ital. Urol. Androl.* 78, 125–129.

Bateman, R.J., Xiong, C., Benzinger, T.L., Fagan, A.M., Goate, A., Fox, N.C., Marcus, D.S. et al. 2012. Clinical and biomarker changes in dominantly inherited Alzheimer's disease. *N. Engl. J. Med.* 367, 795–804.

Bazarian, J.J., Zemlan, F.P., Mookerjee, S., and Stigbrand, T. 2006. Serum S-100B and cleaved-tau are poor predictors of long-term outcome after mild traumatic brain injury. *Brain Inj.* 20, 759–765.

Begaz, T., Kyriacou, D.N., Segal, J., and Bazarian, J.J. 2006. Serum biochemical markers for post-concussion syndrome in patients with mild traumatic brain injury. *J. Neurotrauma* 23, 1201–1210.

Berger, R.P., Adelson, P.D., Pierce, M.C., Dulani, T., Cassidy, L.D., and Kochanek, P.M. 2005. Serum neuron-specific enolase, S100B, and myelin basic protein concentrations after inflicted and noninflicted traumatic brain injury in children. *J. Neurosurg.* 103, 61–68.

Berger, R.P., Bazaco, M.C., Wagner, A.K., Kochanek, P.M., and Fabio, A. 2010. Trajectory analysis of serum biomarker concentrations facilitates outcome prediction after pediatric traumatic and hypoxemic brain injury. *Dev. Neurosci.* 32, 396–405.

Berger, R.P., Dulani, T., Adelson, P.D., Leventhal, J.M., Richichi, R., and Kochanek, P.M. 2006. Identification of inflicted traumatic brain injury in well-appearing infants using serum and cerebrospinal markers: A possible screening tool. *Pediatrics* 117, 325–332.

Berger, R.P., Hayes, R.L., Richichi, R., Beers, S.R., and Wang, K.K. 2012. Serum concentrations of ubiquitin C-terminal hydrolase-L1 and alphaII-spectrin breakdown product 145 kDa correlate with outcome after pediatric TBI. *J. Neurotrauma* 29, 162–167.

Berger, R.P., Pierce, M.C., Wisniewski, S.R., Adelson, P.D., and Kochanek, P.M. 2002. Serum S100B concentrations are increased after closed head injury in children: A preliminary study. *J. Neurotrauma* 19, 1405–1409.

Biberthaler, P., Linsenmeier, U., Pfeifer, K.J., Kroetz, M., Mussack, T., Kanz, K.G., Hoecherl, E.F. et al. 2006. Serum S-100B concentration provides additional information for the indication of computed tomography in patients after minor head injury: A prospective multicenter study. *Shock* 25, 446–453.

Binder, L.I., Frankfurter, A., and Rebhun, L.I. 1985. The distribution of tau in the mammalian central nervous system. *J. Cell Biol.* 101, 1371–1378.

Bloomfield, S.M., McKinney, J., Smith, L., and Brisman, J. 2007. Reliability of S100B in predicting severity of central nervous system injury. *Neurocrit. Care* 6, 121–138.

Blyth, B.J., Farahvar, A., He, H., Nayak, A., Yang, C., Shaw, G., and Bazarian, J.J. 2011. Elevated serum ubiquitin carboxy-terminal hydrolase L1 is associated with abnormal blood-brain barrier function after traumatic brain injury. *J. Neurotrauma* 28, 2453–2462.

Brophy, G.M., Mondello, S., Papa, L., Robicsek, S.A., Gabrielli, A., Tepas, J., III, Buki, A. et al. 2011. Biokinetic analysis of ubiquitin C-terminal hydrolase-L1 (UCH-L1) in severe traumatic brain injury patient biofluids. *J. Neurotrauma* 28, 861–870.

Bulut, M., Koksal, O., Dogan, S., Bolca, N., Ozguc, H., Korfali, E., Ilcol, Y.O., and Parklak, M. 2006. Tau protein as a serum marker of brain damage in mild traumatic brain injury: Preliminary results. *Adv. Ther.* 23, 12–22.

Buttram, S.D., Wisniewski, S.R., Jackson, E.K., Adelson, P.D., Feldman, K., Bayir, H., Berger, R.P., Clark, R.S., and Kochanek, P.M. 2007. Multiplex assessment of cytokine and chemokine levels in cerebrospinal fluid following severe pediatric traumatic brain injury: Effects of moderate hypothermia. *J. Neurotrauma* 24, 1707–1717.

Cardali, S. and Maugeri, R. 2006. Detection of alphaII-spectrin and breakdown products in humans after severe traumatic brain injury. *J. Neurosurg. Sci.* 50, 25–31.

Chabok, S.Y., Moghadam, A.D., Saneei, Z., Amlashi, F.G., Leili, E.K., and Amiri, Z.M. 2012. Neuron-specific enolase and S100BB as outcome predictors in severe diffuse axonal injury. *J. Trauma Acute Care Surg.* 72, 1654–1657.

Chien, D.T., Bahri, S., Szardenings, A.K., Walsh, J.C., Mu, F., Su, M.Y., Shankle, W.R., Elizarov, A., and Kolb, H.C. 2013. Early clinical PET imaging results with the novel PHF-tau radioligand [F-18]-T807. *J. Alzheimers Dis.* 34, 457–468.

Collins, M.W., Lovell, M.R., Iverson, G.L., Cantu, R.C., Maroon, J.C., and Field, M. 2002. Cumulative effects of concussion in high school athletes. *Neurosurgery.* 51, 1175–1179.

Corsellis, J.A., Bruton, C.J., and Freeman-Browne, D. 1973. The aftermath of boxing. *Psychol. Med.* 3, 270–303.

de Kruijk, J.R., Leffers, P., Menheere, P.P., Meerhoff, S., and Twijnstra, A. 2001. S-100B and neuron-specific enolase in serum of mild traumatic brain injury patients. A comparison with health controls. *Acta Neurol. Scand.* 103, 175–179.

Doolan, A.W., Day, D.D., Maerlender, A.C., Goforth, M., and Gunnar, B.P. 2012. A review of return to play issues and sports-related concussion. *Ann. Biomed. Eng.* 40, 106–113.

Egea-Guerrero, J.J., Revuelto-Rey, J., Murillo-Cabezas, F., Munoz-Sanchez, M.A., Vilches-Arenas, A., Sanchez-Linares, P., Dominguez-Roldan, J.M., and Leon-Carrion, J. 2012. Accuracy of the S100beta protein as a marker of brain damage in traumatic brain injury. *Brain Inj.* 26, 76–82.

Farde, L., Halldin, C., Stone-Elander, S., and Sedvall, G. 1987. PET analysis of human dopamine receptor subtypes using 11C-SCH 23390 and 11C-raclopride. *Psychopharmacology* (Berl) 92, 278–284.

Field, M., Collins, M.W., Lovell, M.R., and Maroon, J. 2003. Does age play a role in recovery from sports-related concussion? A comparison of high school and collegiate athletes. *J. Pediatr.* 142, 546–553.

Franz, G., Beer, R., Kampfl, A., Engelhardt, K., Schmutzhard, E., Ulmer, H., and Deisenhammer, F. 2003. Amyloid beta 1–42 and tau in cerebrospinal fluid after severe traumatic brain injury. *Neurology* 60, 1457–1461.

Gatson, J.W., Barillas, J., Hynan, L.S., Diaz-Arrastia, R., Wolf, S.E., and Minei, J.P. 2014. Detection of neurofilament-H in serum as a diagnostic tool to predict injury severity in patients who have suffered mild traumatic brain injury. *J Neurosurg.* 121, 1–7.

Gatson, J.W., Warren, V., Abdelfattah, K., Wolf, S., Hynan, L.S., Moore, C., Diaz-Arrastia, R., Minei, J.P., Madden, C., and Wigginton, J.G. 2013. Detection of beta-amyloid oligomers as a predictor of neurological outcome after brain injury. *J. Neurosurg.* 118, 1336–1342.

Geyer, C., Ulrich, A., Grafe, G., Stach, B., and Till, H. 2009. Diagnostic value of S100B and neuron-specific enolase in mild pediatric traumatic brain injury. *J. Neurosurg. Pediatr.* 4, 339–344.

Govind, V., Gold, S., Kaliannan, K., Saigal, G., Falcone, S., Arheart, K.L., Harris, L., Jagid, J., and Maudsley, A.A. 2010. Whole-brain proton MR spectroscopic imaging of mild-to-moderate traumatic brain injury and correlation with neuropsychological deficits. *J. Neurotrauma* 27, 483–496.

Goyal, A., Failla, M.D., Niyonkuru, C., Amin, K., Fabio, A., Berger, R.P., and Wagner, A.K. 2013. S100b as a prognostic biomarker in outcome prediction for patients with severe traumatic brain injury. *J. Neurotrauma* 30, 946–957.

Hergenroeder, G., Redell, J.B., Moore, A.N., Dubinsky, W.P., Funk, R.T., Crommett, J., Clifton, G.L., Levine, R., Valadka, A., and Dash, P.K. 2008. Identification of serum biomarkers in brain-injured adults: Potential for predicting elevated intracranial pressure. *J. Neurotrauma* 25, 79–93.

Herrmann, M., Curio, N., Jost, S., Grubich, C., Ebert, A.D., Fork, M.L., and Synowitz, H. 2001. Release of biochemical markers of damage to neuronal and glial brain tissue is associated with short and long term neuropsychological outcome after traumatic brain injury. *J. Neurol. Neurosurg. Psychiatry* 70, 95–100.

Herrmann, M., Jost, S., Kutz, S., Ebert, A.D., Kratz, T., Wunderlich, M.T., and Synowitz, H. 2000. Temporal profile of release of neurobiochemical markers of brain damage after traumatic brain injury is associated with intracranial pathology as demonstrated in cranial computerized tomography. *J. Neurotrauma* 17, 113–122.

Honda, M., Tsuruta, R., Kaneko, T., Kasaoka, S., Yagi, T., Todani, M., Fujita, M., Izumi, T., and Maekawa, T. 2010. Serum glial fibrillary acidic protein is a highly specific biomarker for traumatic brain injury in humans compared with S-100B and neuron-specific enolase. *J. Trauma* 69, 104–109.

Ingebrigtsen, T. and Romner, B. 2003. Biochemical serum markers for brain damage: A short review with emphasis on clinical utility in mild head injury. *Restor. Neurol. Neurosci.* 21, 171–176.

Ingebrigtsen, T., Romner, B., Marup-Jensen, S., Dons, M., Lundqvist, C., Bellner, J., Alling, C., and Borgesen, S.E. 2000. The clinical value of serum S-100 protein measurements in minor head injury: A Scandinavian multicentre study. *Brain Inj.* 14, 1047–1055.

Ingebrigtsen, T., Waterloo, K., Jacobsen, E.A., Langbakk, B., and Romner, B. 1999. Traumatic brain damage in minor head injury: Relation of serum S-100 protein measurements to magnetic resonance imaging and neurobehavioral outcome. *Neurosurgery* 45, 468–475.

Johnson, V.E., Stewart, W., and Smith, D.H. 2010. Traumatic brain injury and amyloid-beta pathology: A link to Alzheimer's disease? *Nat. Rev. Neurosci.* 11, 361–370.

Jonsson, H., Johnsson, P., Hoglund, P., Alling, C., and Blomquist, S. 2000. Elimination of S100B and renal function after cardiac surgery. *J. Cardiothorac. Vasc. Anesth.* 14, 698–701.

Kadhim, H.J., Duchateau, J., and Sebire, G. 2008. Cytokines and brain injury: Invited review. *J. Intensive Care Med.* 23, 236–249.

Kosik, K.S. and Finch, E.A. 1987. MAP2 and tau segregate into dendritic and axonal domains after the elaboration of morphologically distinct neurites: An immunocytochemical study of cultured rat cerebrum. *J. Neurosci.* 7, 3142–3153.

Liliang, P.C., Liang, C.L., Weng, H.C., Lu, K., Wang, K.W., Chen, H.J., and Chuang, J.H. 2010. Tau proteins in serum predict outcome after severe traumatic brain injury. *J. Surg. Res.* 160, 302–307.

Lima, D.P., Simao, F.C., Abib, S.C., and de Figueiredo, L.F. 2008. Quality of life and neuropsychological changes in mild head trauma. Late analysis and correlation with S100B protein and cranial CT scan performed at hospital admission. *Injury* 39, 604–611.

Ma, M., Lindsell, C.J., Rosenberry, C.M., Shaw, G.J., and Zemlan, F.P. 2008. Serum cleaved tau does not predict postconcussion syndrome after mild traumatic brain injury. *Am. J. Emerg. Med.* 26, 763–768.

Macedo, R.C., Tomasi, C.D., Giombelli, V.R., Alves, S.C., Bristot, M.L., Locks, M.F., Petronilho, F. et al. 2013. Lack of association of S100β and neuron-specific enolase with mortality in critically ill patients. *Rev. Bras. Psiquiatr.* 35, 267–270.

Matek, J., Vajtr, D., Krska, Z., Springer, D., Filip, M., and Zima, T. 2012. [Protein S100b in differential diagnosis of brain concussion and superficial scalp injury in inebriated patients]. *Rozhl. Chir.* 91, 545–549.

Maugans, T.A., Farley, C., Altaye, M., Leach, J., and Cecil, K.M. 2012. Pediatric sports-related concussion produces cerebral blood flow alterations. *Pediatrics* 129, 28–37.

McKee, A.C., Kosik, K.S., and Kowall, N.W. 1991. Neuritic pathology and dementia in Alzheimer's disease. *Ann. Neurol.* 30, 156–165.

Messe, A., Caplain, S., Paradot, G., Garrigue, D., Mineo, J.F., Soto, A.G., Ducreux, D. et al. 2011. Diffusion tensor imaging and white matter lesions at the subacute stage in mild traumatic brain injury with persistent neurobehavioral impairment. *Hum. Brain Mapp.* 32, 999–1011.

Mondello, S., Gabrielli, A., Catani, S., D'Ippolito, M., Jeromin, A., Ciaramella, A., Bossu, P. et al. 2012. Increased levels of serum MAP-2 at 6-months correlate with improved outcome in survivors of severe traumatic brain injury. *Brain Inj.* 26, 1629–1635.

Mondello, S., Jeromin, A., Buki, A., Bullock, R., Czeiter, E., Kovacs, N., Barzo, P. et al. 2012. Glial neuronal ratio: A novel index for differentiating injury type in patients with severe traumatic brain injury. *J. Neurotrauma* 29, 1096–1104.

Mondello, S., Papa, L., Buki, A., Bullock, M.R., Czeiter, E., Tortella, F.C., Wang, K.K., and Hayes, R.L. 2011. Neuronal and glial markers are differently associated with computed tomography findings and outcome in patients with severe traumatic brain injury: A case control study. *Crit Care* 15, R156.

Mondello, S., Robicsek, S.A., Gabrielli, A., Brophy, G.M., Papa, L., Tepas, J., Robertson, C. et al. 2010. αII-Spectrin breakdown products (SBDPs): Diagnosis and outcome in severe traumatic brain injury patients. *J. Neurotrauma* 27, 1203–1213.

Morganti-Kossmann, M.C., Rancan, M., Otto, V.I., Stahel, P.F., and Kossmann, T. 2001. Role of cerebral inflammation after traumatic brain injury: A revisited concept. *Shock* 16, 165–177.

Mortimer, J.A., van Duijn, C.M., Chandra, V., Fratiglioni, L., Graves, A.B., Heyman, A., Jorm, A.F., Kokmen, E., Kondo, K., and Rocca, W.A. 1991. Head trauma as a risk factor for Alzheimer's disease: A collaborative re-analysis of case-control studies. EURODEM Risk Factors Research Group. *Int. J. Epidemiol.* 20 Suppl 2, S28–S35.

Muller, K., Townend, W., Biasca, N., Unden, J., Waterloo, K., Romner, B., and Ingebrigtsen, T. 2007. S100B serum level predicts computed tomography findings after minor head injury. *J. Trauma* 62, 1452–1456.

Neselius, S., Brisby, H., Theodorsson, A., Blennow, K., Zetterberg, H., and Marcusson, J. 2012. CSF-biomarkers in Olympic boxing: Diagnosis and effects of repetitive head trauma. *PLoS One* 7, e33606.

Neselius, S., Zetterberg, H., Blennow, K., Randall, J., Wilson, D., Marcusson, J., and Brisby, H. 2013. Olympic boxing is associated with elevated levels of the neuronal protein tau in plasma. *Brain Inj.* 27, 425–433.

Neselius, S., Zetterberg, H., Blennow, K., Randall, J., Wilson, D., Marcusson, J., and Brisby, H. 2013. Olympic boxing is associated with elevated levels of the neuronal protein tau in plasma. *Brain Inj.* 27, 425–433.

Nylen, K., Ost, M., Csajbok, L.Z., Nilsson, I., Blennow, K., Nellgard, B., and Rosengren, L. 2006. Increased serum-GFAP in patients with severe traumatic brain injury is related to outcome. *J. Neurol. Sci.* 240, 85–91.

Okonkwo, D.O., Yue, J.K., Puccio, A.M., Panczykowski, D.M., Inoue, T., McMahon, P.J., Sorani, M.D. et al. 2013. GFAP-BDP as an acute diagnostic marker in traumatic brain injury: Results from the prospective transforming research and clinical knowledge in traumatic brain injury study. *J. Neurotrauma* 30, 1490–1497.

Olsson, A., Csajbok, L., Ost, M., Hoglund, K., Nylen, K., Rosengren, L., Nellgard, B., and Blennow, K. 2004. Marked increase of beta-amyloid(1–42) and amyloid precursor protein in ventricular cerebrospinal fluid after severe traumatic brain injury. *J. Neurol.* 251, 870–876.

Ondruschka, B., Pohlers, D., Sommer, G., Schober, K., Teupser, D., Franke, H., and Dressler, J. 2013. S100B and NSE as useful postmortem biochemical markers of traumatic brain injury in autopsy cases. *J. Neurotrauma* 30, 1862–1871.

Ost, M., Nylen, K., Csajbok, L., Ohrfelt, A.O., Tullberg, M., Wikkelso, C., Nellgard, P., Rosengren, L., Blennow, K., and Nellgard, B. 2006. Initial CSF total tau correlates with 1-year outcome in patients with traumatic brain injury. *Neurology* 67, 1600–1604.

Papa, L., Akinyi, L., Liu, M.C., Pineda, J.A., Tepas, J.J., III, Oli, M.W., Zheng, W. et al. 2010. Ubiquitin C-terminal hydrolase is a novel biomarker in humans for severe traumatic brain injury. *Crit Care Med.* 38, 138–144.

Papa, L., Lewis, L.M., Falk, J.L., Zhang, Z., Silvestri, S., Giordano, P., Brophy, G.M. et al. 2012. Elevated levels of serum glial fibrillary acidic protein breakdown products in mild and moderate traumatic brain injury are associated with intracranial lesions and neurosurgical intervention. *Ann. Emerg. Med.* 59, 471–483.

Papa, L., Lewis, L.M., Silvestri, S., Falk, J.L., Giordano, P., Brophy, G.M., Demery, J.A. et al. 2012. Serum levels of ubiquitin C-terminal hydrolase distinguish mild traumatic brain injury from trauma controls and are elevated in mild and moderate traumatic brain injury patients with intracranial lesions and neurosurgical intervention. *J. Trauma Acute Care Surg.* 72, 1335–1344.

Papa, L., Silvestri, S., Brophy, G.M., Giordano, P., Falk, J.L., Braga, C.F., Tan, C.N. et al. 2014. GFAP outperforms S100β in detecting traumatic intracranial lesions on computed tomography in trauma patients with mild traumatic brain injury and those with extracranial lesions. *J. Neurotrauma* 31, 1815–1822.

Park, S.Y., Tournell, C., Sinjoanu, R.C., and Ferreira, A. 2007. Caspase-3- and calpain-mediated tau cleavage are differentially prevented by estrogen and testosterone in beta-amyloid-treated hippocampal neurons. *Neuroscience* 144, 119–127.

Pelinka, L.E., Hertz, H., Mauritz, W., Harada, N., Jafarmadar, M., Albrecht, M., Redl, H., and Bahrami, S. 2005. Nonspecific increase of systemic neuron-specific enolase after trauma: Clinical and experimental findings. *Shock* 24, 119–123.

Pelinka, L.E., Kroepfl, A., Leixnering, M., Buchinger, W., Raabe, A., and Redl, H. 2004. GFAP versus S100B in serum after traumatic brain injury: Relationship to brain damage and outcome. *J. Neurotrauma* 21, 1553–1561.

Pelinka, L.E., Kroepfl, A., Schmidhammer, R., Krenn, M., Buchinger, W., Redl, H., and Raabe, A. 2004. Glial fibrillary acidic protein in serum after traumatic brain injury and multiple trauma. *J. Trauma* 57, 1006–1012.

Piazza, O., Storti, M.P., Cotena, S., Stoppa, F., Perrotta, D., Esposito, G., Pirozzi, N., and Tufano, R. 2007. S100B is not a reliable prognostic index in paediatric TBI. *Pediatr. Neurosurg.* 43, 258–264.

Pike, B.R., Zhao, X., Newcomb, J.K., Posmantur, R.M., Wang, K.K., and Hayes, R.L. 1998. Regional calpain and caspase-3 proteolysis of α-spectrin after traumatic brain injury. *Neuroreport* 9, 2437–2442.

Pineda, J.A., Lewis, S.B., Valadka, A.B., Papa, L., Hannay, H.J., Heaton, S.C., Demery, J.A. et al. 2007. Clinical significance of αII-spectrin breakdown products in cerebrospinal fluid after severe traumatic brain injury. *J. Neurotrauma* 24, 354–366.

Plassman, B.L., Havlik, R.J., Steffens, D.C., Helms, M.J., Newman, T.N., Drosdick, D., Phillips, C. et al. 2000. Documented head injury in early adulthood and risk of Alzheimer's disease and other dementias. *Neurology* 55, 1158–1166.

Pleines, U.E., Morganti-Kossmann, M.C., Rancan, M., Joller, H., Trentz, O., and Kossmann, T. 2001. S-100β reflects the extent of injury and outcome, whereas neuronal specific enolase is a better indicator of neuroinflammation in patients with severe traumatic brain injury. *J. Neurotrauma* 18, 491–498.

Purcell, L. 2009. What are the most appropriate return-to-play guidelines for concussed child athletes? *Br. J. Sports Med.* 43 Suppl 1, i51–5.

Raabe, A., Kopetsch, O., Woszczyk, A., Lang, J., Gerlach, R., Zimmermann, M., and Seifert, V. 2003. Serum S-100B protein as a molecular marker in severe traumatic brain injury. *Restor. Neurol. Neurosci.* 21, 159–169.

Ramlackhansingh, A.F., Brooks, D.J., Greenwood, R.J., Bose, S.K., Turkheimer, F.E., Kinnunen, K.M., Gentleman, S. et al. 2011. Inflammation after trauma: Microglial activation and traumatic brain injury. *Ann. Neurol.* 70, 374–383.

Ross, S.A., Cunningham, R.T., Johnston, C.F., and Rowlands, B.J. 1996. Neuron-specific enolase as an aid to outcome prediction in head injury. *Br. J. Neurosurg.* 10, 471–476.

Savola, O., Pyhtinen, J., Leino, T.K., Siitonen, S., Niemela, O., and Hillbom, M. 2004. Effects of head and extracranial injuries on serum protein S100B levels in trauma patients. *J. Trauma.* 56, 1229–1234.

Setsuie, R. and Wada, K. 2007. The functions of UCH-L1 and its relation to neurodegenerative diseases. *Neurochem. Int.* 51, 105–111.

Siman, R., Toraskar, N., Dang, A., McNeil, E., McGarvey, M., Plaum, J., Maloney, E., and Grady, M.S. 2009. A panel of neuron-enriched proteins as markers for traumatic brain injury in humans. *J. Neurotrauma* 26, 1867–1877.

Skogseid, I.M., Nordby, H.K., Urdal, P., Paus, E., and Lilleaas, F. 1992. Increased serum creatine kinase BB and neuron specific enolase following head injury indicates brain damage. *Acta Neurochir. (Wien.)* 115, 106–111.

Stalnacke, B.M., Bjornstig, U., Karlsson, K., and Sojka, P. 2005. One-year follow-up of mild traumatic brain injury: Post-concussion symptoms, disabilities and life satisfaction in relation to serum levels of S-100B and neurone-specific enolase in acute phase. *J. Rehabil. Med.* 37, 300–305.

Stewart, T.C., Polgar, D., Gilliland, J., Tanner, D.A., Girotti, M.J., Parry, N., and Fraser, D.D. 2011. Shaken baby syndrome and a triple-dose strategy for its prevention. *J. Trauma* 71, 1801–1807.

Su, S.H., Xu, W., Li, M., Zhang, L., Wu, Y.F., Yu, F., and Hai, J. 2014. Elevated C-reactive protein levels may be a predictor of persistent unfavourable symptoms in patients with mild traumatic brain injury: A preliminary study. *Brain Behav. Immun.* 38, 111–117.

Tapia, F.J., Polak, J.M., Barbosa, A.J., Bloom, S.R., Marangos, P.J., Dermody, C., and Pearse, A.G. 1981. Neuron-specific enolase is produced by neuroendocrine tumours. *Lancet* 1, 808–811.

Trojanowski, J.Q., Schuck, T., Schmidt, M.L., and Lee, V.M. 1989. Distribution of tau proteins in the normal human central and peripheral nervous system. *J. Histochem. Cytochem.* 37, 209–215.

Vagnozzi, R., Signoretti, S., Cristofori, L., Alessandrini, F., Floris, R., Isgro, E., Ria, A. et al. 2010. Assessment of metabolic brain damage and recovery following mild traumatic brain injury: A multicentre, proton magnetic resonance spectroscopic study in concussed patients. *Brain* 133, 3232–3242.

Vagnozzi, R., Signoretti, S., Floris, R., Marziali, S., Manara, M., Amorini, A.M., Belli, A. et al. 2013. Decrease in N-acetylaspartate following concussion may be coupled to decrease in creatine. *J. Head Trauma Rehabil.* 28, 284–292.

Vagnozzi, R., Signoretti, S., Tavazzi, B., Floris, R., Ludovici, A., Marziali, S., Tarascio, G., Amorini, A.M., Di Pietro, V., Delfini, R., and Lazzarino, G. 2008. Temporal window of metabolic brain vulnerability to concussion: A pilot 1H-magnetic resonance spectroscopic study in concussed athletes—Part III. *Neurosurgery* 62, 1286–1295.

Vajtr, D., Benada, O., Linzer, P., Samal, F., Springer, D., Strejc, P., Beran, M., Prusa, R., and Zima, T. 2012. Immunohistochemistry and serum values of S-100B, glial fibrillary acidic protein, and hyperphosphorylated neurofilaments in brain injuries. *Soud. Lek.* 57, 7–12.

Vos, P.E., Lamers, K.J., Hendriks, J.C., van Haaren, M., Beems, T., Zimmerman, C., van Geel, W., de Reus, H., Biert, J., and Verbeek, M.M. 2004. Glial and neuronal proteins in serum predict outcome after severe traumatic brain injury. *Neurology* 62, 1303–1310.

Weiner, M.W., Veitch, D.P., Aisen, P.S., Beckett, L.A., Cairns, N.J., Green, R.C., Harvey, D. et al. 2013. The Alzheimer's Disease Neuroimaging Initiative: A review of papers published since its inception. *Alzheimers Dement.* 9, e111–e194.

Whitney, N.P., Eidem, T.M., Peng, H., Huang, Y., and Zheng, J.C. 2009. Inflammation mediates varying effects in neurogenesis: Relevance to the pathogenesis of brain injury and neurodegenerative disorders. *J. Neurochem.* 108, 1343–1359.

Yeo, R.A., Phillips, J.P., Jung, R.E., Brown, A.J., Campbell, R.C., and Brooks, W.M. 2006. Magnetic resonance spectroscopy detects brain injury and predicts cognitive functioning in children with brain injuries. *J. Neurotrauma* 23, 1427–1435.

Yokobori, S., Gajavelli, S., Mondello, S., Mo-Seaney, J., Bramlett, H.M., Dietrich, W.D., and Bullock, M.R. 2013. Neuroprotective effect of preoperatively induced mild hypothermia as determined by biomarkers and histopathological estimation in a rat subdural hematoma decompression model. *J. Neurosurg.* 118, 370–380.

Zemlan, F.P., Jauch, E.C., Mulchahey, J.J., Gabbita, S.P., Rosenberg, W.S., Speciale, S.G., and Zuccarello, M. 2002. C-tau biomarker of neuronal damage in severe brain injured patients: Association with elevated intracranial pressure and clinical outcome. *Brain Res.* 947, 131–139.

Zemlan, F.P., Rosenberg, W.S., Luebbe, P.A., Campbell, T.A., Dean, G.E., Weiner, N.E., Cohen, J.A., Rudick, R.A., and Woo, D. 1999. Quantification of axonal damage in traumatic brain injury: Affinity purification and characterization of cerebrospinal fluid tau proteins. *J. Neurochem.* 72, 741–750.

Zetterberg, H., Hietala, M.A., Jonsson, M., Andreasen, N., Styrud, E., Karlsson, I., Edman, A. et al. 2006. Neurochemical aftermath of amateur boxing. *Arch. Neurol.* 63, 1277–1280.

Zohar, O., Lavy, R., Zi, X., Nelson, T.J., Hongpaisan, J., Pick, C.G., and Alkon, D.L. 2011. PKC activator therapeutic for mild traumatic brain injury in mice. *Neurobiol. Dis.* 41, 329–337.

Zurek, J., Bartlova, L., and Fedora, M. 2011. Hyperphosphorylated neurofilament NF-H as a predictor of mortality after brain injury in children. *Brain Inj.* 25, 221–226.

Zurek, J. and Fedora, M. 2012. The usefulness of S100B, NSE, GFAP, NF-H, secretagogin and Hsp70 as a predictive biomarker of outcome in children with traumatic brain injury. *Acta Neurochir. (Wien.)* 154, 93–103.

7 Neuro-Immune Signaling Circuit in Traumatic Brain Injury

Scott D. Olson

CONTENTS

7.1 INFLAMMATORY REFLEX: CENTRAL NERVOUS SYSTEM CONTROL OF THE PERIPHERAL IMMUNE SYSTEM

Inflammation and immune response are traditionally thought to be a process governed on a cellular level through complicated autocrine and paracrine feedback mechanisms. For example, the standard response to infection consists of local antigen-presenting cells inducing an acute response from the innate immune system (namely, phagocytic cells), which then stimulates T cells and B cells to mount an adaptive immune response. The initial pro-inflammatory response generates an anti-inflammatory reaction during the adaptive immune response from Th2 and Treg activated T cells, which restores homeostasis in a self-regulating cytokine-based system.

Drug design for inflammatory diseases and injuries has focused on targeting this model of inflammation by utilizing neutralizing antibodies against cytokines, prohibiting the synthesis of inflammatory proteins, administering glucocorticoids with broad anti-inflammatory effects, and using other anti-inflammatory drugs and compounds. In studying a potential drug that inhibited the release of inflammatory cytokines from macrophages, Borovikova and colleagues in the Tracey lab stumbled upon a revelatory observation (Borovikova et al., 2000a). They found that if they administered the drug CNI-1493 intracerebroventricularly, they could preserve the anti-inflammatory effects while lowering the dose by up to 6 logs compared to systemic usage in the model of acute inflammation. They went on to describe how this observation required the presence of an intact vagus nerve, with similar effects obtained with local administration of acetylcholine.

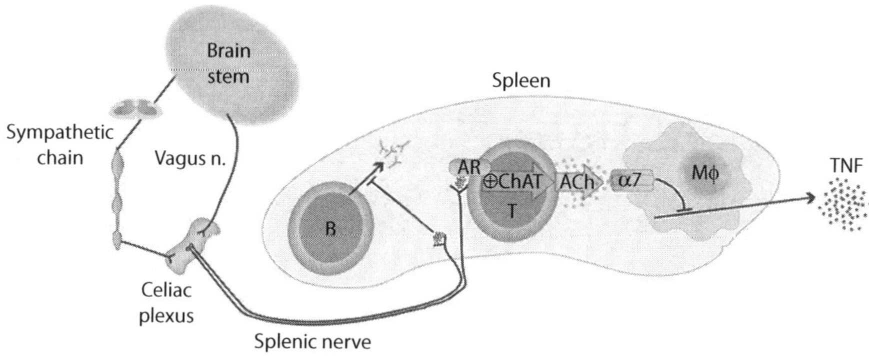

FIGURE 7.1 Rethinking inflammation: neural circuits in the regulation of immunity. (Adapted from Olofsson PS et al. 2012. *Immunol Rev* 248:188–204.)

In the last decade, Tracey's group and others have gone on to better describe a homeostatic neuroimmune signaling system that is often referred to as the *inflammatory reflex*, due to the concerted relationship between detection in the sensory afferent vagus and an active response delivered via the efferent vagus, as summarized in Figure 7.1 (Andersson and Tracey, 2012a,b). The afferent arm of the reflex is not as well understood as the efferent portion at this point, although much progress has been made recently. Studies have found that peripheral inflammation, particularly TNF-α, induces afferent vagal signaling, although this alone does not seem to be sufficient to trigger activation of the system. Direct activation of the reflex occurs when neurons in the brainstem detect increased levels of systemic inflammatory cytokines, making the effective afferent arm relatively short. The motor arm of the efferent vagus results in the release of acetylcholine into the celiac plexus, stimulating the splenic nerve (Diamond and Tracey, 2011). The splenic nerve is catecholaminergic, releasing norepinephrine (NE) into the spleen, where a population of T cells has been found that express both beta-adrenergic receptors (βAR) and choline acetyltransferase (ChAT). In response to NE, the T cells produce and secrete copious amounts of acetylcholine (ACh) that affects splenic macrophage/monocytes expressing the α7 nicotinic acetylcholine receptor (α7nAChR), inhibiting the production and release of TNF-α to restore homeostasis (Rosas-Ballina et al., 2011). Recent studies indicate that a similar signaling circuit, or a branch off of the above pathway, has a similar effect on resident microglia in the brain (Gnatek et al., 2012; Parada et al., 2013; Vijayaraghavan et al., 2013).

7.2 INFLAMMATORY REFLEX IN INJURY AND DISEASE

The role of the inflammatory reflex in injury and disease has been most well studied in models of endotoxemia and sepsis. The classical immune response to infection results in raised systemic levels of TNF-α, IL-1β, and a number of other proinflammatory cytokines. These cytokines, although important to mount an immune response in the absence of modern antibiotics, additionally result in the potentially dangerous physiologic response to infection, including fever, shock, tissue injury, and organ failure. In this model, these cytokines directly act on a set of neurons that use muscarinic acetylcholine receptors for intermediate signaling. Once activated, the inflammatory reflex is able to significantly reduce the levels of systemic cytokines by the innate immune system. This response in endotoxemia appears to be reliable and profound and was used to determine a majority of the elements involved in the reflex.

Tracey's group was groundbreaking in their demonstration that an intact vagus nerve was required for TNF-α inhibition in an animal model of endotoxemia, where the authors found that a vagotomy resulted in systemic rise in TNF-α. Electrical or pharmacological stimulation of the vagus has been shown to reduce inflammatory injuries. The reflex was then tracked to the splenic nerve, with similar observations occurring with splenic nerve ligation and stimulation resulting in increased and decreased TNF-α, respectively. Treatment of the spleen with reserpine or using βAR knockout mice removed the reflex, indicating that norepinephrine is a key component that stimulates βAR. In nude mice, vagus stimulation did not result in a decrease in TNF-α, indicating that T cells are also required (Rosas-Ballina et al., 2011). This observation resulted in the discovery of the specialized ChAT expressing T cells in the spleen that convert norepinephrine to acetylcholine and directly inhibit activated α7nAChR expressing macrophages.

Much of the mechanism of the inflammatory reflex was initially described in models of endotoxemia and sepsis. Tracey and colleagues were able to show that vagal ligation, splenic nerve ligation, and splenectomy all drastically increased the inflammatory response to LPS with decreased survival and greater inflammation-related injuries (Borovikova et al., 2000a,b; Bernik et al., 2002a,b; Wang et al., 2003; Saeed et al., 2005; Huston et al., 2006, 2008; Parrish et al., 2008; Rosas-Ballina et al., 2008). Additionally, they were able to show that broad cholinergic agonists and selective α7nAChR agonists were able to decrease inflammatory injury, while antagonists simulated the effects of disruption to the inflammatory reflex. In experimental models of endotoxemia, several pharmaceuticals have been used to manipulate the inflammatory reflex for therapeutic gain. The AChE inhibitor galantamine was able to reduce serum TNF-α and improve survival (Pavlov et al., 2009). Physostigmine was reported to reduce vascular leakage in a similar model (Peter et al., 2010). In a parallel effort, selective stimulation of α7nAChR using GTS-21 resulted in reduced serum TNF-α and improved overall survival (Pavlov et al., 2007).

There has been some recent interest in dysfunction in vagal activation in early stage amyotrophic lateral sclerosis (ALS) patients that is hypothesized to contribute to early death from cardiovascular complications. A preliminary study found that ALS patients had reduced baseline vagal activity compared to control groups (Linden et al., 1998). A different group found no difference in normal vagus activity between early ALS and normal controls; however, they found reduced vagal stimulation from the baroreflex response (Hilz et al., 2002). The authors concluded that beta blockade therapy may prove beneficial to restore endogenous cardiovascular control. In Parkinson's disease (PD), a novel observation that there is an inverse correlation between smoking and PD (Gorell et al., 1999; Hernan et al., 2002) led to the discovery that nicotine is neuroprotective in animal models. In unrelated studies, two groups were able to show that nicotine and an α7nAChR agonist were able to lower neuroinflammation and reduce neural loss in two different animal models of PD (Park et al., 2007; Stuckenholz et al., 2013). A new study of soluble ChAT potentially provides evidence of cholinergic immune disruptions in Alzheimer's disease and multiple sclerosis (Vijayaraghavan et al., 2013).

Outside of the central nervous system (CNS), disruptions to the inflammatory reflex are being observed in a large number of conditions. Olofsson et al. have done a remarkable job of assimilating a wide range of studies of the inflammatory reflex in different diseases along with documenting cases where vagal nerve stimulation, α7nAChR agonists, or AChE inhibitors have been found to have therapeutic applications (Olofsson et al., 2012). Just one α7nAChR agonist, PNU-282987, has been used in 69 publications to date since its discovery in 2005 (Bodnar et al., 2005). PNU-282987 has been applied to models of drug-related neural injuries (Chipana et al., 2008a,b; Garcia-Rates et al., 2010; Ishida et al., 2011; Gould et al., 2013), liver damage (Hiramoto et al., 2008; Li et al., 2013), gut injury (Kawahara et al., 2011), lung injury (Bregeon et al., 2011), Alzheimer's disease (Del Barrio et al., 2011; Vicens et al., 2013a,b), subarachnoid hemorrhage and stroke (Duris et al., 2011; Hijioka et al., 2012; Krafft et al., 2012; Lafargue et al., 2012), hypertension (Li et al., 2011), burn injuries (Hu et al., 2014), PD (Stuckenholz et al., 2013), and a number of cognitive and behavioral models that may have less to do with the inflammatory reflex than with the cholinergic signaling pathways of the brain. This list is an indication of the rapidly developing body of work describing how perturbations to the neuroimmune cholinergic signaling system can contribute to a wide variety of conditions that have pathology linked to an over- or underactive immune response.

7.3 INFLAMMATORY REFLEX IN TRAUMATIC BRAIN INJURY AND TRAUMA

It is not yet fully understood what the role of the inflammatory reflex is in traumatic brain injury (TBI). Traumatic brain injuries encompass a wide variety of injuries that can affect any number of areas of the brain, including the brainstem. Several studies have sought to measure autonomic nervous system by measuring heart rate variability (HRV). Heart rate is controlled by innervation of the sinoatrial node by the vagus. Action potential in the vagus releases acetylcholine, which delays the next heart contraction to slow heart rate (Campos et al., 2013). Computer analysis of echocardiogram tracings reveals the variability between beats as a function of time or frequency (Pagani et al., 1986). Transformative analysis gives several measures that have been shown to correlate with activity in the vagus nerve in physiological and pharmacological models, demonstrating that the high-frequency (HF) component corresponds to efferent vagus activity and reduced TNF-α and IL-6 (Marsland et al., 2007; Pavlov et al., 2009). The low-frequency (LF) component is generally used to reflect sympathetic input, and the high-frequency component represents parasympathetic activity (Rimoldi et al., 1990; Montano et al., 1994). Often these components are presented as a ratio of LF/HF, and occasionally this ratio is misleadingly also referred to as HRV.

In a number of clinical observations, low ANS activity, measured by HRV, has correlated with increased morbidity and mortality (La Rovere et al., 1998; Schmidt et al., 2005; Bruchfeld et al., 2010). In the context of the inflammatory reflex, low HRV would indicate that the vagus is not actively responding to systemic inflammation, either due to a lack of initiation of the reflex or dysfunction in the efferent response. This would allow inflammation to persist unchecked, as in the endotoxemia models with vagotomy that Tracey's group used to describe the reflex (Borovikova et al., 2000a,b; Huston et al., 2006, 2008; Huston et al., 2007; Rosas-Ballina et al., 2008). More observational research is necessary to determine if and why the inflammatory reflex is seemingly functional in some patients and not others, resulting in drastically different outcomes from seemingly identical initial prognosis.

Several studies have described the effects of general trauma and traumatic brain injury on vagal activity measured by HRV, reporting that in the hours and days following TBI, there is increased vagal activity (Winchell and Hoyt, 1997; Biswas et al., 2000; Keren et al., 2005; Baguley et al., 2006; Colombo et al., 2008; Kox et al., 2008, 2012). In TBI, a recent study was able to show an inverse correlation between the high-frequency component of the HRV and TNF-α in patients suffering from TBI, intracranial hemorrhage, and subarachnoid hemorrhage, demonstrating a link between autonomic vagal activity and inflammation (Kox et al., 2012). This study additionally demonstrated that vagal activity remained overactive 4 days after injury with a decrease over time in both IL-6 and IL-10 and an overall decrease in TNF-α, suggesting an immune system that is frozen in a state that is neither pro- or anti-inflammatory. This effect is likely linked to the increased incidence of infection following TBI (Boddie et al., 2003).

This observation runs counterintuitive to the secondary injuries following TBI resulting from overactive inflammatory responses (Moppett, 2007; Hergenroeder et al., 2008). Increased vagal activity (measured by HRV), should induce a systemic anti-inflammatory response in the previously described model of the inflammatory reflex. On the contrary, increased vagal activity is linked to "immune paralysis," whereby there is an increased occurrence of opportunistic infections in TBI and autonomic nervous system activity observed to be significantly higher in nonsurvivors of blunt and penetrating trauma compared to survivors. In another study it was found that vagal tone was directly related to intracranial pressure, suggesting that vagal activity could be a result of physical compression of the brainstem or another neural element upstream of the vagus (Matsuura et al., 1984). This opens the door for several theories as to why changes to HRV could result in negative outcome in TBI, despite the anti-inflammatory effects of vagal activity.

The first possibility is that there is an early exhaustion of the reflex by overactive afferent and efferent vagus activity. In practice, this could occur by several mechanisms. First, neurons and synapses downstream of the vagus could be damaged by excitotoxicity or have reduced function from downregulation of receptors or signal transduction components. There are reports of changed expression of α7nAChR and AChE following TBI in the CNS, which may be mirrored by changes in the peripheral nervous system. A second possibility is that the responsive leukocyte populations are exhausted by the prolonged activation of a reflex that may normally only be used in bursts. The ChAT expressing T cells may no longer be present or responsive to βAR signaling, or the downstream monocytes and macrophages may have largely mobilized into circulation.

A third possibility is that the systemic immune system is overwhelmed by massively conflicting signals, where a major trauma in an environment that is normally immuno-isolated is leaking alarmins and antigens that stimulate an innate immune response while the inflammatory reflex actively suppresses those same cells, causing them to be minimally activated and unresponsive to other regulatory stimuli. This theory is partially supported by a literature review that found an increase in pneumonia as a result of vagal nerve stimulation (VNS) to explain the increased incidence of infection following TBI (Hall et al., 2014). The authors of the study propose that in the presence of an overactive vagus, macrophages are unable to be stimulated to mount a proper immune response, leading to opportunistic infections. Based on these observations, we can safely conclude that too little vagal activity is bad for acute morbidity and mortality from TBI, and too much is bad for long-term health and secondary complications related to TBI. Physicians may need to dynamically adjust VNS based on systemic processes, or perhaps realize that HRV as a direct indicator of inflammatory reflex activity has some faults.

The above scenarios have been posed assuming that changes in HRV directly correlate with inflammatory reflex activity culminating with changes to splenic effector cells. This assumption is potentially dangerous and flawed. A number of other physiological factors can affect HRV, such as baroreceptor activity and even respiratory patterns (Pomeranz et al., 1985; Ori et al., 1992). Additionally, the vagus nerve innervates a wide variety of organs, with HRV providing an indirect measure of the cumulative activity of the vagus without respect to which vagal branch may be important for immune-modifying functions. In essence, HRV and VNS studies are downgrading the vagus nerve from a complicated highway of neural communication to a two-lane country road of afferent/efferent signaling for simplicity's sake. The full relationship between HRV and splenic activity is difficult to determine in a system without massive perturbations, such as severe trauma, vagotomy, or splenectomy that could indirectly change systemic immune response. Another complication lies in the fact that studies sometimes label different measurements of ANS activity simply as HRV, leading to confusing contradictions where one study will report HRV as being raised (by HF) while another states HRV is lower (by total spectral power).

In trauma and TBI, the inflammatory reflex is potentially subject to a number of disruptions. First, with TBI there is a chance that the reflex itself will be damaged through injury to the brainstem or vagus. As previously stated, intracranial pressure can result in compression to the medulla oblongata and brainstem, resulting in increased vagal activity from mechanical stimulation. Additionally, most cases of TBI occur alongside trauma to other tissues and organs, confounding the sources and effects of cytokines and signals. Animal models of TBI and vagus activity have their own complications. Long-term monitoring of vagus activity through HRV is technically formidable, and implantable biopotential monitors have their own shortcomings (short battery life, hard to properly mount, expensive). Animal models of TBI have a number of technical limitations made to increase their reliability and decrease animal mortality (Xiong et al., 2013). Based on a limited number of studies, there also appears to be a striking temporal component to HRV activity. One study focused on HRV immediately upon admission and found a decrease in vagal activity (Su et al., 2005). Meanwhile, several studies of TBI patients during a late, chronic stage of injury that also report a decrease in vagal activity (King et al., 1997; Keren et al., 2005; Baguley et al., 2009). There is a need for studies that detail and define the roles of HRV, afferent and efferent vagal activity, release of neurotransmitters to the spleen, and the resulting direct effects on the systemic immune system.

7.4 INFLAMMATORY REFLEX–MODIFYING STRATEGIES

A number of groups are studying how direct cervical VNS can be used as a therapeutic tool following TBI. Endogenous HRV activity in clinical studies does not correlate with long-term recovery, even though it results in lower serum TNF-α; other groups have found therapeutic effects from VNS in both short- and long-term TBI studies. A study that utilized proactive VNS stimulation in an animal model for 10 minutes immediately prior to TBI reported a decrease in blood-brain barrier (BBB) permeability with an accompanying decrease in aquaporin expression as a result of VNS (Lopez et al., 2012). Some studies have begun to explore the effects of VNS on the immune system, finding that stimulation 2 hours after TBI caused an increase in ghrelin and an associated decrease in TNF-α that could be prevented with a blocking antibody against ghrelin (Bansal et al., 2012), with similar results reported in sepsis (Fink, 2009). A Houston, Texas-based company (Cyberonics) has patented the use of VNS as a rehabilitation tool, with preliminary data indicating that VNS can improve memory and cognition recovery during the months and years following TBI. They attribute this to improving neural plasticity, primarily studying the effects of VNS on the brain. The company has additional studies ongoing in stroke and Alzheimer's disease.

It has been known for some time now that serum NE increases following TBI, with the amount correlating to morbidity and mortality (Woolf et al., 1988, 1987; Hamill et al., 1987; Woolf et al., 1992). Traditionally, studies have focused on the effects of the neurotransmitter on the brain, however, this observation is consistent with the reports that HRV is increased following TBI. There is a distinct possibility that the two observations are related. It is doubtful that vagus activity alone would be sufficient to account for the increase in serum NE. However, in combination with the overactive vagal activity indicated from HRV measurements, NE released into the serum from damaged neurons could function as a second hit to overstress the immune system in the presence of conflicting pro- and anti-inflammatory cues and induce immune paralysis. The NE saturation is concurrent with oxidative stress, indicating that NE may additionally be causing neurological injury by increasing cerebral metabolism with oxidative injuries from associated free radicals.

Retrospective analysis found that patients taking the pan-β adrenergic blocker propranolol had significantly higher survival following TBI compared to historical controls (Cotton et al., 2007; Riordan et al., 2007; Heffernan et al., 2010). Propranolol is proposed to have therapeutic effects by reducing the brain's metabolism following TBI, and thus avoiding ischemic-reperfusion injury (Heffernan et al., 2010). In experimental animal models, propranolol treatment following TBI resulted in a reduction in edema, blood-brain permeability, and an increase in cerebral perfusion (Ley et al., 2009, 2010, 2012; Zlotnik et al., 2012). These studies have resulted in the initiation of a clinical trial testing the usage of propranolol and clonidine to treat TBI

(Patel et al., 2012). The effects of the "beta-blockers" on TBI are interesting, as inhibiting catecholamine signaling would counteract the raised NE and hyperactive vagus in the hours and days following TBI. However, the retrospective studies also noted a significant increase in complications after TBI, including a number of opportunistic and inflammatory-based injuries, like sepsis, acute respiratory distress syndrome (ARDS), and multiple-organ failure along with an increased length of stay (Cotton et al., 2007; Heffernan et al., 2010; Bukur et al., 2012). Prolonged and systemic β adrenergic antagonists may be ill suited to treat the entirety of the injury associated with TBI; future studies may find that a more specific antagonist or actively titrating a response can maintain the improved survival with less secondary complications.

In an extreme measure, splenectomy has been shown to reduce injuries associated with TBI. Walker was able to show that splenectomized animals had reductions in BBB edema, likely by reducing the secondary inflammatory injury (Walker et al., 2011, 2012). Walker was using cellular therapies to treat TBI and found that splenectomy also removed the effects of cell treatment, indicating that the cell therapy required an intact spleen to work, in a style similar to the inflammatory reflex. Although splenectomy is an extreme measure to decrease the inflammatory injury following TBI, there are a number of scenarios where TBI is accompanied by widespread trauma that may necessitate the removal of the spleen, such as with internal trauma in a car wreck. However, there should be more elegant ways to reduce the involvement of splenic leukocytes in secondary injury.

Acetylcholine is reduced following TBI with a reported change in tissue ACh, AChE, and ACh receptors (Hoffmeister et al., 2011; Bansal et al., 2012; Kelso and Oestreich, 2012; Valiyaveettil et al., 2013). The inflammatory reflex in the current model ends in the spleen, with CNS-based signaling culminating in the release of ACh onto activated splenic macrophages to inhibit the secretion of TNF-α. The effects of NE and ACh on other splenic leukocytes are not yet fully understood. A wide variety of cells, including leukocytes (Sanders, 2012), platelets (Schedel et al., 2011), and even mitochondria (Gergalova et al., 2012) express ACh and NE receptors with their role in signaling and inflammation being mostly speculative. It is known that the reflex can reduce systemic TNF-α, but it is not yet known if it induces any activation of anti-inflammatory cytokines, such as IL-10, or changes or skews leukocyte activation states (e.g., M1 versus M2 or Th1 versus Th2 or Treg). Following injuries there are reports of a massive egress of leukocytes from the spleen, including dramatic intravital microscopy of monocytes and macrophages releasing into circulation from the spleen following stroke (Seifert et al., 2012). It is not yet known whether the cholinergic response prohibits or changes the mobilization of leukocytes following injury.

The inflammatory reflex could also be manipulated through the cholinergic signaling pathway. There are a number of potential targets here that could potentiate the anti-inflammatory reflex, such as activating or overexpressing choline acetyltransferase, inhibiting acetylcholinesterase, agonizing the α7nAChR, or even using an acetylcholine analogue. The most potent AChE inhibitors are potent neurotoxic chemical warfare agents, including sarin, VX, and soman. Potent AChR antagonists are found in snake venom, such as the very selective alpha-bungarotoxin from the Taiwanese banded krait. Moderate AChE inhibitors were used for many years as potent insecticides, and they still are used in some developing countries where they have not yet been banned. Some milder, reversible AChE inhibitors (and some moderate AChE inhibitors with high specificity) have been used recently as a potential treatment for Alzheimer's disease, such as the organophosphate pesticide trichlorfon that is now used pharmaceutically under an alternate name, metrifonate. It is thought that they will help potentiate ACh signaling in the diseased brain, decreasing dementia and other cognitive/behavioral issues. Currently there are few studies using AChE inhibitors to treat TBI.

Nicotine patches have additionally been studied as nAChR agonists, with some promising results. It was found that in mild TBI and cognitive impairment in an elderly population, nicotine patches increased memory and cognitive abilities (Gold et al., 2012; Newhouse et al., 2012; Roh and Evins, 2012; Cooper et al., 2013), an effect that was also seen in animal models (Shin et al., 2012). The therapeutic effect was limited to nonsmokers, perhaps due to the prior disposition of smoker's cells to nicotine. The authors of the study concluded that nicotine patches present some benefit to improve outcome following mild TBI. This observation reinforces the idea that cholinergic stimulation, perhaps limited to nicotinic receptors or even more specifically α7nAChR, may yield therapeutic benefit in treating TBI. Similar to the observations involving changes to HRV and alterations to systemic catecholamines, this therapy is likely to have serious temporal considerations if it is to be truly optimized.

Future therapies may seek to further expand upon Walker et al.'s usage of cellular therapies to manipulate and modify the inflammatory reflex. Mesenchymal stem cells (MSCs) have been widely used to treat a number of injuries where their efficacy is largely derived from the ability of MSC to modulate the inflammatory response. Recent studies indicate that much of this activity occurs systemically through interactions between MSC and leukocytes in the spleen, lung, and circulatory system. There is a growing amount of circumstantial evidence that MSCs may participate in the inflammatory reflex in some capacity. The MSCs have been reported to express βAR (Li et al., 2010a,b; Ishizuka et al., 2012), α7nAChR (Hoogduijn et al., 2009; Schraufstatter et al., 2009), ChAT (Hoogduijn et al., 2009), and even AChE (Hoogduijn et al., 2006, 2009) in addition to a number of inflammatory cytokine receptors (Singer and Caplan, 2011). These receptors and enzymes would allow MSCs to replace ChAT expressing T cells to relay NE-based signals or respond to ACh to assist in reducing systemic inflammation. Additionally, MSCs could act outside of the reflex, responding to non-neurological cues by manipulating splenic T cells and macrophages by modifying ACh signaling.

CONCLUSIONS

The involvement of the inflammatory reflex to TBI is yet to be rigorously defined. Heterogeneity in injury severity and the potential for the brainstem to be directly injured allows for many different potential responses. However, TBIs generally follow a somewhat predictable pattern of primary injury followed in the hours and days later by an inflammation-linked secondary injury. Heart rate variability studies mentioned earlier indicate that the vagus is overactive in this period, which should lead to a potent anti-inflammatory response through the understood mechanism. This inherent contradiction may be explained somewhat by the localization of the injury, where the reflex is being triggered not by systemic inflammation, but instead by local damage. This activates the reflex, inhibiting systemic inflammation but doing little to affect the activated microglia in the microenvironment. Persistent activation of the reflex could be excitotoxic or negative regulating, causing the systemic immune system to undergo "immune paralysis," where it becomes refractory to pro- and anti-inflammatory cues. The exact effects of an overactive vagus nerve on the immune system are yet to be determined; however, studies that intentionally manipulate the reflex and its components indicate that the system has a number of therapeutic targets for TBI.

REFERENCES

Andersson U and KJ Tracey. 2012a. Neural reflexes in inflammation and immunity. *J Exp Med* 209:1057–1068.

Andersson U and KJ Tracey. 2012b. Reflex principles of immunological homeostasis. *Annu Rev Immunol* 30:313–335.

Baguley IJ, RE Heriseanu, KL Felmingham, and ID Cameron. 2006. Dysautonomia and heart rate variability following severe traumatic brain injury. *Brain Inj* 20:437–444.

Baguley IJ, RE Heriseanu, MT Nott, J Chapman, and J Sandanam. 2009. Dysautonomia after severe traumatic brain injury: Evidence of persisting overresponsiveness to afferent stimuli. *Am J Phys Med Rehabil* 88:615–622.

Bansal V, SY Ryu, N Lopez, S Allexan, M Krzyzaniak, B Eliceiri, A Baird, and R Coimbra. 2012. Vagal stimulation modulates inflammation through a ghrelin mediated mechanism in traumatic brain injury. *Inflammation* 35:214–220.

Bernik TR, SG Friedman, M Ochani, R DiRaimo, S Susarla, CJ Czura, and KJ Tracey. 2002a. Cholinergic antiinflammatory pathway inhibition of tumor necrosis factor during ischemia reperfusion. *J Vasc Surg* 36:1231–1236.

Bernik TR, SG Friedman, M Ochani, R DiRaimo, L Ulloa, H Yang, S Sudan, CJ Czura, SM Ivanova, and KJ Tracey. 2002b. Pharmacological stimulation of the cholinergic antiinflammatory pathway. *J Exp Med* 195:781–788.

Biswas AK, WA Scott, JF Sommerauer, and PM Luckett. 2000. Heart rate variability after acute traumatic brain injury in children. *Crit Care Med* 28:3907–3912.

Boddie DE, DG Currie, O Eremin, and SD Heys. 2003. Immune suppression and isolated severe head injury: A significant clinical problem. *Br J Neurosurg* 17:405–417.

Bodnar AL, LA Cortes-Burgos, KK Cook, DM Dinh, VE Groppi, M Hajos, NR Higdon et al. 2005. Discovery and structure-activity relationship of quinuclidine benzamides as agonists of α7 nicotinic acetylcholine receptors. *J Med Chem* 48:905–908.

Borovikova LV, S Ivanova, D Nardi, M Zhang, H Yang, M Ombrellino, and KJ Tracey. 2000a. Role of vagus nerve signaling in CNI-1493-mediated suppression of acute inflammation. *Auton Neurosci* 85:141–147.

Borovikova LV, S Ivanova, M Zhang, H Yang, GI Botchkina, LR Watkins, H Wang, N Abumrad, JW Eaton, and KJ Tracey. 2000b. Vagus nerve stimulation attenuates the systemic inflammatory response to endotoxin. *Nature* 405:458–462.

Bregeon F, F Xeridat, N Andreotti, H Lepidi, S Delpierre, A Roch, S Ravailhe, Y Jammes, and JG Steinberg. 2011. Activation of nicotinic cholinergic receptors prevents ventilator-induced lung injury in rats. *PLoS One* 6:e22386.

Bruchfeld A, RS Goldstein, S Chavan, NB Patel, M Rosas-Ballina, N Kohn, AR Qureshi, and KJ Tracey. 2010. Whole blood cytokine attenuation by cholinergic agonists ex vivo and relationship to vagus nerve activity in rheumatoid arthritis. *J Intern Med* 268:94–101.

Bukur M, T Lustenberger, B Cotton, S Arbabi, P Talving, A Salim, EJ Ley, and K Inaba. 2012. Beta-blocker exposure in the absence of significant head injuries is associated with reduced mortality in critically ill patients. *Am J Surg* 204:697–703.

Campos LA, VL Pereira, Jr., A Muralikrishna, S Albarwani, S Bras, and S Gouveia. 2013. Mathematical biomarkers for the autonomic regulation of cardiovascular system. *Front Physiol* 4:279.

Chipana C, J Camarasa, D Pubill, and E Escubedo. 2008a. Memantine prevents MDMA-induced neurotoxicity. *Neurotoxicology* 29:179–183.

Chipana C, I Torres, J Camarasa, D Pubill, and E Escubedo. 2008b. Memantine protects against amphetamine derivatives-induced neurotoxic damage in rodents. *Neuropharmacology* 54:1254–1263.

Colombo J, WC Shoemaker, H Belzberg, G Hatzakis, P Fathizadeh, and D Demetriades. 2008. Noninvasive monitoring of the autonomic nervous system and hemodynamics of patients with blunt and penetrating trauma. *J Trauma* 65:1364–1373.

Cooper C, R Li, C Lyketsos, and G Livingston. 2013. Treatment for mild cognitive impairment: Systematic review. *Br J Psychiatry* 203:255–264.

Cotton BA, KB Snodgrass, SB Fleming, RO Carpenter, CD Kemp, PG Arbogast, and JA Morris, Jr. 2007. Beta-blocker exposure is associated with improved survival after severe traumatic brain injury. *J Trauma* 62:26–33; discussion 33–35.

Del Barrio L, MD Martin-de-Saavedra, A Romero, E Parada, J Egea, J Avila, JM McIntosh, S Wonnacott, and MG Lopez. 2011. Neurotoxicity induced by okadaic acid in the human neuroblastoma SH-SY5Y line can be differentially prevented by α7 and β2* nicotinic stimulation. *Toxicol Sci* 123:193–205.

Diamond B and KJ Tracey. 2011. Mapping the immunological homunculus. *Proc Natl Acad Sci USA* 108:3461–3462.

Duris K, A Manaenko, H Suzuki, WB Rolland, PR Krafft, and JH Zhang. 2011. α7 nicotinic acetylcholine receptor agonist PNU-282987 attenuates early brain injury in a perforation model of subarachnoid hemorrhage in rats. *Stroke* 42:3530–3536.

Fink MP. 2009. Sepsis, ghrelin, the cholinergic anti-inflammatory pathway, gut mucosal hyperpermeability, and high-mobility group box 1. *Crit Care Med* 37:2483–2485.

Garcia-Rates S, J Camarasa, AI Sanchez-Garcia, L Gandia, E Escubedo, and D Pubill. 2010. The effects of 3,4-methylenedioxymethamphetamine (MDMA) on nicotinic receptors: Intracellular calcium increase, calpain/caspase 3 activation, and functional upregulation. *Toxicol Appl Pharmacol* 244:344–353.

Gergalova G, O Lykhmus, O Kalashnyk, L Koval, V Chernyshov, E Kryukova, V Tsetlin, S Komisarenko, and M Skok. 2012. Mitochondria express α7 nicotinic acetylcholine receptors to regulate Ca2+ accumulation and cytochrome c release: Study on isolated mitochondria. *PLoS One* 7:e31361.

Gnatek Y, G Zimmerman, Y Goll, N Najami, H Soreq, and A Friedman. 2012. Acetylcholinesterase loosens the brain's cholinergic anti-inflammatory response and promotes epileptogenesis. *Front Mol Neurosci* 5:66.

Gold M, PA Newhouse, D Howard, and RJ Kryscio. 2012. Nicotine treatment of mild cognitive impairment: A 6-month double-blind pilot clinical trial. *Neurology* 78:1895; author reply 1895.

Gorell JM, BA Rybicki, CC Johnson, and EL Peterson. 1999. Smoking and Parkinson's disease: A dose-response relationship. *Neurology* 52:115–119.

Gould RW, PK Garg, S Garg, and MA Nader. 2013. Effects of nicotinic acetylcholine receptor agonists on cognition in rhesus monkeys with a chronic cocaine self-administration history. *Neuropharmacology* 64:479–488.

Hall S, A Kumaria, and A Belli. 2014. The role of vagus nerve overactivity in the increased incidence of pneumonia following traumatic brain injury. *Br J Neurosurg* 28(2):181–186.

Hamill RW, PD Woolf, JV McDonald, LA Lee, and M Kelly. 1987. Catecholamines predict outcome in traumatic brain injury. *Ann Neurol* 21:438–443.

Heffernan DS, K Inaba, S Arbabi, and BA Cotton. 2010. Sympathetic hyperactivity after traumatic brain injury and the role of beta-blocker therapy. *J Trauma* 69:1602–1609.

Hergenroeder GW, JB Redell, AN Moore, and PK Dash. 2008. Biomarkers in the clinical diagnosis and management of traumatic brain injury. *Mol Diagn Ther* 12:345–358.

Hernan MA, B Takkouche, F Caamano-Isorna, and JJ Gestal-Otero. 2002. A meta-analysis of coffee drinking, cigarette smoking, and the risk of Parkinson's disease. *Ann Neurol* 52:276–284.

Hijioka M, H Matsushita, H Ishibashi, A Hisatsune, Y Isohama, and H Katsuki. 2012. α7 Nicotinic acetylcholine receptor agonist attenuates neuropathological changes associated with intracerebral hemorrhage in mice. *Neuroscience* 222:10–19.

Hilz MJ, MJ Hecht, F Mittelhamm, B Neundorfer, and CM Brown. 2002. Baroreflex stimulation shows impaired cardiovagal and preserved vasomotor function in early-stage amyotrophic lateral sclerosis. *Amyotroph Lateral Scler Other Motor Neuron Disord* 3:137–144.

Hiramoto T, Y Chida, J Sonoda, K Yoshihara, N Sudo, and C Kubo. 2008. The hepatic vagus nerve attenuates Fas-induced apoptosis in the mouse liver via α7 nicotinic acetylcholine receptor. *Gastroenterology* 134:2122–2131.

Hoffmeister PG, CK Donat, MU Schuhmann, C Voigt, B Walter, K Nieber, J Meixensberger, R Bauer, and P Brust. 2011. Traumatic brain injury elicits similar alterations in α7 nicotinic receptor density in two different experimental models. *Neuromolecular Med* 13:44–53.

Hoogduijn MJ, A Cheng, and PG Genever. 2009. Functional nicotinic and muscarinic receptors on mesenchymal stem cells. *Stem Cells Dev* 18:103–112.

Hoogduijn MJ, Z Rakonczay, and PG Genever. 2006. The effects of anticholinergic insecticides on human mesenchymal stem cells. *Toxicol Sci* 94:342–350.

Hu Q, MH Du, S Hu, JK Chai, HM Luo, XH Hu, L Zhang et al. 2014. PNU-282987 Improves the hemodynamic parameters by alleviating vasopermeability and tissue edema in dogs subjected to a lethal burns shock. *J Burn Care Res* 35(4):e197–e204.

Huston JM, M Gallowitsch-Puerta, M Ochani, K Ochani, R Yuan, M Rosas-Ballina, M Ashok et al. 2007. Transcutaneous vagus nerve stimulation reduces serum high mobility group box 1 levels and improves survival in murine sepsis. *Crit Care Med* 35:2762–2768.

Huston JM, M Ochani, M Rosas-Ballina, H Liao, K Ochani, VA Pavlov, M Gallowitsch-Puerta et al. 2006. Splenectomy inactivates the cholinergic antiinflammatory pathway during lethal endotoxemia and polymicrobial sepsis. *J Exp Med* 203:1623–1628.

Huston JM, H Wang, M Ochani, K Ochani, M Rosas-Ballina, M Gallowitsch-Puerta, M Ashok, L Yang, KJ Tracey, and H Yang. 2008. Splenectomy protects against sepsis lethality and reduces serum HMGB1 levels. *J Immunol* 181:3535–3539.

Ishida S, Y Kawasaki, H Araki, M Asanuma, H Matsunaga, T Sendo, H Kawasaki, Y Gomita, and Y Kitamura. 2011. α7 Nicotinic acetylcholine receptors in the central amygdaloid nucleus alter naloxone-induced withdrawal following a single exposure to morphine. *Psychopharmacology (Berl)* 214:923–931.

Ishizuka T, H Goshima, A Ozawa, and Y Watanabe. 2012. β1-Adrenoceptor stimulation enhances the differentiation of mouse induced pluripotent stem cells into neural progenitor cells. *Neurosci Lett* 525:60–65.

Kawahara R, M Yasuda, H Hashimura, K Amagase, S Kato, and K Takeuchi. 2011. Activation of α7 nicotinic acetylcholine receptors ameliorates indomethacin-induced small intestinal ulceration in mice. *Eur J Pharmacol* 650:411–417.

Kelso ML and JH Oestreich. 2012. Traumatic brain injury: Central and peripheral role of α7 nicotinic acetylcholine receptors. *Curr Drug Targets* 13:631–636.

Keren O, S Yupatov, MM Radai, R Elad-Yarum, D Faraggi, S Abboud, H Ring, and Z Groswasser. 2005. Heart rate variability (HRV) of patients with traumatic brain injury (TBI) during the post-insult sub-acute period. *Brain Inj* 19:605–611.

King ML, SW Lichtman, G Seliger, FA Ehert, and JS Steinberg. 1997. Heart-rate variability in chronic traumatic brain injury. *Brain Inj* 11:445–453.

Kox M, JC Pompe, P Pickkers, CW Hoedemaekers, AB van Vugt, and JG van der Hoeven. 2008. Increased vagal tone accounts for the observed immune paralysis in patients with traumatic brain injury. *Neurology* 70:480–485.

Kox M, MQ Vrouwenvelder, JC Pompe, JG van der Hoeven, P Pickkers, and CW Hoedemaekers. 2012. The effects of brain injury on heart rate variability and the innate immune response in critically ill patients. *J Neurotrauma* 29:747–755.

Krafft PR, O Altay, WB Rolland, K Duris, T Lekic, J Tang, and JH Zhang. 2012. α7 Nicotinic acetylcholine receptor agonism confers neuroprotection through GSK-3β inhibition in a mouse model of intracerebral hemorrhage. *Stroke* 43:844–850.

La Rovere MT, JT Bigger, Jr., FI Marcus, A Mortara, and PJ Schwartz. 1998. Baroreflex sensitivity and heart-rate variability in prediction of total cardiac mortality after myocardial infarction. ATRAMI (Autonomic Tone and Reflexes After Myocardial Infarction) Investigators. *Lancet* 351:478–484.

Lafargue M, L Xu, M Carles, E Serve, N Anjum, KE Iles, X Xiong, R Giffard, and JF Pittet. 2012. Stroke-induced activation of the α7 nicotinic receptor increases *Pseudomonas aeruginosa* lung injury. *FASEB J* 26:2919–2929.

Ley EJ, MA Clond, M Bukur, R Park, M Chervonski, G Dagliyan, DR Margulies, PD Lyden, PS Conti, and A Salim. 2012. β-Adrenergic receptor inhibition affects cerebral glucose metabolism, motor performance, and inflammatory response after traumatic brain injury. *J Trauma Acute Care Surg* 73:33–40.

Ley EJ, R Park, G Dagliyan, D Palestrant, CM Miller, PS Conti, DR Margulies, and A Salim. 2010. In vivo effect of propranolol dose and timing on cerebral perfusion after traumatic brain injury. *J Trauma* 68:353–356.

Ley EJ, J Scehnet, R Park, S Schroff, G Dagliyan, PS Conti, DR Margulies, and A Salim. 2009. The in vivo effect of propranolol on cerebral perfusion and hypoxia after traumatic brain injury. *J Trauma* 66:154–159; discussion 159–161.

Li DJ, RG Evans, ZW Yang, SW Song, P Wang, XJ Ma, C Liu, T Xi, DF Su, and FM Shen. 2011. Dysfunction of the cholinergic anti-inflammatory pathway mediates organ damage in hypertension. *Hypertension* 57:298–307.

Li F, Z Chen, Q Pan, S Fu, F Lin, H Ren, H Han, TR Billiar, F Sun, and Q Li. 2013. The protective effect of PNU-282987, a selective α7 nicotinic acetylcholine receptor agonist, on the hepatic ischemia-reperfusion injury is associated with the inhibition of high-mobility group box 1 protein expression and nuclear factor kappaB activation in mice. *Shock* 39:197–203.

Li H, C Fong, Y Chen, G Cai, and M Yang. 2010a. β2- and β3-, but not β1-adrenergic receptors are involved in osteogenesis of mouse mesenchymal stem cells via cAMP/PKA signaling. *Arch Biochem Biophys* 496:77–83.

Li H, C Fong, Y Chen, G Cai, and M Yang. 2010b. Beta-adrenergic signals regulate adipogenesis of mouse mesenchymal stem cells via cAMP/PKA pathway. *Mol Cell Endocrinol* 323:201–207.

Linden D, RR Diehl, and P Berlit. 1998. Reduced baroreflex sensitivity and cardiorespiratory transfer in amyotrophic lateral sclerosis. *Electroencephalogr Clin Neurophysiol* 109:387–390.

Lopez NE, MJ Krzyzaniak, TW Costantini, J Putnam, AM Hageny, B Eliceiri, R Coimbra, and V Bansal. 2012. Vagal nerve stimulation decreases blood-brain barrier disruption after traumatic brain injury. *J Trauma Acute Care Surg* 72:1562–1566.

Marsland AL, PJ Gianaros, AA Prather, JR Jennings, SA Neumann, and SB Manuck. 2007. Stimulated production of proinflammatory cytokines covaries inversely with heart rate variability. *Psychosom Med* 69:709–716.

Matsuura S, H Sakamoto, Y Hayashida, and M Kuno. 1984. Efferent discharges of sympathetic and parasympathetic nerve fibers during increased intracranial pressure in anesthetized cats in the absence and presence of pressor response. *Brain Res* 305:291–301.

Montano N, TG Ruscone, A Porta, F Lombardi, M Pagani, and A Malliani. 1994. Power spectrum analysis of heart rate variability to assess the changes in sympathovagal balance during graded orthostatic tilt. *Circulation* 90:1826–1831.

Moppett IK. 2007. Traumatic brain injury: Assessment, resuscitation and early management. *Br J Anaesth* 99:18–31.

Newhouse P, K Kellar, P Aisen, H White, K Wesnes, E Coderre, A Pfaff, H Wilkins, D Howard, and ED Levin. 2012. Nicotine treatment of mild cognitive impairment: A 6-month double-blind pilot clinical trial. *Neurology* 78:91–101.

Olofsson PS, M Rosas-Ballina, YA Levine, and KJ Tracey. 2012. Rethinking inflammation: Neural circuits in the regulation of immunity. *Immunol Rev* 248:188–204.

Ori Z, G Monir, J Weiss, X Sayhouni, and DH Singer. 1992. Heart rate variability. Frequency domain analysis. *Cardiol Clin* 10:499–537.

Pagani M, F Lombardi, S Guzzetti, O Rimoldi, R Furlan, P Pizzinelli, G Sandrone et al. 1986. Power spectral analysis of heart rate and arterial pressure variabilities as a marker of sympatho-vagal interaction in man and conscious dog. *Circ Res* 59:178–193.

Parada E, J Egea, I Buendia, P Negredo, AC Cunha, S Cardoso, MP Soares, and MG Lopez. 2013. The Microglial α7-acetylcholine nicotinic receptor is a key element in promoting neuroprotection by inducing heme oxygenase-1 via nuclear factor erythroid-2-related factor 2. *Antioxid Redox Signal* 19:1135–1148.

Park HJ, PH Lee, YW Ahn, YJ Choi, G Lee, DY Lee, ES Chung, and BK Jin. 2007. Neuroprotective effect of nicotine on dopaminergic neurons by anti-inflammatory action. *Eur J Neurosci* 26:79–89.

Parrish WR, M Rosas-Ballina, M Gallowitsch-Puerta, M Ochani, K Ochani, LH Yang, L Hudson et al. 2008. Modulation of TNF release by choline requires α7 subunit nicotinic acetylcholine receptor-mediated signaling. *Mol Med* 14:567–574.

Patel MB, JW McKenna, JM Alvarez, A Sugiura, JM Jenkins, OD Guillamondegui, and PP Pandharipande. 2012. Decreasing adrenergic or sympathetic hyperactivity after severe traumatic brain injury using propranolol and clonidine (DASH After TBI Study): Study protocol for a randomized controlled trial. *Trials* 13:177.

Pavlov VA, M Ochani, LH Yang, M Gallowitsch-Puerta, K Ochani, X Lin, J Levi et al. 2007. Selective α7-nicotinic acetylcholine receptor agonist GTS-21 improves survival in murine endotoxemia and severe sepsis. *Crit Care Med* 35:1139–1144.

Pavlov VA, WR Parrish, M Rosas-Ballina, M Ochani, M Puerta, K Ochani, S Chavan, Y Al-Abed, and KJ Tracey. 2009. Brain acetylcholinesterase activity controls systemic cytokine levels through the cholinergic anti-inflammatory pathway. *Brain Behav Immun* 23:41–45.

Peter C, K Schmidt, S Hofer, M Stephan, E Martin, MA Weigand, and A Walther. 2010. Effects of physostigmine on microcirculatory alterations during experimental endotoxemia. *Shock* 33:405–411.

Pomeranz B, RJ Macaulay, MA Caudill, I Kutz, D Adam, D Gordon, KM Kilborn et al. 1985. Assessment of autonomic function in humans by heart rate spectral analysis. *Am J Physiol* 248:H151–H153.

Rimoldi O, S Pierini, A Ferrari, S Cerutti, M Pagani, and A Malliani. 1990. Analysis of short-term oscillations of R-R and arterial pressure in conscious dogs. *Am J Physiol* 258:H967–H976.

Riordan WP, Jr., BA Cotton, PR Norris, LR Waitman, JM Jenkins, and JA Morris, Jr. 2007. Beta-blocker exposure in patients with severe traumatic brain injury (TBI) and cardiac uncoupling. *J Trauma* 63:503–510; discussion 510–511.

Roh S and AE Evins. 2012. Possible role of nicotine for the treatment of mild cognitive impairment. *Expert Rev Neurother* 12:531–533.

Rosas-Ballina M, M Ochani, WR Parrish, K Ochani, YT Harris, JM Huston, S Chavan, and KJ Tracey. 2008. Splenic nerve is required for cholinergic antiinflammatory pathway control of TNF in endotoxemia. *Proc Natl Acad Sci USA* 105:11008–11013.

Rosas-Ballina M, PS Olofsson, M Ochani, SI Valdes-Ferrer, YA Levine, C Reardon, MW Tusche et al. 2011. Acetylcholine-synthesizing T cells relay neural signals in a vagus nerve circuit. *Science* 334:98–101.

Saeed RW, S Varma, T Peng-Nemeroff, B Sherry, D Balakhaneh, J Huston, KJ Tracey, Y Al-Abed, and CN Metz. 2005. Cholinergic stimulation blocks endothelial cell activation and leukocyte recruitment during inflammation. *J Exp Med* 201:1113–1123.

Sanders VM. 2012. The β2-adrenergic receptor on T and B lymphocytes: Do we understand it yet? *Brain Behav Immun* 26:195–200.

Schedel A, S Thornton, P Schloss, H Kluter, and P Bugert. 2011. Human platelets express functional α7-nicotinic acetylcholine receptors. *Arterioscler Thromb Vasc Biol* 31:928–934.

Schmidt H, U Muller-Werdan, T Hoffmann, DP Francis, MF Piepoli, M Rauchhaus, R Prondzinsky et al. 2005. Autonomic dysfunction predicts mortality in patients with multiple organ dysfunction syndrome of different age groups. *Crit Care Med* 33:1994–2002.

Schraufstatter IU, RG DiScipio, and SK Khaldoyanidi. 2009. α7 Subunit of nAChR regulates migration of human mesenchymal stem cells. *J Stem Cells* 4:203–215.

Seifert HA, AA Hall, CB Chapman, LA Collier, AE Willing, and KR Pennypacker. 2012. A transient decrease in spleen size following stroke corresponds to splenocyte release into systemic circulation. *J Neuroimmune Pharmacol* 7:1017–1024.

Shin SS, ER Bray, and CE Dixon. 2012. Effects of nicotine administration on striatal dopamine signaling after traumatic brain injury in rats. *J Neurotrauma* 29:843–850.

Singer NG and AI Caplan. 2011. Mesenchymal stem cells: Mechanisms of inflammation. *Annu Rev Pathol* 6:457–478.

Stuckenholz V, M Bacher, M Balzer-Geldsetzer, D Alvarez-Fischer, WH Oertel, RC Dodel, and C Noelker. 2013. The α7 nAChR agonist PNU-282987 reduces inflammation and MPTP-induced nigral dopaminergic cell loss in mice. *J Parkinsons Dis* 3:161–172.

Su CF, TB Kuo, JS Kuo, HY Lai, and HI Chen. 2005. Sympathetic and parasympathetic activities evaluated by heart-rate variability in head injury of various severities. *Clin Neurophysiol* 116:1273–1279.

Valiyaveettil M, YA Alamneh, SA Miller, R Hammamieh, P Arun, Y Wang, Y Wei, S Oguntayo, JB Long, and MP Nambiar. 2013. Modulation of cholinergic pathways and inflammatory mediators in blast-induced traumatic brain injury. *Chem Biol Interact* 203:371–375.

Vicens P, D Ribes, L Heredia, M Torrente, and JL Domingo. 2013a. Effects of an α7 nicotinic receptor agonist and stress on spatial memory in an animal model of Alzheimer's disease. *Biomed Res Int* 2013:952719.

Vicens P, D Ribes, L Heredia, M Torrente, and JL Domingo. 2013b. Motor and anxiety effects of PNU-282987, an α7 nicotinic receptor agonist, and stress in an animal model of Alzheimer's disease. *Curr Alzheimer Res* 10:516–523.

Vijayaraghavan S, A Karami, S Aeinehband, H Behbahani, A Grandien, B Nilsson, KN Ekdahl, RP Lindblom, F Piehl, and T Darreh-Shori. 2013. Regulated extracellular choline acetyltransferase activity—The plausible missing link of the distant action of acetylcholine in the cholinergic anti-inflammatory pathway. *PLoS One* 8:e65936.

Walker PA, PA Letourneau, S Bedi, SK Shah, F Jimenez, and CS Jr. 2011. Progenitor cells as remote “bioreactors”: Neuroprotection via modulation of the systemic inflammatory response. *World J Stem Cells* 3:9–18.

Walker PA, SK Shah, F Jimenez, KR Aroom, MT Harting, and CS Cox, Jr. 2012. Bone marrow-derived stromal cell therapy for traumatic brain injury is neuroprotective via stimulation of non-neurologic organ systems. *Surgery* 152:790–793.

Wang H, M Yu, M Ochani, CA Amella, M Tanovic, S Susarla, JH Li et al. 2003. Nicotinic acetylcholine receptor α7 subunit is an essential regulator of inflammation. *Nature* 421:384–388.

Winchell RJ and DB Hoyt. 1997. Analysis of heart-rate variability: A noninvasive predictor of death and poor outcome in patients with severe head injury. *J Trauma* 43:927–933.

Woolf PD, RW Hamill, LA Lee, C Cox, and JV McDonald. 1987. The predictive value of catecholamines in assessing outcome in traumatic brain injury. *J Neurosurg* 66:875–82.

Woolf PD, RW Hamill, LA Lee, and JV McDonald. 1988. Free and total catecholamines in critical illness. *Am J Physiol* 254:E287–E291.

Woolf PD, JV McDonald, DV Feliciano, MM Kelly, D Nichols, and C Cox. 1992. The catecholamine response to multisystem trauma. *Arch Surg* 127:899–903.

Xiong Y, A Mahmood, and M Chopp. 2013. Animal models of traumatic brain injury. *Nat Rev Neurosci* 14:128–142.

Zlotnik A, Y Klin, BF Gruenbaum, SE Gruenbaum, S Ohayon, A Leibowitz, R Kotz et al. 2012. α2 Adrenergic-mediated reduction of blood glutamate levels and improved neurological outcome after traumatic brain injury in rats. *J Neurosurg Anesthesiol* 24:30–38.

8 Soluble Factors and Mechanisms of Action of Mesenchymal Stem Cells in Traumatic Brain Injury and Organ Injury Induced by Trauma

Shibani Pati and Stuart L. Gibb

CONTENTS

8.1 BACKGROUND ON TRAUMATIC BRAIN INJURY AND MESENCHYMAL STEM CELLS

Trauma is the leading cause of death worldwide in all individuals between the ages of 1 and 44, and traumatic brain injury (TBI) is the leading cause of death in pediatric populations. There are currently few if any effective therapies to treat the acute edema and increased intracranial pressures (ICPs) resulting from TBI (Chesnut, 2004; Bazarian et al., 2005; Menon, 2009; Shlosberg et al., 2010). Patients who survive the initial insult of TBI often suffer from significant memory loss and neurocognitive decline that significantly impact their quality of life (Resch et al., 2009; Wrona, 2010). Mesenchymal stem cell (MSC) transplantation has been shown to be a promising therapeutic modality in the treatment of several disorders characterized by vascular instability including TBI (Harvey and Chopp, 2003; Mahmood et al., 2004a,b; Khakoo et al., 2006; Parr et al., 2007; Bernardo et al., 2009; Garcia-Gomez et al., 2010; Myers et al., 2010; Joyce et al., 2010; Lee JW et al., 2011; Lee RH et al., 2011; Pati et al., 2011a,b; Rogers et al., 2011; Uccelli et al., 2011; Wang et al., 2011). Although much phenomenological data have been collected on their use in these conditions, questions regarding their exact mechanism(s) of action remain unanswered. Aside from the TBI, trauma, and hemorrhage associated with the initial injury, victims of traumatic injury often succumb to other conditions in the critical care multiorgan failure setting such as multiple organ failure (MOF) which encompasses renal failure, liver failure, and acute respiratory distress syndrome (ARDS). Mesenchymal stem cells have been shown to have potent therapeutic anti-inflammatory and immunomodulatory effects in a number of these conditions associated with traumatic injury (Picinich et al., 2007). Mechanistically, a number of groups have demonstrated that MSCs regulate vascular stability and inflammation in traumatic injury (Picinich et al., 2007; Lee JW et al., 2011; Lee RH et al., 2011; Oliveri et al., 2014). The mechanisms of action are still not fully known, but over the years it has become clear that these cells have potent biological effects through factors they secrete. Further investigation in this area has revealed that the beneficial effects of MSCs are recapitulated by conditioned culture media (CM) collected from MSCs, thereby suggesting that there are soluble factor(s) produced by MSCs that mediate the observed effects.

The MSCs have been shown to secrete multiple factors that modulate these effects in many disease applications (Picinich et al., 2007; Lee JW et al., 2011; Lee RH et al., 2011; Oliveri et al., 2014). Understanding the mechanisms of action is critical in stem cell therapeutics, because it can lead to the discovery of potential side effects and also potency markers that can be used clinically to track the therapeutic response to treatment, an essential component of successful phase 3 clinical trials and a challenge in general for all stem cell therapies. In general, one can ask the translational question of "why do we need to know the exact mechanisms of action of MSCs if they are working to mitigate outcomes in patients?" The reasons are multifold and focus on determining (1) safety, (2) efficacy, (3) therapeutic optimization, (4) and de-risking late-stage clinical trials from failure. Through mechanistic insight, we can anticipate potential side effects, enhance beneficial effects, decrease deleterious effects, and define therapeutic markers of efficacy that are required for stage 3 clinical trials. Furthermore, learning more about the mechanisms of action of MSCs allows us to use the MSCs as a "discovery platform" from which we can identify novel therapeutics (i.e., conditioned media or soluble factors) for the treatment of diseases like TBI.

8.2 THERAPEUTIC POTENTIAL OF MSCs

Currently there are more than 300 clinical trials using MSCs for a variety of human diseases registered at https://clinicaltrials.gov. These diseases include congestive heart failure, myocardial infarction, graft-versus-host disease, and inflammatory bowel disease. The original excitement about MSCs arose in the late 1990s when MSCs were described as cells that had the potential to differentiate into a number of different cell types under defined culture conditions and also possessed an uncanny ability to home into sites of injury in preclinical models of disease (Picinich et al., 2007). These cell-type MSCs were found to differentiate into included chondrocyte, adipocytes, and osteocytes, which are still considered criteria for the characterization and definition of MSCs (Picinich et al., 2007). However, in the past 15 years it has becomes apparent that the therapeutic potential of the cells is predominantly due to trophic factors (soluble factors, microvesicles, and exosomes) they secrete that modulate their therapeutic effects in disease, as opposed to the concept that they engraft, differentiate, and replace lost or damaged tissue. Detailed tracking studies of MSCs administered by various routes in preclinical models of TBI have shown that few MSCs are found in the brain, and that MSCs do not engraft with most cells being undetectable days after administration (Khakoo et al., 2006; Fischer et al., 2009).

As a therapeutic, MSCs have biological properties that make them an ideal candidate for a cell-based treatment of pathologic edema associated with TBI and other trauma-related conditions, such as acute lung injury (ALI) and ARDS (Aggarwal and Pittenger, 2005; Matthay et al., 2013). Bone marrow–derived MSCs, isolated from the mononuclear fraction of a bone marrow aspirate, are the cells that adhere to plastic tissue culture dishes. The MSCs can be readily expanded to hundreds of millions of cells, proliferate for many passages in culture, and are easily transfectable, allowing for easy ex vivo modification with genes of interest. The MSCs are also relatively nonimmunogenic, and thus appear to be well tolerated when administered in the allogenic setting (Aggarwal and Pittenger, 2005; Picinich et al., 2007), although the mechanism of their immune privilege is not understood and is under intense study. The safety record of MSCs has been highly favorable without any adverse events found in completed clinical trials. No infusions or delayed toxicities have been noted with delivery of autologous or heterologous MSCs.

8.3 MECHANISM OF ACTION OF MSCs AND SOLUBLE FACTORS IN TRAUMATIC BRAIN INJURY

Traumatic brain injury is a heterogeneous disease that is characterized by a number of simultaneously occurring processes that lead to morbidity and mortality. In TBI, blood-brain barrier (BBB) leakage, cerebral edema, neuroinflammation, and neuronal cell death occur simultaneously. Neuroinflammation is initiated by resident astrocytes and microglia that release pro-inflammatory cytokines and chemokines that recruit circulating leukocytes and neutrophils into the brain (Morganti-Kossmann et al., 2001, 2002; Kelley et al., 2007). Neuroinflammation after TBI has been demonstrated to have both acute and chronic phases and has been directly associated with neuronal cell loss and poor outcome (Morganti-Kossmann et al., 2001, 2002; Kelley et al., 2007; Morganti-Kossmann et al., 2007; Schmidt et al., 2007; Hein and O'Banion, 2009). It has been shown that neurons and progenitor cells in selective areas such as the dentate, hilus, and CA3 region of the hippocampus are particularly vulnerable to TBI (Lyeth et al., 1990; Hicks et al., 1993; Colicos and Dash, 1996;

Colicos et al., 1996; Blum et al., 1999; Dash et al., 2002; Raghupathi et al., 2004; Raghupathi and Huh, 2007; Pati et al., 2009; Saatman et al., 2010; Gomez-Pinilla and Gomez, 2011; Sahay et al., 2011). The MSCs have the demonstrated capacity to address a number of targets in TBI including (1) neuroinflammation, (2) neuroprotection, and (3) BBB leakage and cerebral edema. To date there have not been any clinical trials with MSCs in TBI; however, Cox and colleagues have investigated the effects of bone marrow mononuclear cells (BMMNCs) in TBI patients (Cox et al., 2011). The BMMNCs contain a very small percentage of MSCs. Plans are underway to investigate multipotent adult progenitor cells (MAPCs) in TBI in the near future. The MAPCs are similar to MSCs but are thought to be earlier precursors of MSCs (Jiang et al., 2002; Bedi et al., 2013).

It has been demonstrated by a number of groups that the conditioned media (CM) from MSCs have potent biological effects in preclinical models and can recapitulate most of the effects of the MSCs themselves (Lee JW et al., 2011). In TBI, prior preclinical work by many groups has demonstrated that MSCs administered after TBI in preclinical models of injury improve overall neurological function in rats and mice (Harvey and Chopp, 2003; Mahmood et al., 2004a,b, 2008). Although the exact mechanisms of benefit are not well established, they appear to be pleiotropic in nature, with modulation of vascular stability and inflammation being key components. Using a controlled cortical impact model of TBI in mice, MSCs have been shown to reduce BBB permeability, decrease neuroinflammation, and improve overall neurocognitive function when delivered intravenously hours after brain injury; however, very few cells administered intravenously are actually detected in the injured brain. Most (>90%) are trapped in the lungs in a first-pass effect, and it is from their position in these organs that MSCs are thought to secrete soluble factors that mediate recovery in injury (Fischer et al., 2009; Menge et al., 2012) (Figure 8.1).

Mechanistic studies have demonstrated cell-to-cell contact between the pulmonary endothelium and MSCs (Pati et al., 2011a,b; Menge et al., 2012). Soluble factors produced by MSCs are believed to act locally and systemically to decrease vascular instability and inflammation in injury (see Figure 8.2). Menge et al. demonstrated that MSCs prevent pathologic BBB permeability in TBI through the production of tissue inhibitor of metalloproteinase 3 (TIMP3) (Menge et al., 2012), a soluble factor secreted by MSCs and induced by MSCs in the lungs. The TIMP3 was shown to be necessary and sufficient to mediate the protective effects of MSCs on BBB permeability (Menge et al., 2012). Using an established mouse model of TBI, Menge et al. showed that (1) intravenous (IV) administration of MSCs increases lung and circulating serum levels of TIMP3, (2) knockdown of TIMP3 by siRNA in MSCs completely abrogates the effect of MSCs on BBB permeability and endothelial adherens junction (AJ) stability, and (3) IV administration of human recombinant TIMP3 in mice recapitulates the effects of MSCs in vivo by decreasing BBB permeability and preserving AJs in the injured brain (Menge et al., 2012) (Figure 8.2). The TIMP3 is a protein known to inhibit inflammation and MMPs (Onyszchuk et al., 2007); however, its role in neuroprotection or neural recovery after injury has not been established (Chen et al., 2003; Luu et al., 2012).

Recent data by Watanabe et al. demonstrate that another soluble factor produced by MSCs, TSG-6 (TNF-α induced protein 6), has potent effects in recovery after

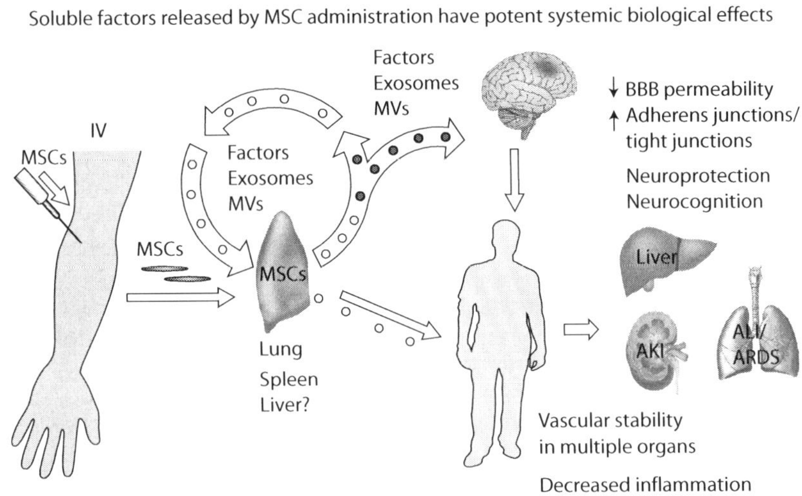

FIGURE 8.1 Intravenous MSCs are primarily trapped in the lungs in a first pass effect. Some MSCs are found in the liver and spleen as well. In the lungs the cells begin to produce circulating factors (growth factors, cytokines, chemokines, exosomes, and microvesicles) which then circulate systemically and mediate potent biological effect in the brain and distant organs, hence leading to repair and recovery after injury.

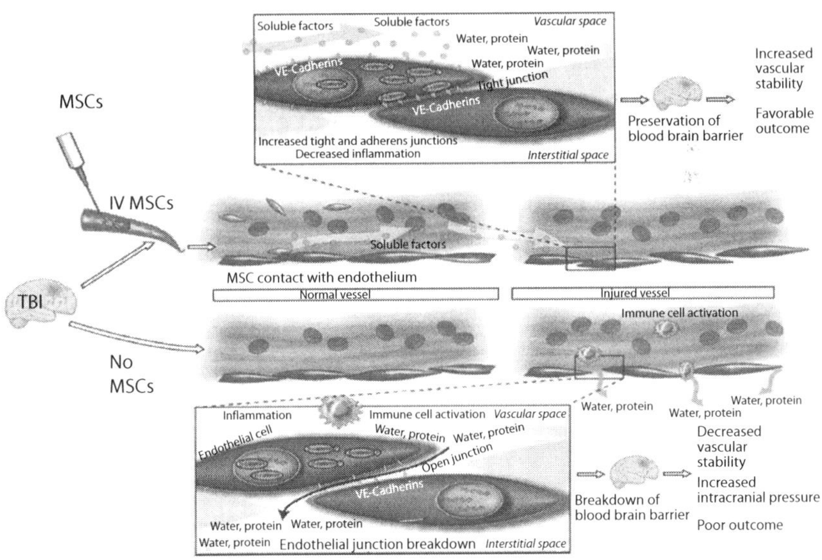

FIGURE 8.2 Systemic model of beneficial effects of mesenchymal stem cells (MSCs) in preventing pathologic edema associated with traumatic brain injury (TBI). Secreted factors (soluble factors, exosomes, and microvesicles) derived from MSC-pulmonary endothelial cell interactions mediate the therapeutic, anti-inflammatory and vascular stabilizing effects associated with MSC infusion.

TBI (Watanabe et al., 2013). The TSG-6 was originally discovered by Prockop and colleagues as a protein secreted by MSCs that had potent effect on recovery after myocardial infarction. In these initial studies, MSCs increased production of the anti-inflammatory protein TSG-6 when administered intravenously after myocardial infarction. Intravenous MSCs, but not MSCs transduced with TSG-6 siRNA, decreased inflammatory responses, reduced infarct size, and improved cardiac function (Lee JW et al., 2009; Lee RH et al., 2009). In TBI, administration of IV mesenchymal stem/stromal cells (MSCs) produced beneficial effects in models of TBI, and the effects were explained by the MSCs being activated to express TSG-6, a multifunctional protein that modulates inflammation mediated by TNF-α (Watanabe et al., 2013). In a mouse model of TBI, they found an initial mild phase of the inflammatory response in TBI for at least 24 hours, followed by a secondary severe response that peaked at 3 days. Intravenous human MSCs or TSG-6 decreased neutrophil extravasation, decreased expression of matrix metalloproteinase-9 by endothelial cells and neutrophils, and diminished subsequent BBB leakage after TBI. Administration of TSG-6 also decreased the lesion size at 2 weeks. Most importantly, the acute administration of TSG-6 alone within 24 hours of TBI was followed by improvement in memory, improvement in depressive-like behavior, and increase in the total number of newly born neurons 6–10 weeks after treatment (Watanabe et al., 2013).

8.4 MECHANISMS OF ACTION OF MSCs AND SOLUBLE FACTORS IN ORGAN (LUNG AND RENAL) INJURY

As mentioned above, secondary consequences of traumatic injury involve other organs including the lungs, kidneys, and liver, all of which have been shown to be therapeutic targets for MSCs, hence arguing for a multipotent therapeutic benefit to using MSCs in traumatic injury.

In lung injury, MSCs have been shown to have potent effects in preclinical models of ALI and ARDS, which are characterized by lung edema and inflammation (Matthay et al., 2010; Lee JW et al., 2011; Lee RH et al., 2011; Lee et al., 2012, 2013; Matthay et al., 2013). The MSCs have been shown to modulate the immune system through the release of anti-inflammatory cytokines, inhibition of inflammatory cytokines, and release of lipid mediators. Considerable work has been conducted in the area of lung injury and MSC-derived soluble factors. The MSCs have been delivered via the lung airspace intratracheally or intravenously. Both routes of delivery have been effective in mitigating ALI and ARDS. In 2003, Ortiz et al. reported that MSC therapy reduced fibrosis in a model of bleomycin-induced lung injury (Ortiz et al., 2003). Further work by this same group demonstrated that the effect was due to a secreted interleukin-1 receptor antagonist (Ortiz et al., 2007). Gupta and associates (Gupta et al., 2007) have shown that MSCs delivered intratracheally inhibit ALI induced by high-dose intra-alveolar delivery of endotoxin. A more recent study by Matthay and colleagues demonstrates that MSCs restore lung epithelial and endothelial permeability, independent of the cells themselves (Matthay et al., 2010). These studies also revealed that MSCs produce the soluble factor—keratinocyte growth factor (KGF)—that enhances the ability of alveolar epithelium to remove fluid and restore lung vascular permeability and extravascular lung water to normal levels. Furthermore, human MSCs have been shown to inhibit alveolar-capillary leak through production of angiopoeitin-1 (Ang-1). Ang-1 was shown to prevent actin stress fiber formation and also claudin-18 disorganization, thereby contributing to pulmonary vascular stability in the lungs (Fang et al., 2010). Taken together, these data suggest that a number of soluble factors produced by MSCs mitigate ALI and ARDS.

In renal injury, soluble factors have also been shown to play a critical role in mediating recovery by MSCs. The role of bone marrow MSCs in the recovery of acute kidney injury (AKI) is well documented. The MSCs have been shown to accelerate recovery from toxic agents (Morigi et al., 2006; Morigi and Benigni, 2013) or ischemia-reperfusion injury (Duffield et al., 2005). The MSCs have also been shown to improve outcomes in chronic renal failure (Choi et al., 2009). After injury to the kidney, nephrons can to some degree recover their functional and structural integrity. It was initially suggested that bone marrow MSCs home into the kidney and differentiate to replace dead epithelial cells (Poulsom et al., 2001, 2002). This hypothesis was subsequently replaced by findings that this is a rare event and that paracrine and endocrine mechanisms are involved in renal repair by MSCs (Duffield et al., 2005; Togel et al., 2005), because MSCs are also undetectable in the kidneys a few days after injury (Hauser et al., 2010). The most compelling support for a paracrine mechanism of MSCs in renal injury comes from experiments showing that administration of conditioned media from MSCs mimicked the effects of MSCs. Bi et al. showed that MSC conditioned media contained factors that limit apoptosis and enhanced proliferation of tubular cells (Bi et al., 2007). Repair was thought to be mediated by release of growth factors such as epidermal growth factor (EGF), hepatocyte growth factor (HGF), insulin-like growth factor-1 (IGF-1), and vascular endothelial growth factor-A (VEGF-A) by MSCs (Morigi et al., 2006; Morigi and Benigni, 2013). Recent studies have suggested that the renal protective effects of MSCs are mediated predominantly by microvesicle transfer between injured renal tubular epithelial cells and MSCs (see below) (Gatti et al., 2011).

8.4.1 Immunomodulatory Effects of MSCs Mediated via Secreted Soluble Factors

There are a number of soluble factors that have been identified as part of the MSC secretome. Many of these factors are produced and also induced by MSCs after administration in various disease models. Many of the factors that have been identified regulate inflammation and endothelial permeability through multiple mechanisms of actions. Immunomodulation is believed to be a key common mechanism of MSC action that enables the cells to display potent effects in multiple disease outcomes. Multiple studies have demonstrated that MSCs possess potent immunosuppressive effects and inhibit the activity of both innate and adaptive immune cells (Aggarwal and Pittenger, 2005; Beyth et al., 2005; Glennie et al., 2005; Corcione et al., 2006). This immunosuppression has been shown to be mediated by contact-dependent and cell contact–independent mechanisms through the release of soluble factors. The long list of candidate mediators released or induced by MSCs include transforming growth factor-β, tumor necrosis factor-α (TNFα)–stimulated gene/protein-6 (TSG-6) (Lee JW et al., 2009; Lee RH et al., 2009), PGE2, indoleamine 2,3-dioxygenase, interleukin-10 (IL-10), and IL-1ra, among others. In a model of sepsis following mice with injury induced by cecal ligation and puncture (CLP) (Nemeth et al., 2009), it was found that bone marrow–derived mouse MSCs, activated by lipopolysaccharide (LPS) or TNFα, secreted PGE2, which reprogrammed alveolar macrophages to secrete IL-10. The MSCs are also major producers of IL-10. The beneficial effect of MSCs on mortality and improved organ function following sepsis was eliminated by macrophage depletion or pretreatment with antibodies to IL-10 or the IL-10 receptor, suggesting an essential role for IL-10 in these experiments. IL-10 is a cytokine secreted predominantly by monocytes that downregulates the expression of T helper 1 cytokines, MHC class II antigens, and costimulatory molecules on macrophages. IL-10 has also been reported to inhibit the rolling, adhesion, and transepithelial migration of neutrophils (Ajuebor et al., 1999). In a model of ALI following intrabronchial *Escherichia coli* endotoxin in mice, Lee and Matthay (Gupta et al., 2007) found that syngeneic MSCs improved survival and lung injury in association with a decrease in macrophage inflammatory protein-2 and TNFα levels in the bronchoalveolar lavage (BAL) fluid and elevated levels of IL-10 in both the plasma and BAL fluids. In bleomycin-induced lung injury and fibrosis in mice, Ortiz et al. (2007) found that mouse MSCs decreased subsequent lung fibrosis and levels of matrix metalloproteinases in part by IL-1rα (IL-1 receptor antagonist) secretion; IL-1rα is a cytokine that competitively competes with IL-1α for IL-1 receptor binding. IL-1α is one of the major inflammatory cytokines in pulmonary edema fluid in patients with ALI/acute respiratory distress syndrome (ARDS) (Geiser et al., 2001).

These studies confirmed the anti-inflammatory effect of MSCs in multiple lung injury experiments in mice due to secreted factors (Ortiz et al., 2003; Rojas et al., 2005; Xu et al., 2007, 2008; Mei et al., 2007). Despite the well-documented anti-inflammatory effects of MSCs, recent literature has also described a dual role for MSCs as immunostimulatory cells (Stagg, 2007). Several investigators have reported that MSCs can upregulate expression of MHC II when exposed to low levels of inflammation and function as antigen-presenting cells stimulating the adaptive immune system (Chan et al., 2006; Stagg et al., 2006). Recent evidence has also shown that MSCs can secrete IL-6 and induce production of IgG by B lymphocytes in an in vitro setting (Rasmusson et al., 2007). In addition, MSCs can prevent neutrophil apoptosis and degranulation in culture without inhibiting their phagocytic or chemotactic capabilities (Raffaghello et al., 2008). Thus, MSCs have more complex effects on the immune system than their classical role as immune suppressor cells. In the future, we have to study the complex and often opposing relationship between the potential immunogenicity of MSCs and their ability to suppress the innate and adaptive immunity to understand the significance of immunomodulation during therapy for lung injury.

8.4.2 Antibacterial Peptides and Factors Secreted by MSCs

One safety concern with MSC-based therapy, particularly in treating disease and conditions with the potential for infection, is their immunosuppressive properties.

This would indeed be a concern in the case of traumatic injury, where the propensity for infection is high. Bacterial pneumonia and sepsis from nonpulmonary causes are two of the most common etiologies of ARDS (Rubenfeld et al., 2005). Given the preponderance of literature that describes the immunosuppressive effect of MSCs, there was a concern that this effect might diminish host defense against infections. However, recent in vivo studies have provided evidence for the beneficial effects of MSCs in the treatment of bacteria-induced sepsis. In the mouse model of sepsis (CLP), IV syngeneic MSCs reduced mortality, improved organ function, and decreased total bacterial counts in blood and peritoneal fluid (Nemeth et al., 2009). Survival benefits in this study were explained in part by the immunomodulatory properties of MSCs mediated by soluble paracrine factors such as IL-10 and PGE2. Other studies have shown similar effects on bacterial infection (Mei et al., 2007; Gonzalez-Rey et al., 2009). Recent work has demonstrated that human bone marrow–derived MSCs can inhibit bacterial growth directly, and their antimicrobial effect is mediated in part through the secretion of an antimicrobial peptide LL-37, which is unregulated upon bacterial stimulation. The secretion of LL-37 by MSCs also improved bacterial clearance in vivo in the mouse model of *E. coli* pneumonia, thereby suggesting balanced effects of the MSCs in immunomodulation of inflammation and response to infection (Krasnodembskaya et al., 2010).

8.5 MITOCHONDRIAL TRANSFER FROM MSCs TO INJURED TARGET CELLS

Recent work has demonstrated that in addition to soluble factor release, MSCs have the capacity to transfer their mitochondria to other cells. Bhattacharya and colleagues demonstrated that in a mouse model of LPS-induced lung injury, MSCs form connexin 43 (Cx43)-containing gap junctional channels (GJCs) with the alveolar epithelia in these mice, releasing mitochondria-containing microvesicles that the epithelia engulf. The presence of BMSC-derived mitochondria in the epithelia resulted in increased alveolar ATP concentrations. The LPS-induced ALI, as indicated by alveolar leukocytosis and protein leak, inhibition of surfactant secretion, and high mortality, was markedly abrogated by the instillation of wild-type mitochondria from MSCs, but not mutant, GJC-incompetent dysfunctional mitochondria. This was the first evidence that MSCs protect against ALI Cx43-dependent alveolar attachment and mitochondrial transfer and may be a mechanism of action of MSCs in multiple therapeutic areas (Islam et al., 2012).

8.6 THERAPEUTIC POTENTIAL OF MSC-DERIVED MICROVESICLES AND EXOSOMES

Cells communicate and exchange information by different ways, including the secretion of soluble factors, the cell-to-cell adhesion contact, and the intercellular exchange of organelles through nanotubular structures. Recent studies have proposed that cell-derived small circular membrane vesicles, called exosomes (Exo) and shedding microvesicles (MVs), represent an additional mechanism of cell-to-cell communication by transfer of membrane components as well as the cytoplasmic content. Exosomes arise from the endosomal membrane cell compartments and are released into the extracellular space after fusion of multivesicular bodies with the plasma membrane (Mathivanan et al., 2009; Thery et al., 2009; Mathivanan et al., 2010; Gyorgy et al., 2011). Exosomes tend to be homogenous in size (30–120 nm) and are released by a p53-regulated exocytosis which is independent of calcium flux. Shedding vesicles or microvesicles originate from direct budding and blebbing of the plasma membrane (Mathivanan et al., 2010). Microvesicles are more heterogenous and range in size from 100 to 1000 nm. They are released by the budding of small cytoplasmic protrusions, which is dependent on cytoskeletal reorganization and calcium flux as well. Exosomes are truly a form of MVs that are usually present together as a mixed population in vitro and in vivo. The shedding of microvesicles from stem cells is considered an active process and is no longer an artifact of cellular damage (Cocucci et al., 2009). It is a process that has been shown to occur in vivo as well.

The MSCs shed significant amounts of MVs containing proteins, RNA, miRNA, lipids, and rarely DNA. Microvesicles (MVs) have been shown to mediate the therapeutic effects of MSCs in lung injury. Zhu, Matthay, and colleagues showed that MVs released by human bone marrow–derived MSCs are effective in restoring lung protein permeability and reducing inflammation in *E. coli* endotoxin-induced ALI in C57BL/6 mice. The intratracheal instillation of MVs improved several indices of ALI at 48 hours. Compared to endotoxin-injured mice, MVs reduced extravascular lung water by 43% and reduced total protein levels in the bronchoalveolar lavage (BAL) fluid by 35%, demonstrating a reduction in pulmonary edema and lung protein permeability. The MVs also reduced the influx of neutrophils and macrophage inflammatory protein-2 levels in the BAL fluid by 73% and 49%, respectively, demonstrating a reduction in inflammation. The KGF siRNA-pretreatment of MSC partially eliminated the therapeutic effects of MVs released by MSCs, suggesting that KGF protein expression was important for the underlying mechanism. These data suggest that microvesicles from MSCs package and harbor a number of the therapeutically active proteins that mediate recovery.

Exosomes have received the most attention in recent years. The secretion of nanovesicles during the maturation of sheep reticulocytes was first described in the 1980s (Harding et al., 1983, 1984; Yu et al., 2014). Investigation into this area revealed that most cell types secrete exosomes—B cells, T cells, dendritic cells, cancer cells, endothelial cells, and MSCs to name a few. They were originally thought to be necessary for the clearance of unneeded proteins from cells (Yu et al., 2014), and the current opinion is that exosomes are specifically mediators of cell-to-cell communication and package in them proteins and genetic material that cells transfer between each other to communicate and mediate biological processes in health and disease. Their exact mechanisms of action remain elusive. As mentioned previously, exosomes are formed by the budding inward of late-stage endosomes and the microvesicular bodies then fuse with the plasma membrane to release to the environment (Yu et al., 2014). They can be secreted in a regulated upon cell surface receptor activation (Saunderson et al., 2008) or constitutive manner. Most exosomes can be sedimented by centrifugation and 100,000×g; however, there are other methods of size fractionation that can produce isolates of exosomes. Other methods include high-performance liquid chromatography (HPLC) for separation. In terms of size, exosomes differ from microvesicles in that they are smaller. Microvesicles are typically between 20 and 1000 nm, and exosomes are considered to be a type of microvesicle (Simpson et al., 2009). Other types of microvesicles include nanoparticles and microparticles.

Exosomes typically are identified by surface markers that are specific to the cells they are derived from and are relatively unique in their lipid, protein, and RNA content. Exosomes have been shown to have specific interactions with targeted recipient cells (Vlassov et al., 2012). The general function of exosomes can range from disposal of unwanted proteins, antigen presentation, genetic exchange, immune responses, angiogenesis, inflammation, spreading of pathogens, and tumor metastases (Thery et al., 2009). The proteomes of exosomes have been analyzed, and there are a number of proteins that are consistently expressed and many that vary (Raimondo et al., 2011). However exosome proteomes account for only a small fraction of the plasma proteome. Similarly, the RNA content or exosomes is similar for many RNAs and also highly variable for a number of them. The MiRNAs are carried by exosomes and may have characteristic signatures in health and disease (Chen et al., 2008).

The MSC-derived exosomes have been studied in the past 5 years. Some of the earliest studies were conducted in cardiovascular disease. The MSCs produce higher amounts of exosomes than other cells. They have common surface markers of CD9 and CD81, and also MSC-specific markers of CD29, CD44, and CD73. Protein content and RNA content do not remain the same for all MSC exosomes. A study attempting to characterize MSC exosomes showed that in three batches of MSCs, 379, 432, and 420 proteins were unique, of which only 154 were common (Lai et al., 2012). These proteins have been characterized, and the potential to modulate a number of biological processes has been assessed. In a mouse model of myocardial infarction–induced ischemia reperfusion injury, outcomes were mitigated by MSC exosome administration (Lai et al., 2010). In a porcine model of myocardial ischemia and reperfusion, IV MSC conditioned media (CM) treatment reduced the infarct size by 50% (Timmers et al., 2007). Size fractionation of the CM indicated that the active component is 30–200 nm in size, which can encompass exosomes and smaller microvesicles. Electron microscopy revealed that the active component is phospholipid vesicles. These results suggested that exosomes can be highly effective as therapeutics in critical illness.

In the area of renal injury, exosomes and microvesicles have been shown to have potent therapeutic effects. It is now known that microvesicles may interact with cells through specific receptor–ligand interactions that mediate their effects (Taraboletti et al., 2002, 2006). Microvesicles transfer profiles of mRNA and of microRNA resulting in genetic exchange between cells (Bruno et al., 2009; Eldh et al., 2010; Herrera et al., 2010; Gatti et al., 2011). This exchange can lead to permanent or semipermanent epigenetic changes in recipient cells, which in many ways may explain the lasting presence of MSC effects even after the cells are not detectable. Cammusi et al. demonstrated that human MSCs release microvesicles that are able to stimulate in vitro proliferation and also apoptosis resistance of tubular epithelial cells (Biancone et al., 2012; Bruno et al., 2012). Furthermore, his group found that MVs accelerate functional and morphological recovery of tubular epithelia in severe combined immunodeficiency (SCID) mice with glycerol-induced AKI (Bruno et al., 2009). Microvesicles were found to be as effective as MSCs in accelerating recovery in AKI, suggesting that the beneficial effect of MSCs is largely due to the microvesicles. RNase treatment of microvesicles abolished both the in vitro and the in vivo effects of the microvesicles, suggesting a mechanism dependent on RNA delivery. Consistent with previous findings in endothelial progenitor cells (EPCs), microvesicles contain a defined subset of RNA transcripts rather than a random sampling of cellular mRNA. Microvesicles released by MSCs contain mRNA representative of the multiple functional properties of MSCs indicating cell specificity of the mRNA content (Bruno et al., 2009). This observation suggests a tightly regulated process of microvesicle formation. It is at present unknown whether mRNA entry in target cells activates translational control mechanisms or translational checkpoints for endogenous transcripts. Although an effective horizontal transfer of mRNA into tubular cells occurs as indicated by the presence of human-specific mRNA and proteins in tubular cells of mice with AKI, it is possible that other cell types are involved in such transfer of genetic material. As suggested by Bonventre (2009), endothelial cells might also be a target for microvesicles, resulting in modulation of renal perfusion, inflammation, and vascular permeability that may affect the recovery of AKI.

Although extensive studies have not been conducted with MSC exosomes and microvesicles in TBI, exosomes and microvesicles have also been shown to have effects and mediate recovery in multiple neurodegenerative diseases such as stroke. Exosomes have been shown to interact with brain parenchymal cells to promote functional recovery. It has been hypothesized that MSCs communicate with brain parenchymal cells via miRNA contained within the exosomes. In one study, rats subjected to middle cerebral artery occlusion (MCAo) were treated with MSCs, and microRNA 133b (miR-133b) levels in the ipsilateral hemisphere increased significantly. In vitro, miR-133b levels increased in exosomes derived from MSCs, which had been exposed to ipsilateral ischemic tissue extracts from rats subjected to MCAo. And miR-133b levels in primary cultured neurons and astrocytes treated with the exosome-enriched fractions released from these MSCs were also increased. However, this increase is significantly diminished by treatment of the astrocytes with exosome-enriched fractions from MSCs transfected with a miR-133b inhibitor. This experiment provided some of the first evidence that MSCs communicate with brain parenchymal cells via exosome-mediated miR-133b transfer, leading to regulation of specific gene expression that enhances neurite outgrowth and functional recovery (Xin et al., 2012). In another study, it was demonstrated that IV injection of MSC-derived exosomes can lead to the increase of axonal density and synaptophysin-positive areas along the ischemic boundary zone of the cortex and striatum and prompt functional recovery confirming that exosomes from MSCs could significantly improve neurologic outcome and contribute to neurovascular remodeling (Xin et al., 2013). In the area of Alzheimer's disease, the accumulation of amyloid (Aβ) peptides in the brain is a characteristic. Furthermore, neprilysin (NEP) is the most important Ab-degrading enzyme in the brain. It has been found that adipose tissue–derived mesenchymal stem cells (ADSCs) secrete functional NEP in association with exosomes. Co-culture of N2a neuroblastoma cells overexpressing human Aβ and ADSCs results in a significant decrease of both Aβ 40 and 42 levels in the culture medium. Moreover, it has been demonstrated that ADSCs express higher levels of NEP than those in bone marrow–derived MSCs. These results suggest that ADSCs may serve as a promising cell source for exosome-based Alzheimer's disease treatments (Katsuda et al., 2013). Figure 8.3 summarizes some of the factors produced by MSCs and their known therapeutic effects.

8.7 RESPONSE OF INJURED TISSUE AND THE INTERPLAY OF MICROVESICLES/EXOSOMES DERIVED FROM INJURED TISSUE AND MSCs

It has been suggested that bone marrow cells have the capacity to produce different cell types from the injured tissue, although the contribution of MSCs to this effect seems very small (Caplan and Dennis, 2006; Li et al., 2009). However, the mechanism underlying stem cell changes after interaction with injured cells is still debated. Differentiation into other cell types as a mechanism of stem cell effect has never been conclusively demonstrated in any experimental setting (Sheldon et al., 2010), and fusion of stem cells with resident injured or noninjured cells has yielded variable results (Sheldon et al., 2010). It has also been suggested (Mause et al., 2010) that stem cell differentiation is related to epigenetic cell changes mediated by signals received from injured cells. These specific signals may be possibly delivered by soluble factors released by injured cells or also by microvesicle-mediated transfer of genetic information (Bruno et al., 2009). In co-culture experiments of MSCs separated from injured lung cells, the expression of genes for lung-specific proteins such as Clara cell-specific protein, surfactant B, and surfactant C were increased in the lung cells (Dooner et al., 2008). Microvesicles released in the cell supernatant of injured cells induced lung-specific gene expression in bone marrow cells. Microvesicles derived from injured cells were shown to contain high levels of lung-specific mRNA and to deliver this mRNA to MSCs. It has been recently reported that conditioned medium from renal tubular epithelial cells initiates differentiation of human MSCs (Baer et al., 2009). Microvesicles derived from injured renal tubular epithelial cells may also induce expression of tubular cell markers in human MSCs. These results suggest that microvesicles derived from injured tissue may potentially mediate changes in stem cells, thereby resulting in a complex interplay between the two populations in physiologic repair of the injured organ.

CONCLUSION

Traumatic brain injury, trauma, and critical care medicine in general have proven to be areas that are truly challenging for targeted therapies and effective drug discovery. Analysis of the multiple failed clinical trials that have been run to date in TBI, for example, reveal that no one therapeutic approach or target will likely address the pleotropic nature of the injury and disease. The MSCs potentially provide a viable option for the treatment of TBI because they therapeutically address multiple endpoints. Although more than hundreds of clinical trials are being conducted to investigate the therapeutic potential of MSCs, little is known about their exact mechanisms of action, but it is possible that a main reason for their positive effect is due to

Disease Area	MSC Bioactive Division	Bioactive Component	Functional Benefit	References
TBI	Soluble protein	TIMP3 TSG-6	Preserves adherens junctions of the BBB Anti-inflammatory	Menge et al. (2012), *STM* Watanabe et al. (2013), *Neurobiol. Dis.*
Stroke	Exosome	miRNA (miRNA-133b) Unknown	Neurite outgrowth Enhanced functional outcome Increased synaptic density Enhances neurogenesis Enhances angiogenesis	Xin et al. (2012), *Stem Cells* Xin et al. (2013) *J. Cereb. Blood Flow and Metab.*
Acute lung injury	Soluble protein	Angiopoietin-1 KGF IL1-R antagonist IL-10 Prostaglandin E2 LL-37 TSG-6	Lung epithelia and endothelial permeability Alveolar fluid transport Vascular Integrity Anti-inflammatory Anti-inflammatory Anti-inflammatory Anti-microbial Anti-inflammatory	Mei et al. (2007), *PLoS Med* Fang et al. (2010), *JBC* Lee et al. (2009), *JNAS* Murakam et al. (2008), *JCI* Ortiz et al. (2007), *PNAS* Gupta et al. (2007), *Journal Immunol.* Nemeth et al. (2009), *Nat Med* Krasnodembskaya et al. (2010), *Stem Cells* Danchuk et al. (2011), *Stem Cell Res. and Ther.*
Acute lung injury	Microvesicle	Includes KGF (Complete list unknown)	Restore lung protein permeability Reduced pulmonary edema Anti-inflammatory	Zhu et al. (2014), *Stem Cells*
Acute alveolar injury	Soluble protein	KGF	Preserves epithelial cell permeability Restores sodium transport	Goolaerts et al. (2014) *Am. J. Physiol. Lung Cell Mol. Physiol.*
Acute renal failure	Exosome	IGF-1R mRNA	Enhanced proximal tubular epithelial cell proliferation	Tomasoni et al. (2013). *Stem Cells and Development*
Acute renal failure	Microvesicle	mRNA	Stimulates tubular epithelial cell proliferation Imparts resistance to apoptosis	Bruno et al. (2012), *PLoS one*
Acute kidney injury	Microvesicles	mRNA	Modulates renal perfusion Anti-inflammatory Enhances vascular permeability	Bonventre et al. (2009), *JASN*
Myocardial infarction	Exosome	Unknown	Reduction in infarc size	Timmers et al. (2007), *Stem Cell Research*

FIGURE 8.3 A summary of the current knowledge on soluble factors known to have therapeutic effects in trauma-related injuries.

the pleiotropic nature of their actions, which are largely mediated through the release of multiple soluble factors, exosomes, and microvesicles (Figure 8.1). It is of great interest from a translational standpoint that naturally occurring, circulating proteins, exosomes, or microvesicles found in blood can be delivered through IV and result in improved outcomes in TBI. From a practical standpoint, there are significant logistical, regulatory, and practical barriers that arise clinically in the translation of stem cell therapies to patients, such as the difficulty of autologous harvest, rejection due to allotransplantation, or the tumorigenic potential of stem cells (Goldring et al., 2011). Thus, the identification and development of a "cell-free therapeutic," a soluble factor(s), microvesicular, or exosome preparation that can recapitulate some of the effects of stem cells in TBI, is highly innovative and could circumvent barriers in translation in the treatment of a disease condition with few therapeutic options.

REFERENCES

Aggarwal S, Pittenger MF. Human mesenchymal stem cells modulate allogeneic immune cell responses. *Blood*. 2005;105:1815–1822.

Ajuebor MN, Das AM, Virag L, Flower RJ, Szabo C, Perretti M. Role of resident peritoneal macrophages and mast cells in chemokine production and neutrophil migration in acute inflammation: Evidence for an inhibitory loop involving endogenous il-10. *J Immunol*. 1999;162:1685–1691.

Baer PC, Bereiter-Hahn J, Missler C, Brzoska M, Schubert R, Gauer S, Geiger H. Conditioned medium from renal tubular epithelial cells initiates differentiation of human mesenchymal stem cells. *Cell Prolif*. 2009;42:29–37.

Bazarian JJ, McClung J, Cheng YT, Flesher W, Schneider SM. Emergency department management of mild traumatic brain injury in the USA. *Emerg Med J*. 2005;22:473–477.

Bedi SS, Hetz R, Thomas C, Smith P, Olsen AB, Williams S, Xue H et al. Intravenous multipotent adult progenitor cell therapy attenuates activated microglial/macrophage response and improves spatial learning after traumatic brain injury. *Stem Cells Transl Med*. 2013;2:953–960.

Bernardo ME, Locatelli F, Fibbe WE. Mesenchymal stromal cells. *Ann NY Acad Sci*. 2009;1176:101–117.

Beyth S, Borovsky Z, Mevorach D, Liebergall M, Gazit Z, Aslan H, Galun E, Rachmilewitz J. Human mesenchymal stem cells alter antigen-presenting cell maturation and induce T-cell unresponsiveness. *Blood*. 2005;105:2214–2219.

Bi B, Schmitt R, Israilova M, Nishio H, Cantley LG. Stromal cells protect against acute tubular injury via an endocrine effect. *J Am Soc Nephrol*. 2007;18:2486–2496.

Biancone L, Bruno S, Deregibus MC, Tetta C, Camussi G. Therapeutic potential of mesenchymal stem cell-derived microvesicles. *Nephrol Dial Transplant*. 2012;27:3037–3042.

Blum S, Moore AN, Adams F, Dash PK. A mitogen-activated protein kinase cascade in the ca1/ca2 subfield of the dorsal hippocampus is essential for long-term spatial memory. *J Neurosci*. 1999;19:3535–3544.

Bonventre JV. Microvesicles from mesenchymal stromal cells protect against acute kidney injury. *J Am Soc Nephrol*. 2009;20:927–928.

Bruno S, Grange C, Collino F, Deregibus MC, Cantaluppi V, Biancone L, Tetta C, Camussi G. Microvesicles derived from mesenchymal stem cells enhance survival in a lethal model of acute kidney injury. *PloS One*. 2012;7:e33115.

Bruno S, Grange C, Deregibus MC, Calogero RA, Saviozzi S, Collino F, Morando L et al. Mesenchymal stem cell-derived microvesicles protect against acute tubular injury. *J Am Soc Nephrol*. 2009;20:1053–1067.

Caplan AI, Dennis JE. Mesenchymal stem cells as trophic mediators. *J Cell Biochem.* 2006;98:1076–1084.

Chan JL, Tang KC, Patel AP, Bonilla LM, Pierobon N, Ponzio NM, Rameshwar P. Antigen-presenting property of mesenchymal stem cells occurs during a narrow window at low levels of interferon-gamma. *Blood.* 2006;107:4817–4824.

Chen X, Ba Y, Ma L, Cai X, Yin Y, Wang K, Guo J et al. Characterization of microRNAs in serum: A novel class of biomarkers for diagnosis of cancer and other diseases. *Cell Res.* 2008;18:997–1006.

Chen XH, Iwata A, Nonaka M, Browne KD, Smith DH. Neurogenesis and glial proliferation persist for at least one year in the subventricular zone following brain trauma in rats. *J Neurotrauma.* 2003;20:623–631.

Chesnut RM. Management of brain and spine injuries. *Crit Care Clin.* 2004;20:25–55.

Choi S, Park M, Kim J, Hwang S, Park S, Lee Y. The role of mesenchymal stem cells in the functional improvement of chronic renal failure. *Stem Cells Dev.* 2009;18:521–529.

Cocucci E, Racchetti G, Meldolesi J. Shedding microvesicles: Artefacts no more. *Trends Cell Biol.* 2009;19:43–51.

Colicos MA, Dash PK. Apoptotic morphology of dentate gyrus granule cells following experimental cortical impact injury in rats: Possible role in spatial memory deficits. *Brain Res.* 1996;739:120–131.

Colicos MA, Dixon CE, Dash PK. Delayed, selective neuronal death following experimental cortical impact injury in rats: Possible role in memory deficits. *Brain Res.* 1996;739:111–119.

Corcione A, Benvenuto F, Ferretti E, Giunti D, Cappiello V, Cazzanti F, Risso M et al. Human mesenchymal stem cells modulate b-cell functions. *Blood.* 2006;107:367–372.

Cox CS, Jr., Baumgartner JE, Harting MT, Worth LL, Walker PA, Shah SK, Ewing-Cobbs L et al. Autologous bone marrow mononuclear cell therapy for severe traumatic brain injury in children. *Neurosurgery.* 2011;68:588–600.

Dash PK, Mach SA, Blum S, Moore AN. Intrahippocampal wortmannin infusion enhances long-term spatial and contextual memories. *Learn Mem.* 2002;9:167–177.

Dooner MS, Aliotta JM, Pimentel J, Dooner GJ, Abedi M, Colvin G, Liu Q, Weier HU, Johnson KW, Quesenberry PJ. Conversion potential of marrow cells into lung cells fluctuates with cytokine-induced cell cycle. *Stem Cells Dev.* 2008;17:207–219.

Duffield JS, Park KM, Hsiao LL, Kelley VR, Scadden DT, Ichimura T, Bonventre JV. Restoration of tubular epithelial cells during repair of the postischemic kidney occurs independently of bone marrow-derived stem cells. *J Clin Invest.* 2005;115:1743–1755.

Eldh M, Ekstrom K, Valadi H, Sjostrand M, Olsson B, Jernas M, Lotvall J. Exosomes communicate protective messages during oxidative stress; possible role of exosomal shuttle rna. *PloS One.* 2010;5:e15353.

Fang X, Neyrinck AP, Matthay MA, Lee JW. Allogeneic human mesenchymal stem cells restore epithelial protein permeability in cultured human alveolar type II cells by secretion of angiopoietin-1. *J Biol Chem.* 2010;285:26211–26222.

Fischer UM, Harting MT, Jimenez F, Monzon-Posadas WO, Xue H, Savitz SI, Laine GA, Cox CS, Jr. Pulmonary passage is a major obstacle for intravenous stem cell delivery: The pulmonary first-pass effect. *Stem Cells Dev.* 2009;18:683–692.

Garcia-Gomez I, Elvira G, Zapata AG, Lamana ML, Ramirez M, Castro JG, Arranz MG, Vicente A, Bueren J, Garcia-Olmo D. Mesenchymal stem cells: Biological properties and clinical applications. *Expert Opin Biol Ther.* 2010;10:1453–1468.

Gatti S, Bruno S, Deregibus MC, Sordi A, Cantaluppi V, Tetta C, Camussi G. Microvesicles derived from human adult mesenchymal stem cells protect against ischaemia-reperfusion-induced acute and chronic kidney injury. *Nephrol Dial Transplant.* 2011;26:1474–1483.

Geiser T, Atabai K, Jarreau PH, Ware LB, Pugin J, Matthay MA. Pulmonary edema fluid from patients with acute lung injury augments in vitro alveolar epithelial repair by an il-1β-dependent mechanism. *Am J Respir Crit Care Med.* 2001;163:1384–1388.

Glennie S, Soeiro I, Dyson PJ, Lam EW, Dazzi F. Bone marrow mesenchymal stem cells induce division arrest anergy of activated t cells. *Blood*. 2005;105:2821–2827.

Goldring CE, Duffy PA, Benvenisty N, Andrews PW, Ben-David U, Eakins R, French N et al. Assessing the safety of stem cell therapeutics. *Cell Stem Cell*. 2011;8:618–628.

Gomez-Pinilla F, Gomez AG. The influence of dietary factors in central nervous system plasticity and injury recovery. *PM R*. 2011;3:S111–S116.

Gonzalez-Rey E, Anderson P, González MA, Rico L, Büscher D, Delgado M. Human adult stem cells derived from adipose tissue protect against experimental colitis and sepsis. *Gut*. 2009;58:929–939.

Gupta N, Su X, Popov B, Lee JW, Serikov V, Matthay MA. Intrapulmonary delivery of bone marrow-derived mesenchymal stem cells improves survival and attenuates endotoxin-induced acute lung injury in mice. *J Immunol*. 2007;179:1855–1863.

Gyorgy B, Szabo TG, Pasztoi M, Pal Z, Misjak P, Aradi B, Laszlo V et al. Membrane vesicles, current state-of-the-art: Emerging role of extracellular vesicles. *Cell Mol Life Sci*. 2011;68:2667–2688.

Harding C, Heuser J, Stahl P. Receptor-mediated endocytosis of transferrin and recycling of the transferrin receptor in rat reticulocytes. *J Cell Biol*. 1983;97:329–339.

Harding C, Heuser J, Stahl P. Endocytosis and intracellular processing of transferrin and colloidal gold-transferrin in rat reticulocytes: Demonstration of a pathway for receptor shedding. *Eur J Cell Biol*. 1984;35:256–263.

Harvey RL, Chopp M. The therapeutic effects of cellular therapy for functional recovery after brain injury. *Phys Med Rehabil Clin N Am*. 2003;14:S143–S151.

Hauser PV, De Fazio R, Bruno S, Sdei S, Grange C, Bussolati B, Benedetto C, Camussi G. Stem cells derived from human amniotic fluid contribute to acute kidney injury recovery. *Am J Pathol*. 2010;177:2011–2021.

Hein AM, O'Banion MK. Neuroinflammation and memory: The role of prostaglandins. *Mol Neurobiol*. 2009;40:15–32.

Herrera MB, Fonsato V, Gatti S, Deregibus MC, Sordi A, Cantarella D, Calogero R, Bussolati B, Tetta C, Camussi G. Human liver stem cell-derived microvesicles accelerate hepatic regeneration in hepatectomized rats. *J Cell Mol Med*. 2010;14:1605–1618.

Hicks RR, Smith DH, Lowenstein DH, Saint Marie R, McIntosh TK. Mild experimental brain injury in the rat induces cognitive deficits associated with regional neuronal loss in the hippocampus. *J Neurotrauma*. 1993;10:405–414.

Islam MN, Das SR, Emin MT, Wei M, Sun L, Westphalen K, Rowlands DJ, Quadri SK, Bhattacharya S, Bhattacharya J. Mitochondrial transfer from bone-marrow-derived stromal cells to pulmonary alveoli protects against acute lung injury. *Nat Med*. 2012;18:759–765.

Jiang Y, Jahagirdar BN, Reinhardt RL, Schwartz RE, Keene CD, Ortiz-Gonzalez XR, Reyes M et al. Pluripotency of mesenchymal stem cells derived from adult marrow. *Nature*. 2002;418:41–49.

Joyce N, Annett G, Wirthlin L, Olson S, Bauer G, Nolta JA. Mesenchymal stem cells for the treatment of neurodegenerative disease. *Regen Med*. 2010;5:933–946.

Katsuda T, Tsuchiya R, Kosaka N, Yoshioka Y, Takagaki K, Oki K, Takeshita F, Sakai Y, Kuroda M, Ochiya T. Human adipose tissue-derived mesenchymal stem cells secrete functional neprilysin-bound exosomes. *Sci Rep*. 2013;3:1197.

Kelley BJ, Lifshitz J, Povlishock JT. Neuroinflammatory responses after experimental diffuse traumatic brain injury. *J Neuropathol Exp Neurol*. 2007;66:989–1001.

Khakoo AY, Pati S, Anderson SA, Reid W, Elshal MF, Rovira, II, Nguyen AT et al. Human mesenchymal stem cells exert potent antitumorigenic effects in a model of Kaposi's sarcoma. *J Exp Med*. 2006;203:1235–1247.

Krasnodembskaya A, Song Y, Fang X, Gupta N, Serikov V, Lee JW, Matthay MA. Antibacterial effect of human mesenchymal stem cells is mediated in part from secretion of the antimicrobial peptide Il-37. *Stem Cells*. 2010;28:2229–2238.

Lai RC, Arslan F, Lee MM, Sze NS, Choo A, Chen TS, Salto-Tellez M et al. Exosome secreted by MSC reduces myocardial ischemia/reperfusion injury. *Stem Cell Res*. 2010;4:214–222.
Lai RC, Tan SS, Teh BJ, Sze SK, Arslan F, de Kleijn DP, Choo A, Lim SK. Proteolytic potential of the MSC exosome proteome: Implications for an exosome-mediated delivery of therapeutic proteasome. *Int J Proteomics*. 2012;2012:971907.
Lee JW, Fang X, Gupta N, Serikov V, Matthay MA. Allogeneic human mesenchymal stem cells for treatment of *E. coli* endotoxin-induced acute lung injury in the ex vivo perfused human lung. *Proc Natl Acad Sci USA*. 2009;106:16357–16362.
Lee JW, Fang X, Krasnodembskaya A, Howard JP, Matthay MA. Concise review: Mesenchymal stem cells for acute lung injury: Role of paracrine soluble factors. *Stem Cells*. 2011;29:913–919.
Lee JW, Krasnodembskaya A, McKenna DH, Song Y, Abbott J, Matthay MA. Therapeutic effects of human mesenchymal stem cells in ex vivo human lungs injured with live bacteria. *Am J Respir Crit Care Med*. 2013;187:751–760.
Lee JW, Zhu Y, Matthay MA. Cell-based therapy for acute lung injury: Are we there yet? *Anesthesiology*. 2012;116:1189–1191.
Lee RH, Oh JY, Choi H, Bazhanov N. Therapeutic factors secreted by mesenchymal stromal cells and tissue repair. *J Cell Biochem*. 2011;112:3073–3078.
Lee RH, Pulin AA, Seo MJ, Kota DJ, Ylostalo J, Larson BL, Semprun-Prieto L, Delafontaine P, Prockop DJ. Intravenous hMSCs improve myocardial infarction in mice because cells embolized in lung are activated to secrete the anti-inflammatory protein TSG-6. *Cell Stem Cell*. 2009;5:54–63.
Li L, Zhang S, Zhang Y, Yu B, Xu Y, Guan Z. Paracrine action mediates the antifibrotic effect of transplanted mesenchymal stem cells in a rat model of global heart failure. *Mol Biol Rep*. 2009;36:725–731.
Luu P, Sill OC, Gao L, Becker S, Wojtowicz JM, Smith DM. The role of adult hippocampal neurogenesis in reducing interference. *Behav Neurosci*. 2012;126:381–391.
Lyeth BG, Jenkins LW, Hamm RJ, Dixon CE, Phillips LL, Clifton GL, Young HF, Hayes RL. Prolonged memory impairment in the absence of hippocampal cell death following traumatic brain injury in the rat. *Brain Res*. 1990;526:249–258.
Mahmood A, Goussev A, Lu D, Qu C, Xiong Y, Kazmi H, Chopp M. Long-lasting benefits after treatment of traumatic brain injury (TBI) in rats with combination therapy of marrow stromal cells (MSCs) and simvastatin. *J Neurotrauma*. 2008;25:1441–1447.
Mahmood A, Lu D, Chopp M. Intravenous administration of marrow stromal cells (MSCs) increases the expression of growth factors in rat brain after traumatic brain injury. *J Neurotrauma*. 2004a;21:33–39.
Mahmood A, Lu D, Chopp M. Marrow stromal cell transplantation after traumatic brain injury promotes cellular proliferation within the brain. *Neurosurgery*. 2004b;55:1185–1193.
Mathivanan S, Ji H, Simpson RJ. Exosomes: Extracellular organelles important in intercellular communication. *J Proteomics*. 2010;73:1907–1920.
Mathivanan S, Simpson RJ. Exocarta: A compendium of exosomal proteins and RNA. *Proteomics*. 2009;9:4997–5000.
Matthay MA, Anversa P, Bhattacharya J, Burnett BK, Chapman HA, Hare JM, Hei DJ et al. Cell therapy for lung diseases. Report from an NIH-NHLBI workshop, November 13–14, 2012. *Am J Respir Crit Care Med*. 2013;188:370–375.
Matthay MA, Goolaerts A, Howard JP, Lee JW. Mesenchymal stem cells for acute lung injury: Preclinical evidence. *Crit Care Med*. 2010;38:S569–573.
Mause SF, Weber C. Microparticles: Protagonists of a novel communication network for intercellular information exchange. *Circ Res*. 2010;107:1047–1057.
Mei SH, McCarter SD, Deng Y, Parker CH, Liles WC, Stewart DJ. Prevention of LPS-induced acute lung injury in mice by mesenchymal stem cells overexpressing angiopoietin 1. *PLoS Med*. 2007;4:e269.

Menge T, Zhao Y, Zhao J, Wataha K, Gerber M, Zhang J, Letourneau P et al. Mesenchymal stem cells regulate blood-brain barrier integrity through TIMP3 release after traumatic brain injury. *Sci Transl Med*. 2012;4:161ra150.

Menon DK. Unique challenges in clinical trials in traumatic brain injury. *Crit Care Med*. 2009;37:S129–S135.

Morganti-Kossmann MC, Rancan M, Otto VI, Stahel PF, Kossmann T. Role of cerebral inflammation after traumatic brain injury: A revisited concept. *Shock*. 2001;16:165–177.

Morganti-Kossmann MC, Rancan M, Stahel PF, Kossmann T. Inflammatory response in acute traumatic brain injury: A double-edged sword. *Curr Opin Crit Care*. 2002;8:101–105.

Morganti-Kossmann MC, Satgunaseelan L, Bye N, Kossmann T. Modulation of immune response by head injury. *Injury*. 2007;38:1392–1400.

Morigi M, Benigni A. Mesenchymal stem cells and kidney repair. *Nephrol Dial Transplant*. 2013;28:788–793.

Morigi M, Benigni A, Remuzzi G, Imberti B. The regenerative potential of stem cells in acute renal failure. *Cell Transplant*. 2006;15 Suppl 1:S111–S117.

Myers TJ, Granero-Molto F, Longobardi L, Li T, Yan Y, Spagnoli A. Mesenchymal stem cells at the intersection of cell and gene therapy. *Expert Opin Biol Ther*. 2010;10:1663–1679.

Nemeth K, Leelahavanichkul A, Yuen PS, Mayer B, Parmelee A, Doi K, Robey PG et al. Bone marrow stromal cells attenuate sepsis via prostaglandin e(2)-dependent reprogramming of host macrophages to increase their interleukin-10 production. *Nat Med*. 2009;15:42–49.

Oliveri RS, Bello S, Biering-Sorensen F. Mesenchymal stem cells improve locomotor recovery in traumatic spinal cord injury: Systematic review with meta-analyses of rat models. *Neurobiol Dis*. 2014;62:338–353.

Onyszchuk G, Al-Hafez B, He YY, Bilgen M, Berman NE, Brooks WM. A mouse model of sensorimotor controlled cortical impact: Characterization using longitudinal magnetic resonance imaging, behavioral assessments and histology. *J Neurosci Methods*. 2007;160:187–196.

Ortiz LA, Dutreil M, Fattman C, Pandey AC, Torres G, Go K, Phinney DG. Interleukin 1 receptor antagonist mediates the antiinflammatory and antifibrotic effect of mesenchymal stem cells during lung injury. *Proc Natl Acad Sci USA*. 2007;104:11002–11007.

Ortiz LA, Gambelli F, McBride C, Gaupp D, Baddoo M, Kaminski N, Phinney DG. Mesenchymal stem cell engraftment in lung is enhanced in response to bleomycin exposure and ameliorates its fibrotic effects. *Proc Natl Acad Sci USA*. 2003;100:8407–8411.

Parr AM, Tator CH, Keating A. Bone marrow-derived mesenchymal stromal cells for the repair of central nervous system injury. *Bone Marrow Transplant*. 2007;40:609–619.

Pati S, Gerber MH, Menge TD, Wataha KA, Zhao Y, Baumgartner JA, Zhao J et al. Bone marrow derived mesenchymal stem cells inhibit inflammation and preserve vascular endothelial integrity in the lungs after hemorrhagic shock. *PloS One*. 2011a;6:e25171.

Pati S, Khakoo AY, Zhao J, Jimenez F, Gerber MH, Harting M, Redell JB et al. Human mesenchymal stem cells inhibit vascular permeability by modulating vascular endothelial cadherin/beta-catenin signaling. *Stem Cells Dev*. 2011b;20:89–101.

Pati S, Orsi SA, Moore AN, Dash PK. Intra-hippocampal administration of the VEGF receptor blocker PTK787/ZK222584 impairs long-term memory. *Brain Res*. 2009;1256:85–91.

Picinich SC, Mishra PJ, Glod J, Banerjee D. The therapeutic potential of mesenchymal stem cells. Cell- & tissue-based therapy. *Expert Opin Biol Ther*. 2007;7:965–973.

Poulsom R, Alison MR, Forbes SJ, Wright NA. Adult stem cell plasticity. *J Pathol*. 2002;197:441–456.

Poulsom R, Forbes SJ, Hodivala-Dilke K, Ryan E, Wyles S, Navaratnarasah S, Jeffery R et al. Bone marrow contributes to renal parenchymal turnover and regeneration. *J Pathol*. 2001;195:229–235.

Raffaghello L, Bianchi G, Bertolotto M, Montecucco F, Busca A, Dallegri F, Ottonello L, Pistoia V. Human mesenchymal stem cells inhibit neutrophil apoptosis: A model for neutrophil preservation in the bone marrow niche. *Stem Cells*. 2008;26:151–162.

Raghupathi R, Huh JW. Diffuse brain injury in the immature rat: Evidence for an age-at-injury effect on cognitive function and histopathologic damage. *J Neurotrauma*. 2007;24: 1596–1608.

Raghupathi R, Mehr MF, Helfaer MA, Margulies SS. Traumatic axonal injury is exacerbated following repetitive closed head injury in the neonatal pig. *J Neurotrauma*. 2004;21:307–316.

Raimondo F, Morosi L, Chinello C, Magni F, Pitto M. Advances in membranous vesicle and exosome proteomics improving biological understanding and biomarker discovery. *Proteomics*. 2011;11:709–720.

Rasmusson I, Le Blanc K, Sundberg B, Ringden O. Mesenchymal stem cells stimulate antibody secretion in human B cells. *Scand J Immunol*. 2007;65:336–343.

Resch JA, Villarreal V, Johnson CL, Elliott TR, Kwok OM, Berry JW, Underhill AT. Trajectories of life satisfaction in the first 5 years following traumatic brain injury. *Rehabil Psychol*. 2009;54:51–59.

Rogers TB, Pati S, Gaa S, Riley D, Khakoo AY, Patel S, Wardlow RD, 2nd et al. Mesenchymal stem cells stimulate protective genetic reprogramming of injured cardiac ventricular myocytes. *J Mol Cell Cardiol*. 2011;50:346–356.

Rojas M, Xu J, Woods CR, Mora AL, Spears W, Roman J, Brigham KL. Bone marrow-derived mesenchymal stem cells in repair of the injured lung. *Am J Respir Cell Mol Biol*. 2005;33:145–152.

Rubenfeld GD, Caldwell E, Peabody E, Weaver J, Martin DP, Neff M, Stern EJ, Hudson LD. Incidence and outcomes of acute lung injury. *N Engl J Med*. 2005;353:1685–1693.

Saatman KE, Creed J, Raghupathi R. Calpain as a therapeutic target in traumatic brain injury. *Neurotherapeutics*. 2010;7:31–42.

Sahay A, Scobie KN, Hill AS, O'Carroll CM, Kheirbek MA, Burghardt NS, Fenton AA, Dranovsky A, Hen R. Increasing adult hippocampal neurogenesis is sufficient to improve pattern separation. *Nature*. 2011;472:466–470.

Saunderson SC, Schuberth PC, Dunn AC, Miller L, Hock BD, MacKay PA, Koch N, Jack RW, McLellan AD. Induction of exosome release in primary b cells stimulated via CD40 and the IL-4 receptor. *J Immunol*. 2008;180:8146–8152.

Schmidt OI, Leinhase I, Hasenboehler E, Morgan SJ, Stahel PF. [The relevance of the inflammatory response in the injured brain.] *Der Orthopade*. 2007;36:248, 250–258.

Sheldon H, Heikamp E, Turley H, Dragovic R, Thomas P, Oon CE, Leek R et al. New mechanism for notch signaling to endothelium at a distance by delta-like 4 incorporation into exosomes. *Blood*. 2010;116:2385–2394.

Shlosberg D, Benifla M, Kaufer D, Friedman A. Blood-brain barrier breakdown as a therapeutic target in traumatic brain injury. Nature reviews. *Neurology*. 2010;6:393–403.

Simpson RJ, Lim JW, Moritz RL, Mathivanan S. Exosomes: Proteomic insights and diagnostic potential. *Expert Rev Proteomics*. 2009;6:267–283.

Stagg J. Immune regulation by mesenchymal stem cells: Two sides to the coin. *Tissue Antigens*. 2007;69:1–9.

Stagg J, Pommey S, Eliopoulos N, Galipeau J. Interferon-gamma-stimulated marrow stromal cells: A new type of nonhematopoietic antigen-presenting cell. *Blood*. 2006;107:2570–2577.

Taraboletti G, D'Ascenzo S, Borsotti P, Giavazzi R, Pavan A, Dolo V. Shedding of the matrix metalloproteinases MMP-2, MMP-9, and MT1-MMP as membrane vesicle-associated components by endothelial cells. *Am J Pathol*. 2002;160:673–680.

Taraboletti G, D'Ascenzo S, Giusti I, Marchetti D, Borsotti P, Millimaggi D, Giavazzi R, Pavan A, Dolo V. Bioavailability of VEGF in tumor-shed vesicles depends on vesicle burst induced by acidic pH. *Neoplasia*. 2006;8:96–103.

Thery C, Ostrowski M, Segura E. Membrane vesicles as conveyors of immune responses. Nature reviews. *Immunology*. 2009;9:581–593.

Timmers L, Lim SK, Arslan F, Armstrong JS, Hoefer IE, Doevendans PA, Piek JJ et al. Reduction of myocardial infarct size by human mesenchymal stem cell conditioned medium. *Stem Cell Res*. 2007;1:129–137.

Togel F, Hu Z, Weiss K, Isaac J, Lange C, Westenfelder C. Administered mesenchymal stem cells protect against ischemic acute renal failure through differentiation-independent mechanisms. *Am J Physiol. Renal Physiol*. 2005;289:F31–F42.0000

Uccelli A, Laroni A, Freedman MS. Mesenchymal stem cells for the treatment of multiple sclerosis and other neurological diseases. *Lancet Neurol*. 2011;10:649–656.

Vlassov AV, Magdaleno S, Setterquist R, Conrad R. Exosomes: Current knowledge of their composition, biological functions, and diagnostic and therapeutic potentials. *Biochim Biophys Acta*. 2012;1820:940–948.

Wang S, Qu X, Zhao RC. Mesenchymal stem cells hold promise for regenerative medicine. *Front Med*. 2011;5:372–378.

Watanabe J, Shetty AK, Hattiangady B, Kim DK, Foraker JE, Nishida H, Prockop DJ. Administration of TSG-6 improves memory after traumatic brain injury in mice. *Neurobiol Dis*. 2013;59:86–99.

Wrona RM. Disability and return to work outcomes after traumatic brain injury: Results from the Washington State Industrial Insurance Fund. *Disabil Rehabil*. 2010;32:650–655.

Xin H, Li Y, Buller B, Katakowski M, Zhang Y, Wang X, Shang X, Zhang ZG, Chopp M. Exosome-mediated transfer of MIR-133b from multipotent mesenchymal stromal cells to neural cells contributes to neurite outgrowth. *Stem Cells*. 2012;30:1556–1564.

Xin H, Li Y, Cui Y, Yang JJ, Zhang ZG, Chopp M. Systemic administration of exosomes released from mesenchymal stromal cells promote functional recovery and neurovascular plasticity after stroke in rats. *J Cereb Blood Flow Metab*. 2013;33:1711–1715.

Xu J, Qu J, Cao L, Sai Y, Chen C, He L, Yu L. Mesenchymal stem cell-based angiopoietin-1 gene therapy for acute lung injury induced by lipopolysaccharide in mice. *J Pathol*. 2008;214:472–481.

Xu J, Woods CR, Mora AL, Joodi R, Brigham KL, Iyer S, Rojas M. Prevention of endotoxin-induced systemic response by bone marrow-derived mesenchymal stem cells in mice. *Am J Physiol Lung Cell Mol Physiol*. 2007;293:L131–L141.

Yu B, Zhang X, Li X. Exosomes derived from mesenchymal stem cells. *Int J Mol Sci*. 2014;15:4142–4157.

9 Use of a Natural Modulator of Inflammation (TSG-6) for the Therapy of Traumatic Brain Injury

Darwin J. Prockop

CONTENTS

9.1 INTRODUCTION

Inflammation is now recognized as making a major contribution to the patho-etiology of many diseases including diseases of the central nervous system (CNS) such as traumatic brain injury (TBI) (Ransohoff and Brown, 2012). The inflammation can be helpful because it clears necrotic and apoptotic cells and other debris. It can also be extremely harmful in that the flood of molecular stimuli of inflammation can initiate a repetitive cycle of tissue destruction. Because of the potential harm generated by excessive inflammation in the sterile environment of most TBIs, a number of anti-inflammatory agents have been tested. The results have been either disappointing or inconclusive to date. Glucocorticoids were used clinically to decrease brain edema but failed in a large clinical trial because of increased mortality (Edwards et al., 2005; Roberts et al., 2004). Also, glucocorticoids were shown to aggravate retrograded memory deficits in a TBI model (Chen et al., 2009). Nonsteroidal anti-inflammatory drugs produced mixed results in models for TBI with some reports indicating improvements (Kovesdi et al., 2012; Thau Zuchman et al., 2012) but others indicating deleterious effects such as worsened cognitive outcomes (Browne et al., 2006). Strategies to reduce inflammation by targeting toll-like receptor (TLR) ligands, TLR receptors, or pro-inflammatory cytokines have also proven ineffective (Rivest, 2011).

We have tested a different strategy to combating excessive inflammation in a model for TBI by using TNF-α stimulated gene/protein 6 (TSG-6), a natural modulator of inflammation. TSG-6 is a multifunctional endogenous protein that is expressed by a variety of cells in response to stimulation by pro-inflammatory cytokines (Fulop et al., 1997; Milner et al., 2006; Szanto et al., 2004). The 35 kDa protein (Blundell et al., 2005) consists primarily of a N-terminal domain similar to the link module of proteoglycans, and a C-terminal domain with sequences similar to complement C1r/C1s, an embryonic sea urchin growth factor Uegf and BMP1 (CUB domain). TSG-6 binds to a large number of components of the extracellular matrix including hyaluronan, heparin, heparan sulfate, thrombospondins-1 and -2, fibronectin, and pentraxin (Baranova et al., 2011; Blundell et al., 2005; Kuznetsova et al., 2005, 2008; Mahoney et al., 2005). These interactions primarily act to stabilize the extracellular matrix.

In addition, TSG-6 modulates inflammatory and immune responses by several interactions, some of which are related to its effects on extracellular matrix but some of which appear to be independent. One of the more complex interactions is that the protein catalytically transfers the heavy chains of inter-α-trypsin inhibitor onto hyaluronan (Rugg et al., 2005). It thereby helps stabilize the extracellular matrix and collapses the polymeric brush of hyaluronan that surrounds many cells (Baranova et al., 2013; Evanko et al., 1999). Transfer of the heavy chains releases the bikunin core of the inter-α-trypsin inhibitor and thereby increases its activity in inhibiting the cascade of proteases released during inflammatory responses (Okroj et al., 2012; Scavenius et al., 2011). In apparently independent interactions, TSG-6 also reduces the migration of neutrophils through endothelial cells by its interactions with the cytokine CXCL8 (Cao et al., 2004; Dyer et al., 2014), forms a tertiary complex with murine mast cell trypases and heparin (Nagyeri et al., 2011), and inhibits FGF-2–induced angiogenesis through an interaction with pentraxin (Leali et al., 2012). In addition, TSG-6 either directly or through a complex with hyaluronan, binds to CD44 on resident macrophages in a manner that decreases TLR2/NF-κB signaling and modulates the initial phase of the inflammatory response of most tissues (Choi et al., 2011; Oh et al., 2010, 2012a,b). TSG-6 thereby reduces the large, second phase of inflammation that is frequently an excessive and deleterious response to sterile injuries (Prockop and Oh, 2012).

These and related observations stimulated interest in the therapeutic potentials of the TSG-6. For example, transgenic mice with localized overexpression of the gene in joints or cartilage had a decreased response to experimentally induced arthritis (Glant et al., 2002; Mindrescu et al., 2002). Conversely, mice with a knockout of the gene had increased susceptibility to proteoglycan-induced arthritis (Szanto et al., 2004). Also, administration of recombinant TSG-6 decreased experimentally induced arthritis in several different models (Bardos et al., 2001; Mindrescu et al., 2000). In addition, the recombinant protein decreased osteoblastogenesis and osteoclast activity (Mahoney et al., 2008, 2011).

Our own interest in the therapeutic potential of the protein was stimulated by our observations that enhanced expression of the protein by adult stem/progenitor cells, referred to as mesenchymal stem/stromal cells (MSCs), explained some of the beneficial effects observed after administration of the cells in animal models for myocardial infarction (Lee et al., 2009), chemical injury to the cornea

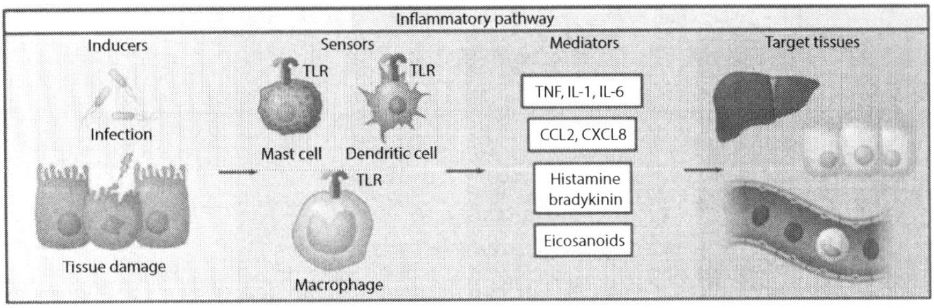

FIGURE 9.1 Summary of the inflammatory pathway as defined by Medzhitov (2010). Stimulator cells in the central nervous system have not been defined but may consist of mast cells or microglia. (Reproduced from Medzhitov R, *Cell* 2010, 140:771–776.)

(Oh et al., 2010, 2012a,b), zymosan-induced peritonitis (Choi et al., 2011), and LPS-induced or bleomycin-induced lung injury (Danchuk et al., 2011; Foskett et al., 2014). In the first three of these models, MSCs with an siRNA knockdown of the TSG-6 gene were ineffective, and recombinant TSG-6 reproduced most of the beneficial effects of MSCs. Of special interest was that the experiments in the model for corneal injury demonstrated that TSG-6 was effective only if administered early in the inflammatory response, i.e., during the first 4 to 6 hours of the initial stages of local cellular stimulation and before the large influx of neutrophils and macrophages that occurred later. In effect, it acted on the "sensor" cells (Figure 9.1) as defined by Medzhitov (2010). The molecular interactions of MSCs and TSG-6 with resident macrophages were defined in detail with experiments in the model for corneal injury (Oh et al., 2012a,b) and the model for peritonitis (Figure 9.2). The sensor cells in the central nervous system have not been identified but may consist of microglia or mast cells (Skaper et al., 2014).

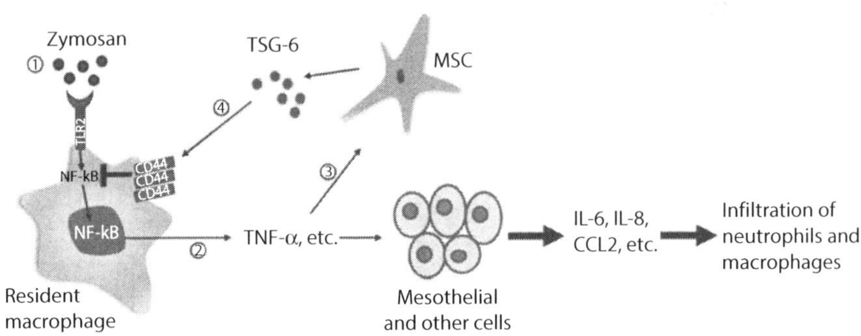

FIGURE 9.2 Anti-inflammatory action of hMSCs mediated mainly through TSG-6. (1) Zymosan activated macrophages via TLR2. (2) Activated NF-B increased the expression of pro-inflammatory cytokines. (3) HMSCs were activated by the pro-inflammatory cytokines to secrete TSG-6. (4) TSG-6 negatively regulated the TLR2-mediated responses through CD44. As indicated, "sensor" cells for the central nervous system that act like resident macrophages have not yet been identified but may be either mast cells or microglia. (Reproduced from Choi H et al., *Blood* 2011, 118(2):330–338.)

9.2 RECENT RESULTS WITH TSG-6 IN A MODEL FOR TRAUMATIC BRAIN INJURY

We recently tested the efficacy of TSG-6 in a mouse model for controlled cortical impact injury (CCI) (see Watanabe et al., 2013). The results will be summarized in brief here.

The administration of recombinant TSG-6 intravenously decreased inflammation in the brain as indicated by a decrease in matrix metalloproteinase 9 (MMP9) and neutrophil infiltration 1 day after the TBI (Figure 9.3). The intravenous administration of TSG-6 also preserved the blood-brain barrier 3 days after the TBI (Figure 9.4). The TSG-6 appeared to be slightly more effective than intravenous administration of 1 million human MSCs. A dose-response assay indicated that administration of 50 µg TSG-6 at 6 hours and again at 24 hours was more effective than a single administration of 50 µg of TSG-6 at 6 hours after TBI (Figure 9.4e). Surprisingly, the administration of TSG-6 within the first 24 hours of TBI had long-term effects on the behavior of the mice. There was an improvement in several Morris water maize tests as long as 40 days after the TBI (Figure 9.5a–d). There was also an improvement in the Y maze working memory test (Figure 9.5e).

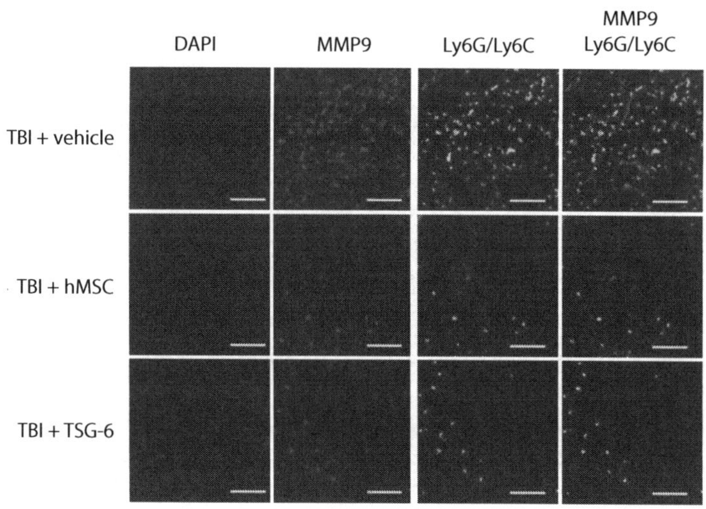

FIGURE 9.3 Neutrophils that infiltrated the brain and blood vessel endothelial cells expressed MMP9 at 24 hours after TBI. hMSCs (10^6 cells/mouse) or TSG-6 protein (50 µg/mouse) was administrated 6 hours after TBI. Representative sections showing double-immunofluorescence labeling of cortical sections from the injured mouse treated with vehicle, hMSC or TSG-6 at 24 hours after TBI. The images were taken from superficial pericontusional areas of cortex. Co-labeling of MMP9 (red) with marker for neutrophils (Ly6G/Ly6C, green). ×20 magnification (scale bars = 100 µm). (Reproduced from Watanabe J et al., *Neurobiol Dis* 2013, 59:86–99.)

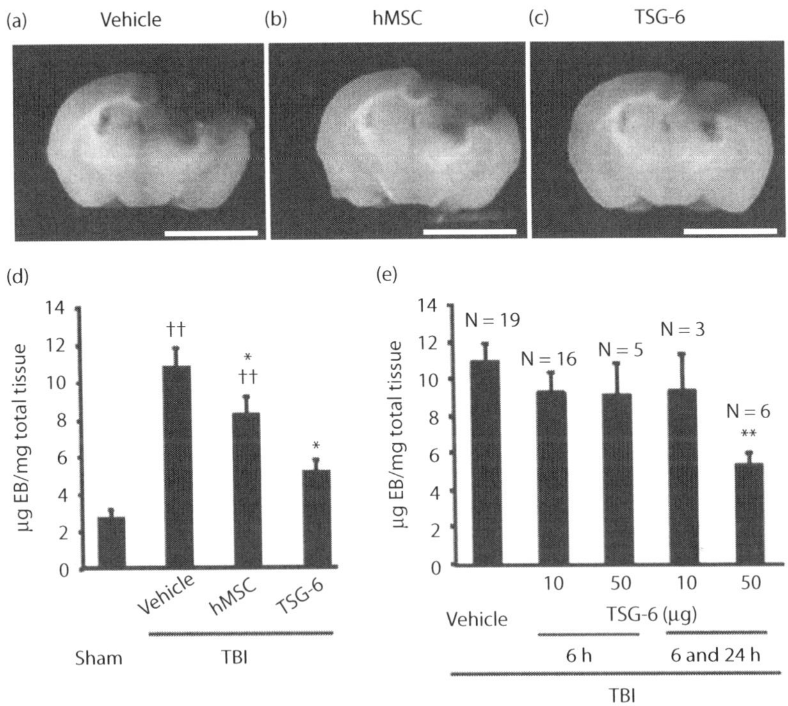

FIGURE 9.4 IV-injected hMSCs or TSG-6 protein decreased blood brain barrier (BBB) permeability in mice 3 days after TBI. (a–d) hMSCs (10^6 cells/mouse) were administrated 6 hours after TBI. TSG-6 protein at a dose of 50 µg/mouse was administered twice at 6 and 24 hours after TBI. (e) In assay, TSG-6 was administrated at 10 or 50 µg/mouse dose once or twice. (a–c) Representative brain slices of the site of cortical contusion injury after administration of vehicle (a), hMSCs (b), or TSG-6 (c) and recovered 3 days after TBI. Darker gray color represents Evans Blue dye extravasation at the site of injury. Scale bars = 5 mm. (d) Quantitative data of Evans Blue level in the ipsilateral cerebral hemisphere tissue of mice from sham operated group ($n = 8$) and injured group treated vehicle ($n = 21$), hMSCs ($n = 14$), or TSG-6 ($n = 6$) at 3 days after CCI. hMSCs (10^6 cells/mouse) were administrated 6 hours after TBI. TSG-6 protein at a dose of 50 µg/mouse was administered twice at 6 and 24 hours after TBI. (e) Dose-dependency and time window of TSG-6 treatment in BBB breakdown following TBI. Numbers of samples are indicated above the columns. Evans Blue dye was extracted from brain at 3 days after CCI. All data are represented as mean ± SEM. ††$P < .01$ versus the sham group, *$P < .05$, **$P < .01$ versus the vehicle group. (Reproduced from Watanabe J et al., *Neurobiol Dis* 2013, 59:86–99.)

9.3 DISCUSSION

We are approaching a consensus that TBI and several other CNS disorders initiate an inflammatory response to sterile injury of the tissues that is more harmful than helpful and compounds the pathology. This conclusion has triggered reexamination of the question of whether agents that suppress inflammation may have important

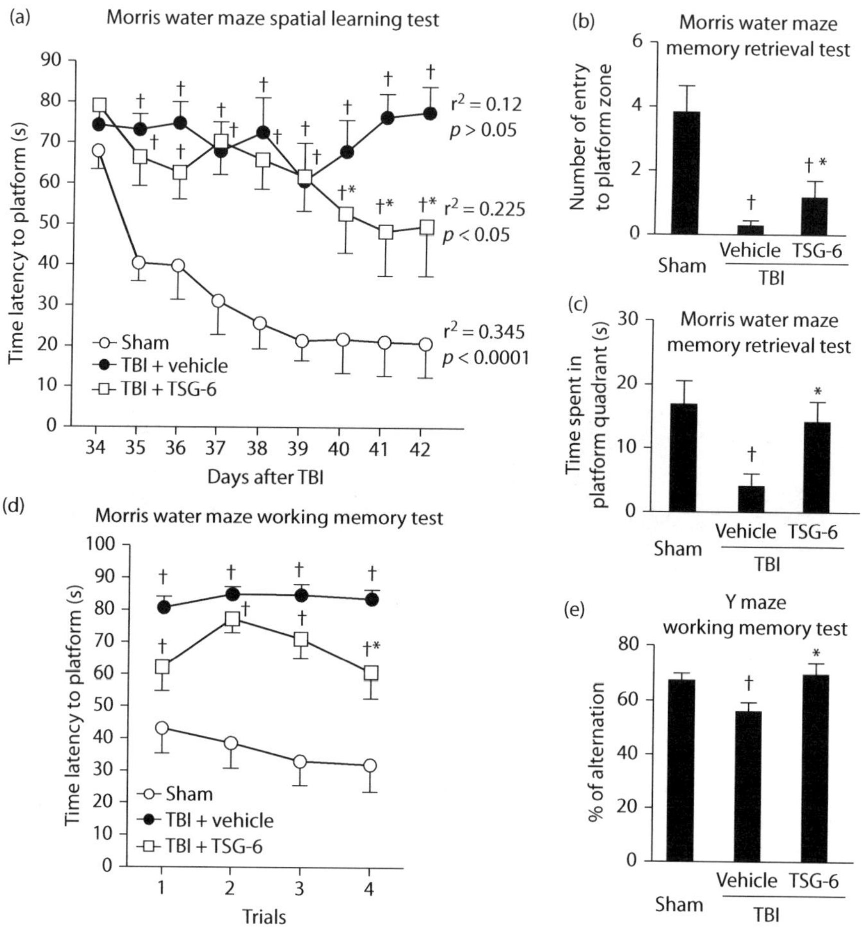

FIGURE 9.5 Protective effect of TSG-6 protein on cognitive function. TSG-6 protein (50 μg/mouse) was administered twice at 6 and 24 hours after TBI. Effect of TSG-6 on learning was assessed in (a) Morris water maze. (b,c) Probe (memory retention) test was performed at 24 h after the last learning session. The defects in working memory were assessed by (d) Morris water maze test and (e) Y-maze spontaneous alternation test. The Morris water maze memory retrieval test was performed 44 days after TBI, the Morris water maze working memory test was performed 48–51 days after TBI, and the Y maize working memory test 32 days after TBI. $n = 9$ or 10 mice/group. All data are represented as mean ± SEM. †$P < .05$ versus the sham group, *$p < .05$ versus the vehicle group. (Reproduced from Watanabe J et al., *Neurobiol Dis* 2013, 59:86–99.)

therapeutic effects. Unfortunately, the long history of the field indicates that the question is not easily answered for several reasons. One reason is the chronic nature of the diseases or, in the case of TBI of brain concussions, the long-term consequences of acute injuries. A second reason is that the readily available animal models do not reflect many aspects of the human disease. Rodents in particular heal peripheral tissues far more readily than man (Prockop et al., 2014), and the same is probably true for the CNS. As a result, it is extremely difficult to carry out laboratory experiments or clinical trials that will generate definitive data.

Within this context, any statements about the superiority of a given therapy must be made cautiously. Trials of standard steroids or nonsteroidal anti-inflammatory agents have been either disappointing or inconclusive, but the data do not rule out the possibility that one or more of such agents administered on the correct schedule may be highly effective. At the other extreme, there are currently several hundred citations in PubMed on the effectiveness of MSCs in animal models for a range of CNS diseases including TBI, stroke, spinal cord injury, parkinsonism, Alzheimer's disease, and cancers. These reports raise a series of additional issues, because there are no standardized protocols to prepare MSCs (Viswanathan et al., 2014), and there are marked differences in the properties of the MSCs from rodents and man (Prockop et al., 2014). We have recently observed that assays for expression of TSG-6 in culture-expanded MSCs may be a useful biomarker to predict the efficacy of human MSCs in suppressing inflammation in a mouse model (Lee et al., 2014). However, the results do not resolve all the issues encountered in clinical trials with MSCs.

Why, therefore, should TSG-6 be pursued as a potential therapy for TBI? One answer is that protein is a natural modulator of inflammation and that it has not to date shown any adverse effects such as those encountered by steroids. Another answer is that its mode of action differs from steroids and most anti-inflammatory agents in that it acts on the initial response of sensors to inflammatory signals (Figures 9.1 and 9.2). It has no effects on the late phase after massive neutrophil and macrophage infiltration that responds to standard anti-inflammatory agents, resolvins (Serhan et al., 2014), and perhaps other drugs. Therefore, TSG-6 may be most effective if used soon after the initial insult or if used in combination with other reagents to shut off a cycle of tissue injury and inflammation that produces further tissue injury and inflammation. And, in the end, we must probably recognize that we are attacking complex problems in trying to develop therapies for TBI and other diseases of the CNS. It is unlikely that any single agent will provide a panacea for all of them. Instead, we probably need an armamentarium of agents and strategies in which TSG-6 may well be an important component.

REFERENCES

Baranova NS, Nileback E, Haller FM, Briggs DC, Svedhem S, Day AJ, Richter RP: The inflammation-associated protein TSG-6 cross-links hyaluronan via hyaluronan-induced TSG-6 oligomers. *J Biol Chem* 2011, 286(29):25675–25686.

Baranova NS, Foulcer SJ, Briggs DC, Tilakaratna V, Enghild JJ, Milner CM, Day AJ, Richter RP: Inter-α-inhibitor impairs TSG-6-induced hyaluronan cross-linking [published online September 4, 2013]. *J Biol Chem* 2013, 288(41):29642–29653.

Bardos T, Kamath RV, Mikecz K, Glant TT: Anti-inflammatory and chondroprotective effect of TSG-6 (tumor necrosis factor-alpha-stimulated gene-6) in murine models of experimental arthritis. *Am J Pathol* 2001, 159(5):1711–1721.

Blundell CD, Almond A, Mahoney DJ, DeAngelis PL, Campbell ID, Day AJ: Towards a structure for a TSG-6 hyaluronan complex by modeling and NMR spectroscopy: Insights into other members of the link module superfamily. *J Biol Chem* 2005, 280(18): 18189–18201.

Browne K et al.: Chronic ibuprofen administration worsens cognitive outcome following traumatic brain injury in rats. *Exp Neurol* 2006, 201:301–307.

Cao TV, La M, Getting SJ, Day AJ, Perretti M: Inhibitory effects of TSG-6 link module on leukocyte-endothelial cell interactions in vitro and in vivo. *Microcirculation* 2004, 11(7):615–624.

Chen X et al.: Glucocorticoids aggravate retrograde memory deficiency associated with traumatic brain injury in rats. *J Neurotrauma* 2009, 26:253–260.

Choi H, Lee RH, Bazhanov N, Oh JY, Prockop DJ: Anti-inflammatory protein TSG-6 secreted by activated MSCs attenuates zymosan-induced mouse peritonitis by decreasing TLR2/NF-kappaB signaling in resident macrophages. *Blood* 2011, 118(2):330–338.

Danchuk S et al.: Human multipotent stromal cells attenuate lipopolysaccharide-induced acute lung injury in mice via secretion of tumor necrosis factor-α-induced protein 6. *Stem Cell Res Ther* 2011, 2(3):27.

Dyer DP, Thomson JM, Hermant A, Jowitt TA, Handel TM, Proudfoot AE, Day AJ, Milner CM: TSG-6 Inhibits neutrophil migration via direct interaction with the chemokine CXCL8 [published online February 4, 2014]. *J Immunol* 2014, 192:2177–2185.

Edwards P et al.: Final results of MRC CRASH, a randomised placebo-controlled trial of intravenous corticosteroid in adults with head injury-outcomes at 6 months. *Lancet* 2005, 365:1957–1959.

Evanko SP, Angello JC, Wight TN: Formation of hyaluronan- and versican-rich pericellular matrix is required for proliferation and migration of vascular smooth muscle cells. *Arterioscler Thromb Vasc Biol* 1999, 19(4):1004–1013.

Foskett AM, Bazhanov N, Ti X, Tiblow A, Bartosh TJ, Prockop DJ: Phase-directed therapy: TSG-6 targeted to early inflammation improves bleomycin-injured lungs. *Am J Physiol Lung Cell Mol Physiol* 2014,306(2):L120–L131.

Fulop C, Kamath RV, Li Y, Otto JM, Salustri A, Olsen BR, Glant TT, Hascall VC: Coding sequence, exon-intron structure and chromosomal localization of murine TNF-stimulated gene 6 that is specifically expressed by expanding cumulus cell-oocyte complexes. *Gene* 1997, 202(1–2):95–102.

Glant TT, Kamath RV, Bardos T, Gal I, Szanto S, Murad YM, Sandy JD, Mort JS, Roughley PJ, Mikecz K: Cartilage-specific constitutive expression of TSG-6 protein (product of tumor necrosis factor α-stimulated gene 6) provides a chondroprotective, but not antiinflammatory, effect in antigen-induced arthritis. *Arthritis Rheum* 2002, 46(8): 2207–2218.

Kovesdi E. et al.: Acute minocycline treatment mitigates the symptoms of mild blast-induced traumatic brain injury. *Front Neurol* 2012, 3:111.

Kuznetsova SA, Day AJ, Mahoney DJ, Rugg MS, Mosher DF, Roberts DD: The N-terminal module of thrombospondin-1 interacts with the link domain of TSG-6 and enhances its covalent association with the heavy chains of inter-α-trypsin inhibitor. *J Biol Chem* 2005, 280(35):30899–30908.

Kuznetsova SA, Mahoney DJ, Martin-Manso G, Ali T, Nentwich HA, Sipes JM, Zeng B, Vogel T, Day AJ, Roberts DD: TSG-6 binds via its CUB_C domain to the cell-binding domain of fibronectin and increases fibronectin matrix assembly. *Matrix Biol* 2008, 27(3):201–210.

Leali D et al.: Long pentraxin 3/tumor necrosis factor-stimulated gene-6 interaction: A biological rheostat for fibroblast growth factor 2-mediated angiogenesis. *Arterioscler Thromb Vasc Biol* 2012, 32(3):696–703.

Lee RH, Pulin AA, Seo MJ, Kota DJ, Ylostalo J, Larson BL, Semprun-Prieto L, Delafontaine P, Prockop DJ: Intravenous hMSCs improve myocardial infarction in mice because cells embolized in lung are activated to secrete the anti-inflammatory protein TSG-6. *Cell Stem Cell* 2009, 5(1):54–63.

Lee RH, Yu JM, Foskett AM, Peltier G, Reneau JC, Bazhanov N, Oh JY, Prockop DJ: TSG-6 as a biomarker to predict efficacy of human mesenchymal stem/progenitor cells (hMSCs) in modulating sterile inflammation in vivo [published online November 10, 2014]. *Proc Natl Acad Sci USA* 2014, 111(47):16766–16771.

Mahoney DJ, Mikecz K, Ali T, Mabilleau G, Benayahu D, Plaas A, Milner CM, Day AJ, Sabokbar A: TSG-6 regulates bone remodeling through inhibition of osteoblastogenesis and osteoclast activation. *J Biol Chem* 2008, 283(38):25952–25962.

Mahoney DJ, Mulloy B, Forster MJ, Blundell CD, Fries E, Milner CM, Day AJ: Characterization of the interaction between tumor necrosis factor-stimulated gene-6 and heparin: Implications for the inhibition of plasmin in extracellular matrix microenvironments. *J Biol Chem* 2005, 280(29):27044–27055.

Mahoney DJ, Swales C, Athanasou NA, Bombardieri M, Pitzalis C, Kliskey K, Sharif M, Day AJ, Milner CM, Sabokbar A: TSG-6 inhibits osteoclast activity via an autocrine mechanism and is functionally synergistic with osteoprotegerin. *Arthritis Rheum* 2011, 63(4):1034–1043.

Medzhitov R: Inflammation 2010: New adventures of an old flame. *Cell*. 2010, 140:771–776.

Milner CM, Higman VA, Day AJ: TSG-6: A pluripotent inflammatory mediator? *Biochem Soc Trans* 2006, 34(Pt 3):446–450.

Mindrescu C, Dias AA, Olszewski RJ, Klein MJ, Reis LF, Wisniewski HG: Reduced susceptibility to collagen-induced arthritis in DBA/1J mice expressing the TSG-6 transgene. *Arthritis Rheum* 2002, 46(9):2453–2464.

Mindrescu C, Thorbecke GJ, Klein MJ, Vilcek J, Wisniewski HG: Amelioration of collagen-induced arthritis in DBA/1J mice by recombinant TSG-6, a tumor necrosis factor/interleukin-1-inducible protein. *Arthritis Rheum* 2000, 43(12):2668–2677.

Nagyeri G, Radacs M, Ghassemi-Nejad S, Tryniszewska B, Olasz K, Hutas G, Gyorfy Z, Hascall VC, Glant TT, Mikecz K: TSG-6 protein, a negative regulator of inflammatory arthritis, forms a ternary complex with murine mast cell tryptases and heparin. *J Biol Chem* 2011, 286(26):23559–23569.

Oh JY, Choi H, Lee RH, Roddy GW, Ylöstalo JH, Wawrousek E, Prockop DJ: Identification of the HSPB4/TLR2/NF-κB axis in macrophage as a therapeutic target for sterile inflammation of the cornea [published online February 22, 2012]. *EMBO Mol Med* 2012a, 4(5):435–448.

Oh JY, Lee RH, Yu JM, Ko JH, Lee HJ, Ko AY, Roddy GW, Prockop DJ: Intravenous mesenchymal stem cells prevented rejection of allogeneic corneal transplants by aborting the early inflammatory response. *Mol Ther* 2012b, 20(11):2143–2152.

Oh JY, Roddy GW, Choi H, Lee RH, Ylostalo JH, Rosa RH, Jr., Prockop DJ: Anti-inflammatory protein TSG-6 reduces inflammatory damage to the cornea following chemical and mechanical injury. *Proc Natl Acad Sci USA* 2010, 107(39):16875–16880.

Okroj M, Holmquist E, Sjolander J, Corrales L, Saxne T, Wisniewski HG, Blom AM: Heavy chains of inter alpha inhibitor (IαI) inhibit the human complement system at early stages of the cascade. *J Biol Chem* 2012, 287(24):20100–20110.

Prockop DJ, Oh JY: Mesenchymal stem/stromal cells (MSCs): Role as guardians of inflammation. *Mol Ther* 2012, 20(1):14–20.

Prockop DJ, Prockop SE, Bertoncello I: Are clinical trials with mesenchymal stem/progenitor cells too far ahead of the science? Lessons from experimental hematology. *Stem Cells* 2014, 32(12):3055–3061.

Ransohoff, R, Brown, M: Innate immunity in the central nervous system. *J Clin Invest* 2012, 122:1164–1171.

Rivest, S: The promise of anti-inflammatory therapies for CNS injuries and diseases. *Expert Rev Neurother* 2011, 11:783–786.

Roberts I et al.: Effect of intravenous corticosteroids on death within 14 days in 10008 adults with clinically significant head injury (MRC CRASH trial): Randomised placebo-controlled trial. *Lancet* 2004, 364:1321–1328.

Rugg MS, Willis AC, Mukhopadhyay D, Hascall VC, Fries E, Fulop C, Milner CM, Day AJ: Characterization of complexes formed between TSG-6 and inter-α-inhibitor that act as intermediates in the covalent transfer of heavy chains onto hyaluronan. *J Biol Chem* 2005, 280(27):25674–25686.

Scavenius C, Sanggaard KW, Nikolajsen CL, Bak S, Valnickova Z, Thogersen IB, Jensen ON, Hojrup P, Enghild JJ: Human inter-α-inhibitor is a substrate for factor XIIIa and tissue transglutaminase. *Biochim Biophys Acta* 2011, 1814(12):1624–1630.

Serhan CN, Chiang N, Dalli J, Levy BD: Lipid mediators in the resolution of inflammation [published online October 30, 2014]. *Cold Spring Harb Perspect Biol* 2014, 30.

Skaper SD, Facci L, Giusti P: Neuroinflammation, microglia and mast cells in the pathophysiology of neurocognitive disorders: A review [published online November 30, 2014]. *CNS Neurol Disord Drug Targets* 2014, 13:1654–1666.

Szanto S, Bardos T, Gal I, Glant TT, Mikecz K: Enhanced neutrophil extravasation and rapid progression of proteoglycan-induced arthritis in TSG-6-knockout mice. *Arthritis Rheum* 2004, 50(9):3012–3022.

Thau Zuchman O et al.: The anti-inflammatory drug carprofen improves long-term outcome and induces gliogenesis after traumatic brain injury. *J Neurotrauma* 2012, 29:375–384.

Viswanathan S, Keating A, Deans R, Hematti P, Prockop D, Stroncek DF, Stacey G, Weiss DJ, Mason C, Rao MS. Soliciting strategies for developing cell-based reference materials to advance mesenchymal stromal cell research and clinical translation [published online March 10, 2014]. *Stem Cells Dev* 2014, 23(11):1157–1167.

Watanabe J, Shetty AK, Hattiangady B, Kim DK, Foraker JE, Nishida H, Prockop DJ: Administration of TSG-6 improves memory after traumatic brain injury in mice [published online July 11, 2013]. *Neurobiol Dis* 2013 59:86–99.

10 Tissue Engineering and Cell-Based Therapeutics for Repair of Myelomeningocele

KuoJen Tsao and Luke R. Putnam

CONTENTS

10.1 INTRODUCTION

Spina bifida, commonly known as myelomeningocele (MMC), is one of the most common birth defects in the United States that portends lifelong childhood paralysis. Affecting 1 in 3000 live births in the United States (Boulet et al., 2008), MMC results in significant long-term morbidity and healthcare costs (Botto et al., 1999). Resulting from the incomplete closure of the neural tube, children with MMC suffer from lower motor and sensory dysfunction due to spinal cord exposure to the intrauterine environment. Clinical manifestations include paralysis, bowel and bladder incontinence, and musculoskeletal deformities, as well as cognitive and cranial nerve disabilities due to hindbrain herniation. With lifelong social and medical care dependence, healthcare costs for children with MMC are 13 times greater than those of other children, with an estimated cost of $200,000,000 annually in the United States alone (*MMWR Morb Mortal Wkly Rep*, 1989; Ouyang et al., 2007).

Children with MMC exhibit a wide spectrum of disease severity and clinical outcomes. The prevalence of MMC is more common in females than males

(Kallen et al., 1994) and has been linked to ethnic and geographic factors with higher prevalence rates in countries such as Great Britain, Pakistan, and Egypt, and lower rates in Japan, Israel, and Finland. Within the United States, MMCs are more commonly diagnosed in eastern and southern regions (Harmon et al., 1995) and among Hispanic populations (Shaw et al., 1994). Of note, the influence of termination rates after prenatal diagnosis is unknown and poorly reported (Cragan et al., 1995) but has been demonstrated to show regional variation (Shurtleff, 2004), which may account for some of the geographic differences in prevalence.

Lifelong disabilities are common and are highly dependent on the level of the neural tube defect (NTD) lesion, with higher spinal cord lesions portending worse outcomes. Approximately 80% of MMC lesions occur in the lumbosacral region (Rintoul et al., 2002), which contains nerve roots that control lower extremity function as well as maintaining bowel, bladder, and sexual function. Given these common disabilities, only half of MMC patients in the United States who survive to adulthood are able to live independently despite intelligence quotient scores usually greater than 80 (Hunt, 1990). Tragically, 30% of MMC patients who progress beyond childhood die from respiratory, urinary, or central nervous system (CNS) complications (Saadai et al., 2012). This wide variation in outcomes suggests that current treatment options are limited and fail to address the underlying causes of neurodevelopmental dysfunction.

10.2 ETIOLOGY

Although spina bifida occurs as an isolated defect in the majority of cases, the etiology of MMC is unclear. Hypothesized to be of multifactorial pathophysiology, genetics, nutrition, and environment factors may all play causative roles in MMC development (Shaw et al., 1994). Maternal influences, including pregestational diabetes, antiepileptics or folic acid inhibitor exposure early in pregnancy, a previously affected pregnancy with the same partner, and inadequate maternal folic acid intake have also been implicated. In 1992, the Centers for Disease Control and Prevention (CDC) released a study demonstrating that women with 400 micrograms of daily folate intake for 3 months prior to conception have a 70%–80% relative reduction of spina bifida in their offspring (CDC, 1992). Subsequently, folate fortification of cereal grains was mandated with a reported 26% decrease in anencephaly and MMC combined (CDC, 2004). Despite better knowledge about MMC and successful interventions to decrease the incidence, MMC remains a common birth defect with significant perinatal morbidity and mortality.

The MMCs are NTDs that result from the derangement of normal embryonic neurulation. The NTDs result from the failure of the anterior and posterior neuropores to close between 25 and 28 days postovulation. Through a coordinated series of events, the neural plate, neural folds, and neural tube develop into the future brain and spinal cord. Depending on the degree of primary and/or secondary neurulation, NTDs can be open or closed defects. Open NTDs (ONTDs) can occur in any area of the CNS as a result of deviations in primary neurulation. These tend to exhibit worse neurological pathophysiology due to the exposed neural elements to the fetal environment. The resulting ONTD has no skin coverage, allowing for α-fetoprotein (AFP) to leak from the cerebrospinal fluid (CSF) into the amniotic fluid. Consequently, an increased AFP level in the maternal serum serves as a prenatal screening biomarker. With the addition of classical intra- and extracranial features on prenatal ultrasonography, many MMCs can be detected prenatally, allowing for potential fetal therapeutic options in the treatment of spina bifida. Conversely, closed NTDs result from secondary neurulation derangements, which are mostly limited to the spine, and are more protected with minimal to no neurologic manifestations. Cranial/cervical manifestations of NTDs include anencephaly, craniorachischisis totalis, encephaloceles, meningoencephaloceles, and MMCs. Distal spinal NTDs include spina bifida aperta, myeloschisis, congenital dermal sinus, lipomyelomeningoceles, MMCs, and caudal agenesis. The MMC or spina bifida cystica aperta is the most severe form of NTD in which the spinal cord protrudes through the spinal canal and is covered only by its meningeal membranes.

10.3 PATHOPHYSIOLOGY

Poor neurologic outcomes associated MMC stem from the "two-hit" hypothesis, first described by Heffez et al. in 1990. The "first hit" is the failure of neurulation and proper neuronal development that occurs early in gestation, while the "second hit" is the ongoing in utero insults to the exposed spinal cord from prolonged exposure to the amniotic environment. This theory was supported by Patten's study in the 1950s that found undamaged neural tissue in early gestation human embryos and fetuses with MMCs (Patten, 1953). Histologic studies showed worsening neural damage to the fetal spinal cord as gestation progressed (Korenromp et al., 1986; Hutchins et al., 1996; Meuli et al., 1997). Clinical observations of fetuses with MMC demonstrated normal leg movements up to 16–17 weeks' gestation and subsequent lost function as the pregnancy progressed (Korenromp et al., 1986). Stiefel et al. described temporal and histologic changes associated with the two-hit hypothesis utilizing a mouse model of naturally occurring lumbosacral NTDs (Stiefel et al., 2007). Failure of the neural tube to close was the only abnormality seen in early gestation. Neuronal connections between the spinal cord and peripheral targets including the hind limbs remained intact. As gestation continued, progressive deterioration of exposed spinal cord contents to nearly complete and irreversible loss by the time of birth was demonstrated. Ongoing damage to the traumatic and/or toxic processes was attributed to exposure within the amniotic cavity during the remainder of the pregnancy, which was consistent with other studies (Heffez et al., 1990, 1993; Meuli et al., 1995; Correia-Pinto et al., 2002).

The other major neurological derangement associated with NTDs is the Arnold-Chiari II malformation (ACM). Present in nearly all MMC patients, ACM represents hindbrain herniation due to the continual CSF leak from the ONTD. As a result, a cerebral pressure gradient and the downward displacement of the cerebellar tonsils through the foramen magnum develops, ultimately leading to hydrocephalus. Motor, cranial-nerve, and cognitive functions can all be affected by this anatomic and pathophysiologic derangement (Saadai and Farmer, 2012). Historically, approximately 90% of patients with MMC and hydrocephalus require surgical decompression of CSF with a ventriculoperitoneal (VP) shunt or other cerebral fluid shunting procedure. The propensity for shunt placement correlates with the anatomic location of the defect. Almost 100% of patients with thoracic level lesions require postnatal shunts, whereas those with lumbar and sacral level lesions are 88% and 68%, respectively (Rintoul et al., 2002). Although shunts may be necessary, they are associated with their own complications. Nearly half of all patients who undergo shunt placement suffer from shunt complications within the first year of placement due to mechanical or infectious causes (Caldarelli et al., 1996). This combination of MMC pathophysiology and the surgical morbidity associated with shunting emphasizes the detrimental effects of hindbrain herniation and hydrocephalus.

Traditional treatment for MMC is initiated after birth. Despite early postnatal intervention with closure of the NTD, the nerve damage that occurs during gestation is irreversible. Repair of the MMC generally occurs 24–72 hours postpartum and aims to preserve viable neural tissue, reconstitute a normal anatomic environment, and minimize the risk of infection. Shunt procedures occur dependent on the degree of hydrocephalus. Although progression may be halted, the ACM still exists, and the sequelae of hindbrain herniation with subsequent risks of shunting remain. Despite aggressive postnatal management, conventional treatment results in 14%–35% 5-year mortality (Worley et al., 1996). Unfortunately, traditional postnatal treatment only addresses the symptoms associated with MMCs with attempts to preserve function and anatomy only.

10.4 FETAL SURGERY

Recognition of the therapeutic limitations of conventional treatment and the natural history of disease progression during gestation precipitated interest in advancing therapy before birth. If MMC pathophysiology results from progress neurological damage throughout pregnancy, perhaps intervention prior to the "second hit" could prevent or ameliorate irreversible insults. Early coverage of the exposed spinal cord could protect it from ongoing exposure to amniotic fluid and uterine trauma as well as prevention of continual CSF leakage. Animal-model experiments in fetal surgery focused on a surgically created MMC lesions and subsequent in utero repair (Heffez et al., 1990; Meuli et al., 1995; Paek et al., 2000; Bouchard et al., 2003). Investigators demonstrated similar neurologic outcomes in their animal models as found in human newborns with MMC. However, if in utero repair was performed prior to irreversible neural damage, live births remained neurologically intact. Although fetal surgery could not reverse the initial non-neurulation of the spinal cord, early intervention would prevent the progressive neurologic deterioration and improve postnatal neurologic outcomes.

After promising results in animal models, fetal surgery for MMC advanced to human subjects and culminated in the Management of Myelomeningocele Study (MOMS) sponsored by the National Institutes of Health, published in 2011 (Adzick et al., 2011). With a primary outcome of death and need for VP shunt at 1 year of life, the randomized, controlled trial also evaluated neurologic function at 30 months. Predicted neurologic function based on the level of the MMC lesion was compared between prenatal surgery to conventional postnatal care. Utilizing an open surgical technique (maternal laparotomy and hysterotomy), the MMC defects were repaired

TABLE 10.1
Inclusion and Major Exclusion Criteria for MOMS Trial

Inclusion Criteria	Exclusion Criteria
1. Singleton pregnancy	1. Fetal anomaly unrelated to MMC
2. MMC with upper boundary located between T1 and SI	2. Severe kyphosis
3. Evidence of hindbrain herniation	3. Risk of preterm birth a. Short cervix b. Previous preterm birth
4. Gestational age of 19.0–25.6 weeks at randomization	4. Placental abruption
5. Normal karyotype	5. Body Mass Index of 35 or more
6. U.S. residency	6. Contraindication to surgery
7. Maternal age 18 years or older	a. Previous hysterotomy in the active uterine segment

with a dural closure, followed by closure of the paraspinal myofascial flaps and the fetal skin, when possible. Of the more than 1000 pregnant women screened, 183 (61%) met strict study criteria and underwent randomization (Table 10.1). Shortly after a fourth interim analysis, the data and safety monitoring committee recommended termination of the trial after 158 randomized women demonstrated significant benefits found in the prenatal surgery group. This included decreased mortality and need for VP shunts at 12 months (relative risk [RR] 0.70; 95% confidence interval (CI) 0.58–0.84; $p < .001$) as well as improved hindbrain herniation in fetuses who underwent prenatal repair (64% versus 96%; RR 0.67; 95% CI: 0.56–0.81; $p < .001$). Additionally, neurologic function was significantly better in the prenatal surgery group. More children walked independently after fetal surgery (42% versus 21%; RR 2.01; 95% CI 1.16–3.48), with better than expected motor function relative to the level of lesion. In fact, 32% of fetal surgery patients demonstrated two or more levels better than predicted compared to nonfetal surgery patients at 12%. This suggested that not only does fetal surgery prevent ongoing neurological damage, but it potentially could reverse some of the sequelae of the "second hit" with earlier intervention in the in utero environment.

Although the results of the MOMS trial were encouraging and offered a viable alternative to traditional treatment, fetal surgery is not the "cure-all" for MMCs. Prenatal repair is limited in its application with strict eligibility criteria. In addition to significant maternal morbidity, the majority of prenatal repair patients still suffered from significant neurological morbidity with 40% requiring cerebral shunts and only 42% walking independently. In addition, with less than 20% of screened patients eligible for fetal surgery, most patients do not qualify and do not benefit from this treatment modality.

10.5 ENGINEERED TISSUE APPROACHES FOR REPAIR OF MYELOMENINGOCELE

Investigational goals pursuing tissue engineering approaches of MMC repair have focused on the creation of barrier constructs overlying neuronal tissue to prevent tethering of the spinal cord, providing neurotropic factors to regenerate neural tissue

(Fauza et al., 2008; Saadai et al., 2011, 2013; Herrera et al., 2012; Brown et al., 2014) as well as prevention of ongoing CSF leakage that may lead to hindbrain herniation (Eggink et al., 2005, 2006; Fontecha et al., 2009; Watanabe et al., 2010; Fontecha et al., 2011; Watanabe et al., 2011; Peiro et al., 2013). Utilizing the surgically created sheep model as well as retinoic acid–induced rat model for MMC, several types of scaffolds have been utilized as viable alternatives to skin closure to protection of exposed neural elements. Endogenous materials such as collagen or gelatin, synthetic materials include silicone or polypropylene with high-density polyethylene, and allogeneic materials like decellularized small intestinal submucosa have all been evaluated (Eggink et al., 2005, 2006, 2008; Fontecha et al., 2009, 2011; Peiro et al., 2013; Watanabe et al., 2010, 2011).

Eggrink et al. evaluated different scaffolds using the surgical sheep model (Eggink et al., 2005, 2006, 2008). Utilizing collagen-based as well as small intestinal submucosa, they demonstrated that coverage of the MMC defect was possible with almost complete coverage of the lesion. Scaffolds were well incorporated into native tissue with minimal histological damage to the underlying spinal cord. There were no significant differences among type of matrix or neurological outcomes (Eggink et al., 2006, 2008).

Fontecha et al. demonstrated that tissue and defect coverage only could be adequate to preserve neurological function (Fontecha et al., 2009, 2011). Utilizing a silicone-based patch, the investigators secured the scaffold to the fetal defect with surgical sealant. All untreated animals were unable to walk and had sphincter incontinence. The lesions remained open with histological damage to the spinal cord, and a severe hindbrain herniation. All covered animals were able to walk, maintain sphincter continence, and showed almost complete closure of the defect with regeneration of several soft tissue layers with minimum hindbrain herniation (Fontecha et al., 2009). These results suggested that a formal three-layer closure of the MMC may not be necessary with a fetal repair approach.

Watanabe et al. introduced a gelatin-based scaffold as a replacement for dura mater in fetal MMC repair (Watanabe et al., 2010). Initial studies included in vitro scaffold creation which was subsequently applied to the retinoic acid-induced rat model of MMC using an open fetal surgical technique (Watanabe et al., 2010, 2011). Utilizing gelatin-hydrogel constructs of sheets and sponges with and without basic fibroblast growth factor (bFGF) over the defect in the retinoic acid-induced fetal rat MMC model, the scaffold demonstrated therapeutic potential with evidence of native fetal tissue migration, epidermal ingrowth, and neovascularization within and around the constructs. Overlying the scaffold was an extracellular matrix consisting of type 1 collagen and hyaluronic acid. Within the sponges, there was hyaluronic acid with nonepithelial cells. The constructs promoted underlying neovascularization with epidermal ingrowth, which were significantly more prominent with bFGF than without (Watanabe et al., 2010). This offered potential for manipulation of wound healing and restorative process if trophic mediators could be exogenously provided or endogenously promoted.

This strategic approach was then applied to the surgical sheep model, where 5 cm MMC defects, encompassing 5 lumbar vertebral, were created at 70 days' gestation. At 100 days' gestation, gelatin-based composites of sponges and sheets were utilized to cover the defect without additional skin closure. Pregnancies were continued until 140 days' gestation, where fetuses were harvested and evaluated. They demonstrated that gelatin scaffolds induced wound healing features that mimicked the phases of proliferation and remodeling with migration of local fibroblasts, reepithelialization from the wound edges, and neovascularization (Watanabe et al., 2014). Scaffolds treated with bFGF promoted improved coverage of the MMC defect compared to scaffolds without bFGF, with more mature, consistent granulation tissue and epithelial tissue across the entire defect. On histological sectioning, scaffolds containing bFGF demonstrated better preservation of the spinal cord, with less associated damage to the white matter compartment.

These studies have demonstrated the feasibility to cover the MMC defect and potentially preserve any residual neurological function. However, significant or complete rescue or regeneration of the spinal cord in MMC had not been demonstrated. Attempts to protect exposed spinal cord, while potentially augmenting neural regeneration, would be the natural progression. Whereas synthetic scaffolds have been utilized to protect neural elements as well as prevent tethered cord by replacing the dura, biodegradable scaffolds may offer some advantages with potential regeneration of neuronal cells, while maintaining the protective aspects of the scaffold. In addition, biological scaffolding offers the potential for cell-based therapies by providing direct exposure to the spinal cord.

Herrera et al. compared a cellulose-based, scaffold repair only with a standard three-layer technique with the fetal sheep (Herrera et al., 2012). By placing a cellulose patch on the spinal cord and then closing the skin, there was neo-dura mater formation, separation of nervous tissue from adjacent muscles, and preservation of the posterior funiculus and gray matter. Although there were no functional assessments, this offered a simplified surgical technique with potential regenerative properties.

Saadai et al. placed biodegradable nanofibrous scaffolds above and below the closed dura mater (Saadai et al., 2011). Utilizing the sheep MMC model, scaffolds, consisting of poly-L-lactic acid, were incorporated into surrounding tissue without evidence of an inflammatory response or foreign-body reaction. This offered the potential for delivery of growth factors, drugs or stem cells, as well as neural cell infiltration, which potentially provide greater potential for functional neurologic recovery.

Autologous materials, such as amniotic membranes (AMs), have also been used in experimental MMC repair (Brown et al., 2014). Derived from the cells of the developing fetus, the AM has been shown to demonstrate anti-inflammatory and epithelialization-promoting properties (Liu et al., 2010; Mamede et al., 2012). Its application in wound healing and tissue replacement in ophthalmology is well documented (Mermet et al., 2007; Kersey and Vivian, 2008; Liu et al., 2010; Mamede et al., 2012). Such properties would seem ideal for repair of MMC. Brown et al. hypothesized that the utilization of autologous AM would provide superior outcomes compared to routine skin closure of MMC defects (Brown et al., 2014). Utilizing the fetal sheep model, autologous AM was placed between spinal cord and skin closure with the chorion side on the exposed neural elements. Results with this strategy were mixed. The AM patches remained intact in all cases but failed to incorporate native tissue. There were minimal amounts of scar tissue with histological preservation of spinal cord and neural architecture. Compared to skin-only repairs, the AM patches failed to promote skin healing over the patch. However, skin-only sheep demonstrated significant destruction of the spinal cord tissue with loss of all neurons. Despite these suboptimal results, AM remains a promising adjunct to MMC repair with its properties of immune system modulation, inhibition of endothelial cell proliferation, and promotion of angiogenesis (Hao et al., 2000; Li et al., 2005).

10.6 STEM CELLS AS AN ADJUNCT FOR REPAIR OF MMC

Currently, the clinical options for spina bifida therapy include conventional postnatal care and prenatal surgery in a select group of patients. Despite improvement in outcomes, the vast majority of patients does not qualify or do not benefit from fetal surgery (Adzick et al., 2011). As biologic alternatives are investigated to cover the MMC defect, the potential role of cellular-based therapies to augment prenatal therapy to improve MMC-associated neuronal damage has grown. Stem cells have emerged as a promising therapeutic modality for several neurodegenerative or neuro-injury conditions due to their reparative function (Liu et al., 2000; Cao et al., 2002; Blesch and Tuszynski, 2004; Setoguchi et al., 2004; Koda et al., 2005; Sykova et al., 2006). Naturally, this strategy has evolved to potential therapeutic interventions for MMC (Lee et al., 2004; Li et al., 2012; Saadai et al., 2013). The role of stem cell–based therapies remains promising, especially within the in utero environment that may provide a natural immunotolerance (Merianos et al., 2008). Different strategies and cell lines have been investigated for prenatal therapies of spina bifida.

10.6.1 Human Embryonic Stem Cells

Utilizing the chick embryo model of NTDs, injected human embryonic stem cells (hESCs) into the amniotic cavity have been shown to promote closure of the defect. Sim et al. demonstrated the chronological changes of reclosure of surgically induced spinal NTDs of chick embryos were associated with the length of the defect (Sim et al., 2000). The reclosure capacity of NTDs was directly associated with the proliferative activity of the neural tube, suggesting that this process could be augmented if cell dynamics could also be manipulated. Lee et al. reported that intra-amniotic injection of hESCs in chick embryos could enhance reclosure capacity of surgically induced spinal NTDs (Lee et al., 2004, 2006). The neural tubes of chick embryos were longitudinally opened for a length of 6 somites at the wing bud level. The hESCs were injected into the amniotic cavity at 24 hours. The sizes of NTDs at reincubation days 3, 5, and 7 were compared between control and hESC-injected groups. Lengths of defects were significantly smaller in the hESC cell group than in control and vehicle groups from postoperative day 3–7, demonstrating enhanced reclosure of the NTDs. Although it was speculated that the enhancement of reclosure capacity may be due to cell replacement, the investigators demonstrated that the hESCs were not inside neural tube tissue. Cells were shown to be on the NTD surface and adjacent surface ectoderm, beyond the level of the embryo surface. These findings suggested the mechanism of therapy was delivery of molecules that promoted repair of structure or the regaining of function and a bridging effect by filling the injury site with injected hESCs to protect injured tissue to promote regeneration. These were some of the first studies that suggested the possibility of cellular-based therapies in the prenatal intervention of human NTDs.

10.6.2 Neural Stem Cells

Neural stem cells (NSCs) are an integral part of neural development. As a primordial cell line of the nervous system, NSCs promote neural development properties in response to various CNS derangements (Snyder et al., 1992; Whittemore and Snyder, 1996; Teng et al., 2002; Lu et al., 2003; Imitola et al., 2004). These uncommitted cells have been demonstrated to self-renew, differentiate into neural cell lineages in different regions and contexts within developing or regenerating sites of the CNS. In addition, NSCs have been shown to integrate with host progenitors and progeny while exhibiting repair potential (Snyder et al., 1992; Teng et al., 2002; Lu et al., 2003).

Fauza et al. utilized the fetal sheep MMC model to evaluated murine NSCs as an adjunctive therapy to fetal repair of spina bifida (Fauza et al., 2008). Comparing three groups of sheep (MMC without repair, surgical MMC repair, and surgical repair with NSCs), animals without repair demonstrated lower survival compared to the other two groups. The incidence of major paralysis (inability to walk) was more significant for nonrepaired fetuses than the other two repaired groups with the least paraparesis in the NSC group. Subjective analysis of muscular contractions and tactile function favored those animals repaired and treated with NSCs. Despite their xenologous origin and lack of host immunosuppression, all animals in the NSC group showed cell engraftment with NSC density highest in areas of most damaged spinal cord, particularly regions directly exposed to the amniotic fluid environment. Although demonstrating engraftment, many of the NSCs exhibited features of undifferentiation in vivo by retaining their native morphology and immunoreactivity of immature neuroepithelial stem cell markers. Nestin-positive cells were only found in this group and not in non-NSC infused sheep. In addition, only donor NSCs expressed the neurotrophic factors glial cell line-derived neurotrophic factor (GDNF) and brain-derived neurotrophic factor (BDNF). Both GDNF and BDNF have been associated with supportive and protective function of spinal cord motor neurons and promotion of axonal growth (Yan et al., 1992; Vejsada et al., 1998). These findings suggested that NSC transplantation may not differentiate in order to produce cell-mediated neural repair. Rather, NSCs maintain their undifferentiated state but produce neurotrophic elements that enhance host repair mechanisms.

10.6.3 Induced Pluripotent Stem Cells

Induced pluripotent stem cells (iPSCs) are reprogrammed somatic cells that become pluripotent after induction and ectopic expression of specific transcription factors. Similarly to hESCs, this cell line can also differentiate into all three embryonic germ layers (Takahashi et al., 2006, 2007). The IPSCs offer several advantages including obviating the need for embryos and oocytes for research. Patient-specific iPSCs alleviates challenges with graft rejection, while providing the ability to restore organ function as well as tissue reconstruction in response to injury or disease. Park et al. demonstrated the ability to recapitulate iPSCs from disease-specific cells lines of Mendelian or complex inheritances (Park et al., 2008). Investigation with iPSCs have included generation of spinal motor neurons and glia from amyotrophic lateral sclerosis (Dimos et al., 2008), iPSC-derived dopamine neurons derived from Parkinson's disease (Soldner et al., 2009), motor neurons from spinal muscular atrophy (Ebert et al., 2009), and derivation of iPSCs from persons with familial dysautonomia (Lee et al., 2009). Because of their direct influence on neural tube closure, transplanted neural crest cells could be reprogrammed to derive specific NTD iPSCs, which could assist in closing fetal NTDs.

Neural crest stem cells (NCSCs) are a multipotent population of cells located on the dorsal side of the developing neural tube. Transplantation of NCSCs has been shown to improve outcomes after neural injury (Biernaskie et al., 2007; Hu et al., 2010; Sieber-Blum, 2010). The therapeutic effects of NCSCs have been attributed to combinations of functions including angiogenesis, myelination, cell replacement, and modulation of wound healing (Sieber-Blum et al., 2006; Fernandes et al., 2008). Saadai et al. was able to seed iPSC-NSCSs onto nanofiber scaffolds (Saadai et al., 2013). Utilizing the sheep MMC model, human iPSCs derived from skin fibroblasts were differentiated into NCSCs. After incorporation into hydrogel, constructs were surgically transplanted between neural elements and skin closure. After demonstrating >95% viability after transplantation, iPSC-NCSCs differentiated into neurites and expressed neuronal markers β-III tubulin and neurofilament markers. This study was the first to demonstrate survival and integration of human iPSC-NCSCs into a fetal sheep model despite lack of host immunosuppression. In addition, cells were able to differentiate into a neural lineage with mature axonal neurofilament markers.

10.6.4 Mesenchymal Stem Cells

Mesenchymal stem cells (MSCs) are a multipotent cell line that can be induced into vascular, muscle, intestinal, bone as well as neuronal tissue (Prockop et al., 2010). The transformative ability into neurons has been investigated in neurodegenerative diseases and spinal cord injury (Hofstetter et al., 2002; Pisati et al., 2007; Torrente and Polli, 2008). Li et al. demonstrated that transplanted MSCs into retinoic acid–induced MMC in rats survived, grew, and expressed markers of neurons, glia, and myoblasts. Surrounding spinal cord tissue also expresses neurotrophic elements and evidence of reduced spinal cord apoptosis. This suggested that MSCs may not only have direct effects on neuronal regeneration, but may also have a paracrine effect on host elements. The transplanted MSCs did not transdifferentiate and produced neurotrophic factors. These findings suggested a neuroprotective role of MSCs in addition to cellular replacement.

As the field of stem cell therapy expands, multipotent stromal cells derived from placental tissues and amniotic fluid provide a prenatal strategy for the treatment of MMC. Both sources have shown to be reliable sources for cell harvest (Fauza, 2004; Gonzalez et al., 2007). Placenta-derived tissues can be procured from chorionic villous sampling as early as 10 weeks' gestation with minimal risk to the mother or fetus, allowing time for culture expansion prior to fetal repair of MMC, which typically occurs during mid to late second trimester (Akolekar et al., 2015).

Placenta-derived MSCs may have greater regenerative potential compared to other sources of MSCs (Gonzalez et al., 2007). Placenta-derived MSCs had higher levels of immunomodulatory cytokines compared to bone marrow–derived or adipose-derived stromal stem cells (Gonzalez et al., 2007). Portmann-Lanz et al. characterized MSC from the chorion, amnion, and villous stroma as well as different periods of gestation (Portmann-Lanz et al., 2006). Although all three sources demonstrated potential for mesenchymal differentiation, chorion-derived cells exhibited better neurogenic differentiation capacity than amnion-derived cells with more neuroprotective effect. Gestational age did not appear to influence these results. However, Lankford et al. suggested that early gestation placenta-derived MSCs could be a therapeutic option and expandable from a small tissue mass via chorion villous sampling (Lankford et al., 2015). Immunophenotyping by flow cytometry demonstrated that early placenta-derived MSCs were positive for MSC markers, and were negative for hematopoietic and endothelial markers. Although they displayed tri-lineage differentiation, early placental MSCs were found to express developmental transcription factors as well as neural-related structural proteins. However, term-gestation MSCs did not express this immunophenotype. In addition, placental MSCs demonstrated compatibility with synthetic and biologic delivery vehicles, such as collagen, fibrin hydrogels, and nanofiber scaffolds made of poly-L-lactic acid. The unique properties of early gestation placenta-derived cells supported the approach of cell-based therapy to augment fetal surgery in the prenatal environment by enhancing immunomodulation and providing neuroprotection.

Revisiting early approaches with stem cell with chick embryos, Dionigi et al. injected amniotic MSCs to elicit in utero coverage of MMC defects (Dionigi et al., 2015). Utilizing the rat model, three groups were created: no manipulation, intrauterine injection of saline, and volume-matched, intrauterine injection of amniotic MSCs. There were no differences in reclosure or coverage seen in the nonmanipulation or saline groups. However, fetuses treated with MSCs demonstrated partial or complete coverage of the MMC, suggesting that a transamniotic technique for stem cell therapy could be an option for prenatal therapy of MMC.

Pennington et al. found similar time-dependent stem cell profiles (Pennington et al., 2015). Utilizing amniotic fluid sampling from a retinoic acid–derived MMC rat model, fetuses with MMCs demonstrated a higher proportion of MSC and NSC compared to normal fetuses. Although gestational age did not affect the relative proportion of stem cells, the study demonstrated that amniotic fluid was a viable cell source at a time of potential diagnostic amniocentesis if cell-based therapies were to be utilized in fetal therapy.

10.7 SUMMARY

In the last 30 years, significant clinical advances have been made in the treatment of spina bifida. With the results of the MOMS trial, fetal surgery for repair of MMC is a viable option. However, this therapy has a limited application and incurs a significant level of morbidity for child and mother. Novel approaches with tissue engineering and cell-based therapies may bridge the gap for the majority of children, whose only option is conventional treatment directed at relief of symptoms. As the field has matured, scaffolds that not only offer mechanical protection in the fetal environment, but also provide neuroregenerative potential and cell-based therapies are being investigated. Multipotent stromal cells, derived from placenta and amniotic fluid, appear to be of most interest. Stromal cells from these fetal sources have been demonstrated to be the most reliable in harvesting and producing autologous stem cells, which could be expanded for potential fetal intervention (Fauza, 2004; Gonzalez et al., 2007).

REFERENCES

Adzick, N.S. et al., A randomized trial of prenatal versus postnatal repair of myelomeningocele. *N Engl J Med*, 2011. 364(11): p. 993–1004.

Akolekar, R. et al., Procedure-related risk of miscarriage following amniocentesis and chorionic villus sampling: A systematic review and meta-analysis. *Ultrasound Obstet Gynecol*, 2015. 45(1): p. 16–26.

Biernaskie, J. et al., Skin-derived precursors generate myelinating Schwann cells that promote remyelination and functional recovery after contusion spinal cord injury. *J Neurosci*, 2007. 27(36): p. 9545–9559.

Blesch, A. and M.H. Tuszynski, Gene therapy and cell transplantation for Alzheimer's disease and spinal cord injury. *Yonsei Med J*, 2004. 45 (Suppl): p. 28–31.

Botto, L.D., Moore, C.A., Khoury, M.J., and Erickson, J.D. Neural-tube defects. *N Engl J Med*, 1999. 341: p. 1509–1519.

Bouchard, S. et al., Correction of hindbrain herniation and anatomy of the vermis after in utero repair of myelomeningocele in sheep. *J Pediatr Surg*, 2003. 38(3): p. 451–458; discussion 451–458.

Boulet, S.L. et al., Trends in the postfortification prevalence of spina bifida and anencephaly in the United States. *Birth Defects Res A Clin Mol Teratol*, 2008. 82(7): p. 527–532.

Brown, E.G. et al., In utero repair of myelomeningocele with autologous amniotic membrane in the fetal lamb model. *J Pediatr Surg*, 2014. 49(1): p. 133–137; discussion 137–138.

Bruner, J.P. et al., Fetal surgery for myelomeningocele and the incidence of shunt-dependent hydrocephalus. *JAMA*, 1999. 282(19): p. 1819–1825.

Caldarelli, M., C. Di Rocco, and F. La Marca, Shunt complications in the first postoperative year in children with meningomyelocele. *Childs Nerv Syst*, 1996. 12(12): p. 748–754.

Cao, Q., R.L. Benton, and S.R. Whittemore, Stem cell repair of central nervous system injury. *J Neurosci Res*, 2002. 68(5): p. 501–510.

Centers for Disease Control and Prevention (CDC). Recommendations for the use of folic acid to reduce the number of cases of spina bifida and other neural tube defects. *MMWR Recomm Rep*, 1992. 41(RR-14): p. 1–7.

Centers for Disease Control and Prevention (CDC). Spina bifida and anencephaly before and after folic acid mandate—United States, 1995–1996 and 1999–2000. *MMWR Morb Mortal Wkly Rep*, 2004. 53: p. 362–365.

Correia-Pinto, J. et al., In utero meconium exposure increases spinal cord necrosis in a rat model of myelomeningocele. *J Pediatr Surg*, 2002. 37(3): p. 488–492.

Cragan, J.D. et al., Surveillance for anencephaly and spina bifida and the impact of prenatal diagnosis—United States, 1985–1994. *MMWR CDC Surveill Summ*, 1995. 44(4): p. 1–13.

Dimos, J.T. et al., Induced pluripotent stem cells generated from patients with ALS can be differentiated into motor neurons. *Science*, 2008. 321(5893): p. 1218–1221.

Dionigi, B. et al., Partial or complete coverage of experimental spina bifida by simple intra-amniotic injection of concentrated amniotic mesenchymal stem cells. *J Pediatr Surg*, 2015. 50(1): p. 69–73.

Ebert, A.D. et al., Induced pluripotent stem cells from a spinal muscular atrophy patient. *Nature*, 2009. 457(7227): p. 277–280.

Economic burden of spina bifida—United States, 1980–1990. *MMWR Morb Mortal Wkly Rep*, 1989. 38(15): p. 264–267.

Eggink, A.J. et al., In utero repair of an experimental neural tube defect in a chronic sheep model using biomatrices. *Fetal Diagn Ther*, 2005. 20(5): p. 335–340.

Eggink, A.J. et al., Histological evaluation of acute covering of an experimental neural tube defect with biomatrices in fetal sheep. *Fetal Diagn Ther*, 2006. 21(2): p. 210–216.

Eggink, A.J. et al., Delayed intrauterine repair of an experimental spina bifida with a collagen biomatrix. *Pediatr Neurosurg*, 2008. 44(1): p. 29–35.

Fauza, D., Amniotic fluid and placental stem cells. *Best Pract Res Clin Obstet Gynaecol*, 2004. 18(6): p. 877–891.

Fauza, D.O. et al., Neural stem cell delivery to the spinal cord in an ovine model of fetal surgery for spina bifida. *Surgery*, 2008. 144(3): p. 367–373.

Fernandes, K.J., J.G. Toma, and F.D. Miller, Multipotent skin-derived precursors: Adult neural crest-related precursors with therapeutic potential. *Philos Trans R Soc Lond B Biol Sci*, 2008. 363(1489): p. 185–198.

Fontecha, C.G. et al., Inert patch with bioadhesive for gentle fetal surgery of myelomeningocele in a sheep model. *Eur J Obstet Gynecol Reprod Biol*, 2009. 146(2): p. 174–179.

Fontecha, C.G. et al., Fetoscopic coverage of experimental myelomeningocele in sheep using a patch with surgical sealant. *Eur J Obstet Gynecol Reprod Biol*, 2011. 156(2): p. 171–176.

Gonzalez, R. et al., Pluripotent marker expression and differentiation of human second trimester mesenchymal stem cells. *Biochem Biophys Res Commun*, 2007. 362(2): p. 491–497.

Hao, Y. et al., Identification of antiangiogenic and antiinflammatory proteins in human amniotic membrane. *Cornea*, 2000. 19(3): p. 348–352.

Heffez, D.S. et al., The paralysis associated with myelomeningocele: Clinical and experimental data implicating a preventable spinal cord injury. *Neurosurgery*, 1990. 26(6): p. 987–992.

Heffez, D.S. et al., Intrauterine repair of experimental surgically created dysraphism. *Neurosurgery*, 1993. 32(6): p. 1005–1010.

Herrera, S.R. et al., Comparison between two surgical techniques for prenatal correction of meningomyelocele in sheep. *Einstein (Sao Paulo)*, 2012. 10(4): p. 455–461.

Hofstetter, C.P. et al., Marrow stromal cells form guiding strands in the injured spinal cord and promote recovery. *Proc Natl Acad Sci USA*, 2002. 99(4): p. 2199–2204.

Hu, Y.F. et al., Epidermal neural crest stem cell (EPI-NCSC)—Mediated recovery of sensory function in a mouse model of spinal cord injury. *Stem Cell Rev*, 2010. 6(2): p. 186–198.

Hunt, G.M., Open spina bifida: Outcome for a complete cohort treated unselectively and followed into adulthood. *Dev Med Child Neurol*, 1990. 32(2): p. 108–118.

Hutchins, G.M. et al., Acquired spinal cord injury in human fetuses with myelomeningocele. *Pediatr Pathol Lab Med*, 1996. 16(5): p. 701–712.

Imitola, J. et al., Directed migration of neural stem cells to sites of CNS injury by the stromal cell-derived factor 1α/CXC chemokine receptor 4 pathway. *Proc Natl Acad Sci USA*, 2004. 101(52): p. 18117–18122.

Kallen, B. et al., International study of sex ratio and twinning of neural tube defects. *Teratology*, 1994. 50(5): p. 322–331.

Kersey, J.P. and A.J. Vivian, Mitomycin and amniotic membrane: A new method of reducing adhesions and fibrosis in strabismus surgery. *Strabismus*, 2008. 16(3): p. 116–118.

Koda, M. et al., Hematopoietic stem cell and marrow stromal cell for spinal cord injury in mice. *Neuroreport*, 2005. 16(16): p. 1763–1767.

Korenromp, M.J. et al., Early fetal leg movements in myelomeningocele. *Lancet*, 1986. 1(8486): p. 917–918.

Lankford, L. et al., Early gestation chorionic villi-derived stromal cells for fetal tissue engineering. *World J Stem Cells*, 2015. 7(1): p. 195–207.

Lee, D.H. et al., Enhancement of re-closure capacity by the intra-amniotic injection of human embryonic stem cells in surgically induced spinal open neural tube defects in chick embryos. *Neurosci Lett*, 2004. 364(2): p. 98–100.

Lee, D.H. et al., Reclosure of surgically induced spinal open neural tube defects by the intraamniotic injection of human embryonic stem cells in chick embryos 24 hours after lesion induction. *J Neurosurg*, 2006. 105(2 Suppl): p. 127–133.

Lee, G. et al., Modelling pathogenesis and treatment of familial dysautonomia using patient-specific iPSCs. *Nature*, 2009. 461(7262): p. 402–406.

Li, H. et al., Immunosuppressive factors secreted by human amniotic epithelial cells. *Invest Ophthalmol Vis Sci*, 2005. 46(3): p. 900–907.

Li, H. et al., Therapeutic potential of in utero mesenchymal stem cell (MSCs) transplantation in rat foetuses with spina bifida aperta. *J Cell Mol Med*, 2012. 16(7): p. 1606–1617.

Liu, J. et al., Update on amniotic membrane transplantation. *Expert Rev Ophthalmol*, 2010. 5(5): p. 645–661.

Liu, S. et al., Embryonic stem cells differentiate into oligodendrocytes and myelinate in culture and after spinal cord transplantation. *Proc Natl Acad Sci USA*, 2000. 97(11): p. 6126–6131.

Lu, P. et al., Neural stem cells constitutively secrete neurotrophic factors and promote extensive host axonal growth after spinal cord injury. *Exp Neurol*, 2003. 181(2): p. 115–129.

Mamede, A.C. et al., Amniotic membrane: From structure and functions to clinical applications. *Cell Tissue Res*, 2012. 349(2): p. 447–458.

Merianos, D., T. Heaton, and A.W. Flake, In utero hematopoietic stem cell transplantation: Progress toward clinical application. *Biol Blood Marrow Transplant*, 2008. 14(7): p. 729–740.

Mermet, I. et al., Use of amniotic membrane transplantation in the treatment of venous leg ulcers. *Wound Repair Regen*, 2007. 15(4): p. 459–464.

Meuli, M. et al., In utero surgery rescues neurological function at birth in sheep with spina bifida. *Nat Med*, 1995. 1(4): p. 342–347.

Meuli, M. et al., The spinal cord lesion in human fetuses with myelomeningocele: Implications for fetal surgery. *J Pediatr Surg*, 1997. 32(3): p. 448–452.

Moore, C.A. et al., Elevated rates of severe neural tube defects in a high-prevalence area in northern China. *Am J Med Genet*, 1997. 73(2): p. 113–118.

Ouyang, L. et al., Health care expenditures of children and adults with spina bifida in a privately insured U.S. population. *Birth Defects Res A Clin Mol Teratol*, 2007. 79(7): p. 552–558.

Paek, B.W. et al., Hindbrain herniation develops in surgically created myelomeningocele but is absent after repair in fetal lambs. *Am J Obstet Gynecol*, 2000. 183(5): p. 1119–1123.

Park, I.H. et al., Disease-specific induced pluripotent stem cells. *Cell*, 2008. 134(5): p. 877–886.

Patten, B.M., Embryological stages in the establishing of myeloschisis with spina bifida. *Am J Anat*, 1953. 93(3): p. 365–395.

Peiro, J.L. et al., Single-access fetal endoscopy (SAFE) for myelomeningocele in sheep model I: Amniotic carbon dioxide gas approach. *Surg Endosc*, 2013. 27(10): p. 3835–3840.

Pennington, E.C. et al., The impact of gestational age on targeted amniotic cell profiling in experimental neural tube defects. *Fetal Diagn Ther*, 2015. 37(1): p. 65–69.

Pisati, F. et al., Induction of neurotrophin expression via human adult mesenchymal stem cells: Implication for cell therapy in neurodegenerative diseases. *Cell Transplant*, 2007. 16(1): p. 41–55.

Portmann-Lanz, C.B. et al., Placental mesenchymal stem cells as potential autologous graft for pre- and perinatal neuroregeneration. *Am J Obstet Gynecol*, 2006. 194(3): p. 664–673.

Prockop, D.J. et al., Evolving paradigms for repair of tissues by adult stem/progenitor cells (MSCs). *J Cell Mol Med*, 2010. 14(9): p. 2190–2199.

Rintoul, N.E. et al., A new look at myelomeningoceles: Functional level, vertebral level, shunting, and the implications for fetal intervention. *Pediatrics*, 2002. 109(3): p. 409–413.

Saadai, P. and D.L. Farmer, Fetal surgery for myelomeningocele. *Clin Perinatol*, 2012. 39(2): p. 279–288.

Saadai, P. et al., Prenatal repair of myelomeningocele with aligned nanofibrous scaffolds—A pilot study in sheep. *J Pediatr Surg*, 2011. 46(12): p. 2279–2283.

Saadai, P. et al., Human induced pluripotent stem cell-derived neural crest stem cells integrate into the injured spinal cord in the fetal lamb model of myelomeningocele. *J Pediatr Surg*, 2013. 48(1): p. 158–163.

Setoguchi, T. et al., Treatment of spinal cord injury by transplantation of fetal neural precursor cells engineered to express BMP inhibitor. *Exp Neurol*, 2004. 189(1): p. 33–44.

Shaw, G.M. et al., Epidemiologic characteristics of phenotypically distinct neural tube defects among 0.7 million California births, 1983–1987. *Teratology*, 1994. 49(2): p. 143–149.

Shurtleff, D.B., Epidemiology of neural tube defects and folic acid. *Cerebrospinal Fluid Res*, 2004. 1(1): p. 5.

Shurtleff, D.B. et al., Meningomyelocele: Management in utero and post natum. *Ciba Found Symp*, 1994. 181: p. 270–280; discussion 280–286.

Sieber-Blum, M., Epidermal neural crest stem cells and their use in mouse models of spinal cord injury. *Brain Res Bull*, 2010. 83(5): p. 189–193.

Sieber-Blum, M. et al., Characterization of epidermal neural crest stem cell (EPI-NCSC) grafts in the lesioned spinal cord. *Mol Cell Neurosci*, 2006. 32(1–2): p. 67–81.

Sim, K.B. et al., Chronological changes of re-closure capacity in surgically induced spinal open neural tube defects of chick embryos. *Neurosci Lett*, 2000. 292(3): p. 151–154.

Snyder, E.Y. et al., Multipotent neural cell lines can engraft and participate in development of mouse cerebellum. *Cell*, 1992. 68(1): p. 33–51.

Soldner, F. et al., Parkinson's disease patient-derived induced pluripotent stem cells free of viral reprogramming factors. *Cell*, 2009. 136(5): p. 964–977.

Stiefel, D., A.J. Copp, and M. Meuli, Fetal spina bifida in a mouse model: Loss of neural function in utero. *J Neurosurg*, 2007. 106(3 Suppl): p. 213–221.

Sykova, E. et al., Bone marrow stem cells and polymer hydrogels—Two strategies for spinal cord injury repair. *Cell Mol Neurobiol*, 2006. 26(7–8): p. 1113–1129.

Takahashi, K. and S. Yamanaka, Induction of pluripotent stem cells from mouse embryonic and adult fibroblast cultures by defined factors. *Cell*, 2006. 126(4): p. 663–676.

Takahashi, K. et al., Induction of pluripotent stem cells from adult human fibroblasts by defined factors. *Cell*, 2007. 131(5): p. 861–872.

Teng, Y.D. et al., Functional recovery following traumatic spinal cord injury mediated by a unique polymer scaffold seeded with neural stem cells. *Proc Natl Acad Sci USA*, 2002. 99(5): p. 3024–3029.

Torrente, Y. and E. Polli, Mesenchymal stem cell transplantation for neurodegenerative diseases. *Cell Transplant*, 2008. 17(10–11): p. 1103–1113.

Vejsada, R. et al., Synergistic but transient rescue effects of BDNF and GDNF on axotomized neonatal motoneurons. *Neuroscience*, 1998. 84(1): p. 129–139.

Watanabe, M., A.G. Kim, and A.W. Flake, Tissue engineering strategies for fetal myelomeningocele repair in animal models. *Fetal Diagn Ther*, 2015. 37(3): p. 197–205.

Watanabe, M. et al., A tissue engineering approach for prenatal closure of myelomeningocele with gelatin sponges incorporating basic fibroblast growth factor. *Tissue Eng Part A*, 2010. 16(5): p. 1645–1655.

Watanabe, M. et al., A tissue engineering approach for prenatal closure of myelomeningocele: Comparison of gelatin sponge and microsphere scaffolds and bioactive protein coatings. *Tissue Eng Part A*, 2011. 17(7–8): p. 1099–1110.

Whittemore, S.R. and E.Y. Snyder, Physiological relevance and functional potential of central nervous system-derived cell lines. *Mol Neurobiol*, 1996. 12(1): p. 13–38.

Worley, G., J.M. Schuster, and W.J. Oakes, Survival at 5 years of a cohort of newborn infants with myelomeningocele. *Dev Med Child Neurol*, 1996. 38(9): p. 816–822.

Yan, Q., J. Elliott, and W.D. Snider, Brain-derived neurotrophic factor rescues spinal motor neurons from axotomy-induced cell death. *Nature*, 1992. 360(6406): p. 753–755.

Index

D

E

M

N

R

S

V

W

Y